IUTAM SYMPOSIUM ON
NON-LINEAR SINGULARITIES IN DEFORMATION AND FLOW

IUTAM Symposium on
Non-linear Singularities in Deformation and Flow

Proceedings of the IUTAM Symposium
held in Haifa, Israel,
17-21 March 1997

Edited by

D. DURBAN
Faculty of Aerospace Engineering,
Israel Institute of Technology,
Technion, Haifa, Israel

and

J.R.A. PEARSON
Fluid Mechanics Department,
Schlumberger Cambridge Research,
Cambridge, U.K.

KLUWER ACADEMIC PUBLISHERS
DORDRECHT / BOSTON / LONDON

Library of Congress Cataloging-in-Publication Data

ISBN 0-7923-5349-8

Published by Kluwer Academic Publishers,
P.O. Box 17, 3300 AA Dordrecht, The Netherlands.

Sold and distributed in North, Central and South America
by Kluwer Academic Publishers,
101 Philip Drive, Norwell, MA 02061, U.S.A.

In all other countries, sold and distributed
by Kluwer Academic Publishers,
P.O. Box 322, 3300 AH Dordrecht, The Netherlands.

Printed on acid-free paper

Printed in the Netherlands

CONTENTS

Numerical Methods

Capillary Breakup and Instabilities

Cusps and Contact Lines

Applications

Preface

The IUTAM Symposium on "Non-Linear Singularities in Deformation and Flow" took place from March 17 to 21, 1997, at the Technion in Haifa, Israel, with 70 participants from 12 countries.

The leitmotif of this Symposium brought together scientists working on singularity-dominated local fields in various branches of continuum mechanics, covering traditional solid and liquid behaviour as well as that of more complex non-linear materials; non-linearities arise either from the constitutive equations for the material or from the presence of interfaces or both. The scientific committee invited speakers who presented 34 papers in 12 sessions. Topics covered in the lectures included near tip fields of cracks, notches and wedges; flow around corners, wedges and cones; interfacial phenomena; moving contact lines in multiphase systems; cusps in fluid interfaces and shocks and localization. There was a general consensus among the participants that singularities induced by non-linearities provide a challenging and currently important area of research in mechanics, engineering and applied mathematics. Presentation and discussions during the symposium initiated further studies of problems in these interesting areas.

This volume contains 30 full length papers, submitted by the lecturers after the symposium and reviewed to the standards of international scientific periodicals.

It is our pleasure to acknowledge the efficient and tireless help of Mrs. Alice Goodman and Mr. Gideon Wachsman of the Faculty of Aerospace Engineering at the Technion.

<table>
<tr><td>David Durban</td><td style="text-align:right">Anthony Pearson</td></tr>
<tr><td>Haifa</td><td style="text-align:right">Cambridge</td></tr>
</table>

April 1998

International Scientific Committee

C. Atkinson (UK)
G.I. Barenblatt (USA)
H.-C. Chang (USA)
D. Durban (Israel, Chairman)
E.J. Hinch (UK)
J.-T. Jeong (Korea)
D.D. Joseph (USA)
H.K. Moffatt (UK)
J.R.A. Pearson (UK, Chairman)
E. Sanchez-Palencia (France)
I. Vardoulakis (Greece)

Local Organizing Committee

D. Durban (Chairman)
I. Frankel
D. Givoli
B. Greenberg
Y. Tirosh
A. Goodman (Secretary)

Sponsors of the IUTAM Symposium on Non-Linear Singularities in Deformation and Flow

International Union of Theoretical and Applied Mechanics (IUTAM)
The Institute of Advanced Studies in Mathematics at the Technion
Technion-Israel Institute of Technology
The S. Neaman Institute for Advanced Studies in Science and Technology
Faculty of Aerospace Engineering, Technion
International Association for Computational Mechanics (IACM)
Kluwer Academic Publishers

LENGTH SCALES, ASYMPTOTICS AND NON-LINEAR SINGULARITIES

J.R.A. PEARSON

23 Chaucer Road, Cambridge CB2 2EB, England

1. Introduction

This short introductory paper will try to give a common perspective to many of the problems and techniques covered in later papers. We shall draw attention to those common features that lead to singularities, and to any similarities in the techniques used to deal with them. Continuum mechanical theories for elasto-plastic solids, for materials with memory and for purely viscous or ideal fluids have tended to be treated separately leading to unnecessary and unhelpful specialisation. The mechanics of porous or fractured media have developed as almost independent subjects, while wave propagation is often studied in isolation. It is hoped that this symposium will help to restore the universality inherent in continuum mechanics.

It is well known that singular solutions exist for many of the sets of equations (mathematical models) describing the bulk continuum mechanics of deformable and fluid materials. In the case of linear models, these simple singular solutions have fundamental mathematical (rather than physical) significance and can be used in a formal fashion to provide general solutions to geometrically-complex boundary-value problems. In cases with smooth boundaries, superposition of these isolated singular solutions can lead to physically acceptable (everywhere) smooth solutions. However, when boundaries have corners or include finite cracks, these singularities (specifically in the stress tensor) remain as localised features of the full solution. Examples are provided later in this volume (Banks-Sills, Matveyenko *et al.*, Leguillon).

The question then arises as to how these physically unacceptable singularities can best be removed (regularised) so as to make the solutions correspond to physical reality. There are several ways in which this can be done.

1

D. Durban and J.R.A. Pearson (eds.),
IUTAM Symposium on Non-Linear Singularities in Deformation and Flow, 1-20.

1. Smooth the boundaries. This is consistent with the notion of having at least two widely separated length scales in the problem: a large or outer length scale, in terms of which the singular nature of the boundary geometry is an obvious approximation and a smaller inner length scale in terms of which the boundaries become smooth. However, if the sharpness of the corners or crack tip correspond to the structural (e.g. crystallite or even molecular) length scale of the material in question, then the continuum constitutive equations used are usually inadequate to describe the material.

2. Alter the boundary conditions or bulk constitutive laws near sharp corners so as to distribute the stresses. For example, in fracture mechaniucs this has been done by the introduction of a process zone near the crack tip. In the moving contact line problem, both relaxation of the no-slip boundary condition and a bulk shear-rate-dependent viscosity have been used.

Some examples are given, mainly as reminders of earlier work, in the next section relating to flow of an ideal fluid, an inertialess (incompressible) viscous fluid and a linear elastic solid, all with prescribed boundaries.

If, however, non-linearities are introduced,

(i) as with acceleration (inertia) terms in a Newtonian fluid, yielding the Navier-Stokes equations, or

(ii) by having a non-linear relation for the Cauchy stress tensor in terms of the rate-of-deformation tensor, yielding a non-Newtonian or elastic fluid, or

(iii) by having a deformable boundary between the chosen material and some other material (an insubstantial gas or Newtonian fluid),

then the nature and significance of singular flow solutions are much less obvious.

Two well-known examples of the first category are recalled in § 3 for the Navier-Stokes equations, based on the inviscid and inertia-free limits respectively. The first arises in the boundary-layer theory for flow over an infinite flat plate; the second for slow flow past a sphere or cylinder. The singularities involved are more apparent than real; they refer to approximations obtained by using reduced forms of the Navier-Stokes equations which cease to be valid at the singular points, and which can be removed by considering more terms in the the defining equation to get a correct (uniformly valid) solution. These provide early examples of matched asymptotic expansions.

For the case of elastic fluids, an example of the second category, asymptotic matching can be necessary even for inertialess steady flow, as Renardy (this volume) discusses for flow around a 270° corner.

Cusps formed by flows driven near interfaces provide an example for the third category (Jeong and Shikhmurzaev in this volume).

Fracture mechanics (which involves a solid deformable boundary) introduces a variety of specialised issues. For the simplest version of the theory, the equilibrium crack in a linear elastic medium introduces a single stress singularity at the crack tip, whose magnitude defines the value of the stress intensity function. Propagation of such a fracture requires the stress intensity function to reach a critical value, the fracture toughness, a constitutive characteristic of the material in question. Such failure of the material involves non-linear behaviour in some viscinity of the crack tip. Further complexity is introduced if the crack propagation is driven by a fluid flow into the crack. The issues are discussed in section § 4.

Another class of complex singularities is associated with steadily-moving contact lines. Here the interface between two fluids that intersects a solid surface is highly deformable: this introduces non-linearity. The velocity-gradient field (and hence the stress field) is inevitably singular at the contact line, which introduces a velocity discontinuity. The singular solution provided by traditional viscous fluid mechanics (Newtonian fluids and a no-slip boundary condition) leads to a singular total force, at and near the contact line, which is wholly unacceptable physically. Resolution of this singularity leads to a reconsideration of the boundary conditions, and of the meaning of contact angle. The use of surface and interfacial tensions (or energies) implies two-dimensional continuum concepts, and any attempt to analyse the molecular sources of these forces inevitably introduces the notion of a very short length scale, at which interfaces become interfacial layers, and the contact line a diffuse cylindrical region. Further details are given in § 5.

A quite different type of singularity arises when unsteady flows involving free surfaces are concerned. One class is associated with extending filaments and a related one with thin films. As we have seen, in the continuum limit the interfaces can sustain surface forces. The interplay between surface and bulk forces can lead to instability and rupture in a finite time. Analysis of such problems leads to asymptotic expansions associated with widely-spaced length and time scales (see papers by Lister *et al.*, Oron & Bankoff, Halpern & Grotberg, Chauhan *et al.* in this volume).

2. Linear Models

Irrotational flow of an ideal (incompressible inviscid) fluid is governed by the harmonic equation $\nabla^2 \phi = 0$, where ϕ is the velocity potential. In the absence of body forces, the only relevant component of the stress tensor is the isotropic pressure given by

$$\frac{p}{\rho} = \frac{\partial \phi}{\partial t} - \frac{1}{2}\nabla\phi.\nabla\phi + F(t)$$

The fundamental source term solution

$$\phi = \frac{m}{4\pi r}$$

and its derivative dipole (double source) solution

$$\phi = \frac{\mu \cos\theta}{4\pi r^2}$$

can be used to give a general solution for flow outside prescribed boundaries, in terms of either a distribution of sources or a distribution of double sources over the boundaries, i.e.,

$$\phi_P = -\frac{1}{4\pi} \int \frac{1}{r}\left(\frac{\partial\phi}{\partial n} + \frac{\partial\phi'}{\partial n'}\right) dS = \frac{1}{4\pi} \int (\phi - \phi')\frac{\partial}{\partial n}\left(\frac{1}{r}\right) dS$$

where r has origin at P within the fluid, $\phi' = 1/r$ and n and n' are the inward and outward normals at the surface S. See, for example, Lamb (1958, § 58) for details. In planar flow, the full power of complex variable theory can be used to derive detailed solutions for particular cases including flow past sharp corners (Schwartz-Christoffel method; see Lamb 1953,§ 73).

If vorticity $\omega = \nabla \times \mathbf{u}$ is present, two new types of singularity, the vortex line $\omega = \delta(\mathbf{r} - \mathbf{r}_0)\mathbf{k}$ where $\mathbf{r} = x\mathbf{i} + y\mathbf{j}$, $\mathbf{i}.\mathbf{k} = \mathbf{j}.\mathbf{k} = \mathbf{i}.\mathbf{j} = 0$, and the vortex sheet, $\omega = \mathbf{k}\delta(n)$ where $\mathbf{n}$ is normal to the sheet and $\mathbf{k}.\mathbf{n} = 0$ (Lamb 1953, §§ 148, 151; Saffman 1992, §§ 2.2, 2.3) can be made use of, e.g. using the Biot-Savart law (Saffman 1992, § 11.1).

For slow flow of an incompressible linear viscous (Newtonian) fluid, the point source singularity is still relevant; a further fundamental flow field due to a point force acting on the fluid of viscosity μ is the Stokeslet where $\psi = Fr\sin^2\theta/8\pi\mu$ is the Stokes stream function in spherical polar coordinates aligned with the point force $\mathbf{F}$ (Batchelor 1967, § 4.9; Happel & Brenner 1973, § 4-12).

For flow past a corner, the planar singular solution has been given by Moffatt (1964). The total force associated with the corner is finite. A more severe singularity, leading to a singular force, is provided by a semi-infinite flat plate sliding at an angle over another fixed plate (Batchelor 1967, § 4.8); this will be referred to later in connection with moving contact lines. These planar examples use the Lagrangian stream function χ which obeys the biharmonic equation $\nabla^2\nabla^2\chi = 0$.

Lastly, in the case of a linear elastic solid fundamental singular solutions are provided by a concentrated force and a concentrated couple (see, e.g., Muskhelishvili 1953, § 57 for the plane case). Again corners in the boundary or discontinuities in the boundary loading can be dealt with.

3. Navier-Stokes Equation

This well-known equation, representing the balance between acceleration, pressure and viscous forces in a flowing linear viscous liquid, is most conveniently written, for a uniform incompressible fluid, in the dimensionless form

$$\frac{D\mathbf{u}}{Dt} = -\nabla p + \frac{1}{Re}\nabla^2\mathbf{u}, \qquad \nabla.\mathbf{u} = 0$$

(gravity having been absorbed into the reduced pressure p) where the Reynolds number

$$Re = \rho V L/\mu = V L/\nu$$

represents the balance between characteristic inertial forces ρV^2 and characteristic viscous forces $\mu V/L$; here ρ is the fluid density, μ the fluid viscosity, V a characteristic velocity and L a characteristic length.

3.1. BOUNDARY-LAYER THEORY

In many real flows Re is very large, and for many practical purposes the ideal fluid theory mentioned in § 2 above, which corresponds to putting $Re^{-1} = 0$, is a useful approximation to the velocity distribution in much of the flow field. It will be noted that this simpler, Euler, equation is of lower order than the full Navier-Stokes equation and so fewer boundary conditions on confining boundaries are needed to specify a solution. If it is further required that there can be no discontinuity in tangential velocity (no slip) at any interface between the moving fluid and some other substantial material (usually taken to be a rigid solid or other linear viscous fluid), then viscous forces must become relevant at least in some neighbourhood of the interface; vorticity is generated locally and when convected away can profoundly affect the rest of the flow field.

A classic and highly singular example is given by flow past a fixed semi-infinite flat plate aligned parallel to a uniform steady flow $V\hat{\mathbf{x}}$ with its leading edge normal to the oncoming flow. Using a planar (x,y) coordinate system, with the plate at $y = 0$, $x \geq 0$, a (Blasius) similarity solution of the form

$$\psi = (2\nu V x)^{1/2}f(\eta), \quad \eta = (V/2\nu x)^{1/2}y$$

with ψ the two-dimensional stream function satisfying the ordinary differential equation

$$f''' + ff'' = 0$$

with boundary conditions

$$f(0) = f'(0) = 0, \quad f'(\infty) = 1$$

provides an asymptotically valid solution of the boundary-layer equations (see, e.g., Rosenhead 1963, chap.V).

An important point to note is that there is no length scale L defined in the problem; there is only a local Reynolds number $Re_x = 2Vx/\nu$ in terms of which the dimensionless velocities become

$$u_x = f'(\eta), \quad u_y = (Re_x)^{1/2}\{\eta f'(\eta) - f(\eta)\}$$

and the coordinates become Re_x and $(Re_x)^{1/2}y/x$.

As $Re_x \to \infty$ with x held fixed, so the boundary layer thickness y/x and outflow velocity u_y tend to zero as $(Re_x)^{-1/2}$. The two-term non-linear boundary layer equation for f is obtained from the full Navier-Stokes equations by neglecting terms that are $O(Re_x^{-1/2})$ compared to those that are retained. Higher-order terms in an asymptotic expansion in Re_x can in principle be obtained.

If we hold V/ν fixed and let $x \to 0$, i.e. as we approach the leading edge of the plate, then Re_x first becomes $O(1)$ and then $o(1)$. Thus the solution becomes singular at $x = 0$. Partial regularisation is achieved (Alden 1948, and more systematically by Kaplun 1954 - see Rosenhead 1963, §§ V.13,18) by changing to parabolic coordinates centred on the leading edge, though a full solution requires matching of the boundary-layer solution to a solution of the linear Stokes equations in the neighbourhood ($Re_x \ll 1$) of the edge.

It is also worth noting that the influence of the boundary-layer flow on the outer Euler flow, where $y = O(x)$, is effectively an $O(Re_x^{1/2})$ outflow at the plate itself; the vorticity created in the boundary layer has significant effects only if there is boundary-layer separation. (For the case of a finite flat plate ending at $x = L$ the two vortex sheets shed from each plate surface cancel one another.) This outflow term could be interpreted as a new constitutive boundary condition for the outer "bulk" flow.

All of this highly singular behaviour can be attributed to the assumed infinitesimal thickness of an infinite flat plate. If a very small radius of curvature l_1 at the nose and a small maximum thickness l_2 ($l_1 \ll l_2 \ll L$)[1] are introduced, then three Reynolds numbers can be defined, all of different

---

[1]For an elliptic plate $Ll_1 = l_2^2$

order, with consequential complication of asymptotic behaviour in the flow field.

3.2. SLOW FLOW PAST BLUFF BODIES

This is typified by uniform flow $V\mathbf{i}$ past a sphere or cylinder of radius a. Here $Re = Va/\mu \ll 1$ and solutions can be sought to the dimensionless Stokes equations

$$\nabla.\mathbf{u} = 0, \quad \nabla^2\mathbf{u} = \nabla p$$

with $\mathbf{u} = \mathbf{0}$ on $r = 1$ and $\mathbf{u} \to \mathbf{i}$ as $r \to \infty$; p has been non-dimensionalised with $\mu V/a$.

In the case of the sphere a solution can be given that satisfies both the no-slip boundary condition on the sphere and the correct boundary condition at infinity. However as $r \to \infty$, it can readily be shown by considering the magnitude of various terms that the Stokes equations are an inadequate representation of the Navier-Stokes equations. This anomaly (the Stokes paradox) and a further one (the Whitehead paradox) which arises when an iterative solution of the inhomogeneous Stokes equations (treating inertial terms as a perturbation) is attempted was resolved by Oseen (1910) who pointed out that more realistic solutions could be obtained if the alternative linearisation

$$\frac{\partial \mathbf{u}}{\partial x} + \nabla p = \nabla^2 \mathbf{u}$$

is used in the far field, where now the lengthscale is, as in boundary-layer theory, ν/V; p is here non-dimensionalised with ρV^2. The boundary condition becomes a double source plus a Stokeslet as $r \to 0$ in the Oseen coordinates, i.e. is singular. Pairs of asymptotic expansions based on the Stokes and Oseen limits can be obtained (full details are given in Van Dyke 1964, Ch.VIII) from which the drag can be calculated to any order in Re.

In the case of the cylinder, the paradox is more obvious, because there is no solution of the Stokes equations satisfying no-slip on the cylinder that is compatible with uniform flow at infinity. Asymptotic matching is essential even to get the leading order term, $O(1/\ln Re)$, to the drag. (Again see Van Dyke 1964 for details.) The inner solution for the velocity field, dominated by the two-dimensional Stokeslet, is logarithmically singular at infinite distance from the cylinder.

4. Fracture Mechanics

An interesting way to describe an equilibrium crack is as the limiting case of an elliptic hole in a large plate when the aspect ratio, b/a, of the ellipse semi-axes goes to zero. There are in such a geometry 4 lengths that can be defined: b, a, the plate thickness c and the plate size d. If $d > a \gg c$ a plane stress approximation usually is appropriate for the large-scale deformation problem; if $c \gg a$, then plane strain will be appropriate at the intermediate scale of the hole tending to a crack. (In either case the limit $b/a \to 0$ yields a geometry analogous to that of the finite flate plate discussed in connection with boundary-layer theory.)

Conventional linear elastic fracture mechanics refers to the $r^{-1/2}$ singularity in stress that arises, in any large-scale stress and deformation solution, as the crack tip, thought of as at $r = 0$, is approached. Because solids are much more inhomogeneous that liquids at a small scale, the regularisation of this unphysical singularity (other than the J-integral approach of Rice) introduces more complexity on any relevant inner length scale than corresponding fluid flow singularities. In consequence, there has been relatively little attempt to seek dimensionless groups and matched asymptotic expansions to probe crack-tip behaviour. The Dugdale-Barenblatt boundary condition (see Barpi *et al.*, this volume, for further discussion) applied along an extended fracture surface is an architypal constitutive alteration of boundary conditions (method 2, §1 above) which preserves the linearity of the bulk Hookean properties. Recourse to an elasto-plastic model for bulk behaviour introduces a non-linearity (cf. means (ii), §1 above) that removes the singularity. Note that a local constitutive (continuum) length scale is obtained from $(K/\sigma_Y)^2$ where K is the fracture toughness and σ_Y the yield stress, both critical quantities.

Even these macroscopic (continuum) concepts are sometimes insufficient to cover the reality of fracture propagation, e.g. when the granularity of a porous medium or the "internal damage" of a microfractured medium is involved. The additional singular characteristics of isolated dislocations (see Vigdergauz, this volume) can be employed to simulate the effect of microscopic structural characteristics. Slepyan (this volume) discusses rate effects.

Progressively loaded cracks are subject to catastrophic instability when the propagation limit (stress intensity factor = fracture toughness) is reached according to static theories. Inertia terms or viscoelasticity (cf. means (i) or (ii) in §1 above) can be introduced to regularise this temporal singularity.

The most elaborate nesting of length scales arises in hydraulic fracturing, when a viscous fluid is pumped into a propagating fracture in saturated porous rock (Ben Naceur 1989). The viscous flow in the fracture is modelled according to two-dimensional lubrication flow, and when coupled to linear elastic fracture mechanics yields a strongly non-linear set of equations. Propagation speed is controlled by the rate of pumping. Consideration of the global problem shows that in many practical situations the fracture toughness can be neglected and so the parameters controlling the process are fluid viscosity μ (in the case of a Newtonian fluid), propagation speed V, fracture length l and an effective elastic modulus $E' = E/(1-\nu)$. These yield a single dimensionless parameter $lE'/\mu V$ which is typically extremely large; alternatively the characteristic length $\mu V/E'$ is extremely small. This suggests that a relevant intermediate approximation is obtained by looking for a universal power-law solution based on $l \to \infty$: this has been shown (SCR Geomechanics Group 1994) to have a negative singularity in fluid pressure at the tip of order $x^{-1/3}$ and a fracture width that grows as $x^{2/3}$. Whatever the far-field (closure) stress σ_0, there will be some point x_{lag} at which the fluid pressure falls below the vapour pressure (in the case of an effectively impermeable rock) of the fracturing fluid and so the limit of fluid penetration will lag the tip of the fracture; this effectively resolves the singularity in the intermediate asymptotic approximation for stress. A still smaller inner region will be governed by the effects of fracture toughness. Further details are given in a later paper by Detournay.

Commercial fracturing fluids are often best described by power-law viscous behaviour; for each power-law index there corresponds a specific power describing the pressure singularity at the tip. A corresponding theory has been given for porous materials subject to fluid leak-off (Lenoach 1995).

5. Contact Lines

When two fluids, or a fluid and a solid, are in contact, significant interfacial energies are involved. In continuum mechanics, these are asociated with an infinitely thin interfacial layer; the equivalent finite force/unit length, best exemplified by the interfacial tension between two liquids, can be regarded as introducing a singular stress at the interface. This tension is a constitutive function characteristic of the two fluids.

When the interface between two fluids meets a solid surface, a three-phase line is present; it is clear from the previous paragraph that a concentrated force is exerted on the solid along this contact line. This causes no difficulty in analysis of the static situation, except that a further constitutive parameter, the contact angle, has to be specified; this provides a unique local two-dimensional solution for the interface geometry near the contact line, which can be matched to a global solution.

If however the contact line is assumed to move[2], then the traditional specification of the fluid-mechanical problem (Navier-Stokes equations in both bulk fluids, no-slip of either fluid at the solid surface, continuity of stress across the fluid/fluid interface) leads to a "nasty" singularity at the

---

[2]E.g., by withdrawing a vertical plate through the interface of two otherwise static liquids (of different density) lying one above the other.

contact line: the velocity gradient tensor has an r^{-1} singularity and so the integrated local interfacial force is also singular. Clearly this has to be resolved.

One possibility is to assume that the contact line cannot move ; however this is not consistent with experiment, and so some local alteration to the constitutive description of the flowing fluids near the contact line has to be made. The commonest is to relax the no-slip condition, e.g. by making the wall slip a very weak linear function of the wall shear stress, a not wholly unphysical proposal. (This is adequate for resolving the associated Taylor scraper problem.)

This leaves a further difficulty: the contact angle is observed to vary with contact-line speed over the solid surface; any singular fluid pressure behaviour at the contact line will have singular effects on the interface shape there, assuming constant interfacial tension, and so the whole meaning of a contact angle has to be carefully addressed. At a small enough length scale, molecular effects become important, and so the question arises of whether the very local molecular effects have global effects or not. It is tempting to retain the bulk constitutive description for both fluids and to embed all molecular effects into constitutive complexity for the boundary/interfacial conditions. Shikhmurzaev (1997) has offered a thermodynamically-based theory which allows of a velocity discontinuity and an interfacial stress, at both the fluid/fluid and fluid/solid interfaces, dependent on a local interfacial density obeying a surface equation of state containing a relaxation time. This subsumes the Young-Laplace equation defining static contact angle and provides a unique definition for the dynamic case, as well as resolving the stress singularity.

Even when the local contact-line problem has been resolved, there are very interesting questions that arise when global problems are addressed. The magnitudes of the "outer" dimensionless groups, capillary number, Reynolds number (or Weber number), and any formed from the parametric values associated with local contact-line dynamics, determine what asymptotic schemes are appropriate.

The issues are discussed in later papers by Davis, Billingham, Shikhmuzaev and Chang. The whole problem cannot be said to have been resolved: it presents an opportunity for experimenters and modellers alike.

6. Discussion

There are other forms of non-linear singularity that have not been mentioned above. Shocks in wave propagation provide a rich source of examples; they are not addressed here at all because they need and deserve a full volume on their own. Localisation of deformation in granular and other complex media is another well-known example, which is only just touched upon in this volume (Ayzenberg & Slepyan).

The basic linear continuum rheological models introduced above (Hookean solid and Newtonian liquid) together with their time-dependent viscoelastic analogues and non-linear elastoplastic or elastoviscoplastic extensions are all deterministic. The materials are usually taken to be homogeneous and isotropic; most of the physically significant singularities discussed in this volume are associated with interfaces between different materials. A different approach to real materials is provided by the introduction of random discontinuities (microcracks, small inclusions of separate material) within their bulk. Analysis of behaviour shifts to mesoscale mechanics and/or statistical approaches (Dyskin, Tirosh and Puyesky & Frankl in this volume), which underlie larger-scale single-phase continuum models.

Further interesting problems arise when computational techniques involving spatial and temporal discretisation are used. The question then arises as to whether a particular physical length or time scale lies above or below the discrete value chosen for numerical computation. (Ponthot & Belytschko address dynamic crack propagation numerically later in this volume.)

To conclude, it is worth repeating that large differences in length scale are often at the root of apparently singular behaviour in real materials; when bulk behaviour is being analysed, many of the smallest-scale effects have to be incorporated into boundary or discontinuity conditions, which are themselves constitutive in character (Vardoulakis & Exadaktylos provide such an example in this volume); only asymptotic or exceedingly refined computational meshes can substitute for this in resolving the singularity.

References

Alden, H.L. (1948) Second approximation to the laminar boundary layer flow over a flat plate, *J.Math.Phys.* **27**, 91-104.

Batchelor, G.K. (1967) *An Introduction to Fluid Mechanics,* Cambridge University Press.

Ben Naceur, K. (1989) Ch.3 in M.J. Economides & K.G. Nolte (eds.), *Reservoir Stimulation,* Prentice Hall, Eaglewood Cliffs, N.J.

SCR Geomechanics Group (1994) The crack tip region in hydrodynamic fracturing, *Proc.Roy.Soc.A* **447**, 39-48.

Kaplun, S. (1954) The rôle of co-ordinate systems in boundary-layer theory, *Z.angew.Math.Phys.* **5**, 111-135.

Happel, J. & Brenner, H. (1973) *Low Reynolds Number Hydrodynamics,* Noordhoff, Leyden.

Lamb, Sir H. (1953) *Hydrodynamics,* Cambridge University Press.

Lenoach, B. (1995) The crack tip solution for hydraulic fracturing in a permeable solid, *J.Mech.Phys.Solids* **43**, 1025-43.

Moffatt, H.K. (1964) Viscous and resistive eddies near a sharp corner, *J.Fluid Mech.* **18**, 1.

Muskhelishvili, N.I. (1953) *Some Basic Problems of the Mathematical Theory of Elesticity,* Noordhoff, Groningen.
Oseen, C.W. (1910) Uber die Stokes'sche Formel, und übereine verwandte Aufgabe in der Hydrodynamik, *Ark.Math.Astronom.Fys.* **6**, no.29.
Saffman, P.G. (1992) *Vortex Dynamics,* Cambridge University Press.
Shikhmurzaev, Yu.D. (1997) Moving contact lines in liquid/liquid/solid systems, *J.Fluid Mech.* **334**, 211-249.
Rosenhead, L. (1963) (ed.) *Laminar Boundary Layers,* Clarendon Press, Oxford.
Van Dyke, M. (1964) *Perturbation Methods in Fluid Mechanics,* Academic Press, New York.

HIGH WEISSENBERG NUMBER ASYMPTOTICS AND CORNER SINGULARITIES IN VISCOELASTIC FLOWS

M. RENARDY
Department of Mathematics
Virginia Tech
Blacksburg, VA 24061-0123, USA

Abstract

Flows of viscoelastic fluids at high Weissenberg number exhibit sharp stress boundary layers and complicated singularities at corners of the domain. The paper reviews recent progress in understanding the analytical nature of high Weissenberg number flows and associated singularities.

1. Introduction

Flows of viscoelastic fluids at high Weissenberg number often exhibit singular behavior which is hard to resolve numerically. These singularities are most pronounced if the upper convected Maxwell model is used as a constitutive theory, and consequently this model has turned out to be the most difficult to compute with, even though it is in a sense the "simplest" model of viscoelastic fluids. Corner singularities, such as the reentrant corners in contraction flows, are an example where the Weissenberg number is effectively infinite, since the stresses and velocity gradients at the corner have a singularity. Even in smooth geometries, sharp layers form along boundaries and separating streamlines. Examples include flow between eccentric cylinders [3] and flow past a sphere [4].

Problems such as these clearly suggest the development of asymptotic approaches to high Weissenberg number flows in order to understand the mathematical nature of singularities and boundary layers. These insights can then be used to guide the further development of numerical approaches. While the asymptotics of high Reynolds number flows is well-developed, the systematic study of high Weissenberg number flows has begun only in the last few years [1-2], [5-11]. Obviously, the high Weissenberg number behavior of viscoelastic fluids is highly dependent on the constitutive model chosen. The upper convected Maxwell (UCM) model offers a convenient starting point because of its mathematical simplicity and also because it is the model for which singular behavior at high Weissenberg number appears to be more pronounced than for other popular models. On the other hand, the UCM model is generally not a realistic description of the behavior of real fluids

13

D. Durban and J.R.A. Pearson (eds.),
IUTAM Symposium on Non-Linear Singularities in Deformation and Flow, 13-20.
© 1999 *Kluwer Academic Publishers. Printed in the Netherlands.*

at high Weissenberg numbers, and it is therefore important to investigate how the ideas can be extended to other models.

In this paper, we review progress which has been made in this field. Section 2 considers the "naive" limit of high Weissenberg number, which is the analogue of the Euler equations as a high Reynolds number limit of the Navier-Stokes equations. Rather curiously, there is actually a connection between the high Weissenberg number limit and the Euler equations. In Section 3, we look at stress boundary layers, and Section 4 investigates reentrant corner singularities.

2. The high Weissenberg number limit of the UCM model and the Euler equations

We consider creeping flows of the upper convected Maxwell fluid. The equations are, in dimensionless form:

$$\operatorname{div} \mathbf{T} - \nabla p = \mathbf{0},$$

$$\operatorname{div} \mathbf{v} = 0,$$

$$W[\frac{\partial \mathbf{T}}{\partial t} + (\mathbf{v} \cdot \nabla)\mathbf{T} - (\nabla \mathbf{v})\mathbf{T} - \mathbf{T}(\nabla \mathbf{v})^T] + \mathbf{T} = \nabla \mathbf{v} + (\nabla \mathbf{v})^T. \tag{1}$$

Here $\mathbf{v}$ denotes the velocity, p the pressure and $\mathbf{T}$ the extra stress tensor.

The "naive" limit of infinite Weissenberg number simply consists in setting W equal to infinity in (1). This reduces the last equation to

$$\frac{\partial \mathbf{T}}{\partial t} + (\mathbf{v} \cdot \nabla)\mathbf{T} - (\nabla \mathbf{v})\mathbf{T} - \mathbf{T}(\nabla \mathbf{v})^T = \mathbf{0}. \tag{2}$$

We now consider this equation in conjunction with the first two equations of (1). To make further progress, we assume that $\mathbf{T}$ is a rank one tensor; it can be shown [10] that at leading order this is indeed the case in two-dimensional flows at high Weissenberg number. We thus set

$$\mathbf{T} = \rho \mathbf{u}\mathbf{u}^T, \tag{3}$$

where ρ is a scalar and $\mathbf{u}$ is a vector. We normalize ρ and $\mathbf{u}$ in such a way that the relation

$$\operatorname{div}(\rho \mathbf{u}) = 0. \tag{4}$$

Inserting (3) into the first equation of (1) and utilizing (4), we find that

$$\rho(\mathbf{u} \cdot \nabla)\mathbf{u} = \nabla p. \tag{5}$$

We note that (4) and (5) are the steady Euler equations.

The Euler equations need to be complemented by an appropriate equation of state. For this, we need to use the constitutive equation (2). Inserting (3) into (2), we obtain

$$-(\frac{\partial \rho}{\partial t} + (\mathbf{v} \cdot \nabla \rho))\mathbf{u}\mathbf{u}^T + \mathbf{y}\mathbf{u}^T + \mathbf{u}\mathbf{y}^T, \tag{6}$$

where

$$\mathbf{y} = \frac{\partial}{\partial t}(\rho\mathbf{u}) + (\mathbf{v} \cdot \nabla)(\rho\mathbf{u}) - (\rho\mathbf{u} \cdot \nabla)\mathbf{v}$$

$$= \frac{\partial}{\partial t}(\rho\mathbf{u}) - \nabla \times (\mathbf{v} \times (\rho\mathbf{u})). \tag{7}$$

Obviously, we satisfy this equation if we set

$$\frac{\partial \rho}{\partial t} + (\mathbf{v} \cdot \nabla)\rho = 0, \tag{8}$$

and

$$\mathbf{y} = \mathbf{0}. \tag{9}$$

Since $\operatorname{div}(\rho\mathbf{u}) = 0$, we can write $\rho\mathbf{u} = \nabla \times \mathbf{a}$, and (9) can be satisfied by setting

$$\mathbf{v} \times (\rho\mathbf{u}) = \frac{\partial \mathbf{a}}{\partial t} + \nabla\psi, \tag{10}$$

where the choice of ψ must be such that $(\partial\mathbf{a}/\partial t) + \nabla\psi$ is perpendicular to $\mathbf{u}$. We can now regard (8) and (10) as equations from which $\mathbf{v}$ can be determined if ρ and $\mathbf{u}$ are known. The solution is unique as long as $(\mathbf{u} \cdot \nabla)\rho \neq 0$. In general, however, the $\mathbf{v}$ obtained in this fashion will not satisfy the divergence condition, i.e. the second equation of (1). This condition therefore imposes a consistency requirement on ρ and $\mathbf{u}$. After some calculation (see [10]) one finds that the condition $\operatorname{div}\mathbf{v} = 0$ is equivalent to

$$[\frac{\partial}{\partial t} + (\mathbf{v} \cdot \nabla)](\rho^2 \operatorname{div}\mathbf{u}) = 0, \tag{11}$$

and in view of (8), we find that

$$[\frac{\partial}{\partial t} + (\mathbf{v} \cdot \nabla)]\operatorname{div}\mathbf{u} = 0. \tag{12}$$

This is satisfied in particular if $\operatorname{div}\mathbf{u}$ is a function of ρ.

We can therefore find solutions for the infinite Weissenberg number limit if we solve the steady Euler equations

$$\rho(\mathbf{u} \cdot \nabla)\mathbf{u} = \nabla p, \quad \operatorname{div}(\rho\mathbf{u}) = 0, \tag{13}$$

with the "equation of state"

$$\operatorname{div}\mathbf{u} = \phi(\rho), \tag{14}$$

where ϕ is an arbitrary function.

In particular, the incompressible Euler equations ($\phi(\rho) = 0$) arise as a special case. We can therefore apply known solutions of these equations, especially potential flow solutions, to high Weissenberg number flows. More generally, if $\phi(\rho) = A\rho^{-1/2}$, then we can set $\rho^{1/2}\mathbf{u} = \nabla q$ and we obtain the equation $\Delta q = A/2$.

Just like for high Reynolds number flows of Newtonian fluids, the Euler equations do not suffice to give a description of fluid behavior. One of the main reasons for this is that solutions of the Euler equations are highly nonunique; even more so in the viscoelastic case because of the arbitrariness of the function ϕ in (14). The question thus arises which of the many possible solutions are actually relevant in a given flow situation. This question cannot be answered without including the terms of lower order which we neglected in (2). In [10], this is pursued to derive a necessary condition for flows with closed streamlines.

3. Stress boundary layers

The boundary layers observed in viscoelastic fluids are quite different from Newtonian boundary layers at high Reynolds number. In the Newtonian case, boundary layers develop because there are rapid variations in the fluid velocity near the boundary. In the viscoelastic case, on the other hand, the velocity profile may be quite smooth, but there are rapid variations in the stresses. Indeed, boundary layers of very much the same nature are found even if the velocity field is prescribed (and smooth), and the stress integration from the constitutive law is considered on its own.

We can understand these stress boundary layers from a physical point of view by considering the fact that viscoelastic fluids have memory. Let us consider a flow near a solid wall with a shear rate varying smoothly along the wall. A particle on the wall does not move because of the no-slip boundary condition. Hence such a particle will only "see" the shear rate at one point regardless how long the memory of the fluid is. On the other hand, if the shear rate is high, then particles even at a small distance to the wall will move a long way within one relaxation time, and hence their stress response is determined by a variable shear rate history. We note that the analysis in the previous section was based on the equation (2), where the term $(\mathbf{v} \cdot \nabla)\mathbf{T}$ was retained as contributing to the leading order balance. On a rigid wall, however, this term is forced to vanish, and the stresses are determined by the viscometric balance instead (i.e. by neglecting this term).

Near a wall, one therefore expects a very rapid variation of stresses in the direction perpendicular to the wall. This suggests a boundary layer analysis, which has been introduced in [8]. We shall consider the equations for steady creeping flow of an upper convected Maxwell fluid in two space dimensions, in dimensionless form, which we restate in component notation:

$$W\left[u\frac{\partial T_{11}}{\partial x} + v\frac{\partial T_{11}}{\partial y} - 2\frac{\partial u}{\partial x}T_{11} - 2\frac{\partial u}{\partial y}T_{12}\right] + T_{11} = 2\frac{\partial u}{\partial x},$$

$$W\left[u\frac{\partial T_{12}}{\partial x} + v\frac{\partial T_{12}}{\partial y} - \frac{\partial u}{\partial y}T_{22} - \frac{\partial v}{\partial x}T_{11}\right] + T_{12} = \frac{\partial u}{\partial y} + \frac{\partial v}{\partial x},$$

$$W\left[u\frac{\partial T_{22}}{\partial x} + v\frac{\partial T_{22}}{\partial y} - 2\frac{\partial v}{\partial x}T_{12} - 2\frac{\partial v}{\partial y}T_{22}\right] + T_{22} = 2\frac{\partial v}{\partial y},$$

$$\frac{\partial u}{\partial x} + \frac{\partial v}{\partial y} = 0,$$

$$\frac{\partial^2}{\partial x \partial y}(T_{11} - T_{22}) + (\frac{\partial^2}{\partial y^2} - \frac{\partial^2}{\partial x^2})T_{12} = 0. \tag{15}$$

We place the boundary at $y = 0$, i.e. the flow domain is the half-plane $y > 0$, and we are interested in solutions for small y and large W. We assume that the nondimensionalization is such that the shear rate $\partial u/\partial y$ is of order one and varies smoothly in the x-direction on a length scale of order one. On the wall, we have viscometric stresses, i.e. $T_{22} = 0$, T_{12} is of order one and T_{11} is of order W. We expect the term $(\mathbf{v} \cdot \nabla)\mathbf{T}$ to enter into the balance when it becomes of the same order of magnitude as the terms which contribute to the viscometric stresses, i.e. when, for instance, $u(\partial T_{11}/\partial x)$ is comparable to $T_{12}(\partial u/\partial y)$. Since T_{11} is of order W, this is the case when u is of order $1/W$, i.e. at a distance of order $1/W$ from the wall. This suggests the following scalings:

$$y = z/W, \ u = \tilde{u}(x,z)/W, \ v = \tilde{v}(x,z)/W^2$$

$$T_{11} = W\tilde{T}_{11}(x,z), \ T_{12} = \tilde{T}_{12}(x,z), \ T_{22} = \tilde{T}_{22}(x,z)/W. \tag{16}$$

We insert (16) into (15), keep only the leading order terms in W and suppress the tildes. This leads to the following set of boundary layer equations

$$u\frac{\partial T_{11}}{\partial x} + v\frac{\partial T_{11}}{\partial z} - 2\frac{\partial u}{\partial x}T_{11} - 2\frac{\partial u}{\partial z}T_{12} + T_{11} = 0,$$

$$u\frac{\partial T_{12}}{\partial x} + v\frac{\partial T_{12}}{\partial z} - \frac{\partial u}{\partial z}T_{22} - \frac{\partial v}{\partial x}T_{11} + T_{12} = \frac{\partial u}{\partial z},$$

$$u\frac{\partial T_{22}}{\partial x} + v\frac{\partial T_{22}}{\partial z} - 2\frac{\partial v}{\partial x}T_{12} - 2\frac{\partial v}{\partial z}T_{22} + T_{22} = 2\frac{\partial v}{\partial z},$$

$$\frac{\partial u}{\partial x} + \frac{\partial v}{\partial z} = 0,$$

$$\frac{\partial^2 T_{11}}{\partial x \partial z} + \frac{\partial^2 T_{12}}{\partial z^2} = 0. \tag{17}$$

In [8], it is shown how these equations can be put into a simpler form by introducing a suitable transformation of variables. This simplified form has the added advantage that is extends to the case where the boundary is curved.

The boundary layer equations are still a difficult system of nonlinear partial differential equations, even after simplification. This raises the question what can be said about solutions. Two particular cases have been addressed (see [8]): First, one can find similarity solutions depending on the similarity variable $\xi = y/x^\alpha$. Such similarity solutions play a role in the reentrant corner problem, discussed in the next section. Second, it is possible to consider the linearized equations governing small perturbations of parallel shear flow with a uniform shear rate. For these linearized equations, it is shown in [8] that it is possible to impose an arbitary periodic modulation of the shear rate and find a corresponding solution of the boundary layer equations. Remarkably, these linearized solutions have no boundary layer behavior in the velocities, but only in the stresses.

It is of interest to consider boundary layers for other constitutive models. This issue is taken up in [11], where the Phan-Thien Tanner (PTT) and Giesekus models are considered. They differ from the UCM model by adding an additional term to the constitutive equation, which is quadratic in the stresses. Specifically, the PTT model adds a term proportional to $(\operatorname{tr}\mathbf{T})\mathbf{T}$ to the left hand side of the constitutive relation (i.e. the third equation of (1)), while the Giesekus model adds a term proportional to $\mathbf{T}^2$. Despite the apparent similarity of the models, their viscometric functions are quite different, and this difference affects the boundary layer analysis. The result is that for the PTT model the thickness of the boundary layer is of order $W^{-1/3}$, while for the Giesekus model it is of order $W^{-1/2}$. For either model, the boundary layer therefore steepens much less rapidly with increasing Weissenberg number than it does for the UCM model. This observation agrees with the computational experience.

4. The reentrant corner singularity

Newtonian flows in domains with corners are well understood: The leading contribution to the singularity at the corner is determined entirely by the Stokes equation and can be obtained by separating variables for the biharmonic equation in polar coordinates. For corners between rigid walls, there is a crucial difference between the cases where the angle is less than 180 degrees and where it is greater (the "reentrant" case). If the angle is less, then the velocity gradient and hence the stress vanish at the corner, if the angle is greater, they are infinite.

In the viscoelastic case, the case of corners less than 180 degrees is simple. Since the velocity gradient and stress are zero at the corner, the fluid simply behaves Newtonian. The reentrant case, one the other hand, corresponds to the infinite Weissenberg number limit; the behavior is nonlinear and dependent on the constitutive model.

The separation of variables solution for the Newtonian case has the form $r^{\alpha} f(\theta)$, and it is tempting to try a similar ansatz for velocities and stresses in viscoelastic flows. Such an attempt, however, fails. The reason for this is the formation of boundary layers on the walls, which occur for the reasons pointed out in the last section. This feature already occurs in the simpler problem where the velocity field is considered given (e.g. the Newtonian velocity field) and only the stress integration is considered. This problem was first investigated in [5]. Besides the occurrence of boundary layers, the analysis in [5] showed that the integration of stresses along streamlines becomes unstable downstream from the corner. This leads to a catastrophic amplification of numerical errors, which can easily produce spurious stresses which are orders of magnitude larger than any of the true stresses at the same distance from the corner. This downstream instability is likely to play a major role in the breakdown of the numerical simulations. A strategy to avoid this amplification of errors was discussed in [6].

It is instructive to compare the UCM model with other models. In [9], the PTT model is considered. Not surprisingly, the stresses for the PTT model are less singular. For Newtonian kinematics, the velocity gradient near a 270 degree corner has a singularity proportional to $r^{-0.455}$. The stress integration for the UCM model yield stresses which behave like $r^{-0.74}$ away from the walls, compared to $r^{-0.33}$ for the PTT model. Another significant difference is in the boundary layers near the walls, which, as discussed in the previous section, are much sharper for the UCM model than for the PTT model. One of the consequences of this is that, for the PTT model, stress relaxation becomes important before the downstream instability becomes damaging, and indeed no spurious stress growth was found in [9]. Interestingly, however, the downstream instability seems to lead to a real stress maximum downstream from the corner for the PTT fluid; this was not found for the UCM model.

The full equations for the UCM model near a reentrant corner were considered in [1],[2], and [7]. Davies and Devlin [1] derive formal series solutions which are valid away from the walls; however, it remains to be resolved how their series solutions might be matched to physically appropriate behavior at the wall. Hinch [2] derives a solution, which, in the context of Section 2 above, can be interpreted as a potential flow solution of the Euler equation. With the notations as in Section 2, we set $\mathbf{u} = \mathbf{v}$, $\phi(\rho) = 0$. We take $\rho^{1/2}\mathbf{u} = \nabla\phi$, where $\phi = r^{2/3}\cos(2\theta/3)$ is the potential for a 270 degree corner, with associated streamfunction $\psi = r^{2/3}\sin(2\theta/3)$. We satisfy all the equations in Section 2 by choosing ρ as an arbitrary function $\rho = g(\psi)$. The stress singularity is proportional to $\rho\mathbf{u}\mathbf{u}^T$, i.e. to $r^{-2/3}$. To determine the function $g(\psi)$, Hinch argues that there must be a transition region near the wall, where the stresses change over to viscometric behavior. This transition must occur when the stretch rate is of order one. On this basis, it is found that $g(\psi) = \psi^{-8/3}$. The true streamfunction is then $\tilde{\psi} = \frac{3}{7}\psi^{7/3} = \frac{3}{7}r^{14/9}\sin^{7/3}(2\theta/3)$. This predicts a velocity gradient proportional to $r^{-4/9}$.

The Hinch solution breaks down near the walls; indeed it yields a zero velocity gradient at the wall, which is not consistent with expected behavior. To repair this deficiency, a matched asymptotic solution is required, in which Hinch's solution serves as the "outer" solution. The inner solution near the wall can be obtained as a similarity solution of the boundary layer equations discussed in the previous section [7]. Such similarity solutions for the upstream wall were found numerically in [7]; the downstream problem remains to be solved.

5. References

1. A.R. Davies and J. Devlin, On corner flows of Oldroyd-B fluids, *J. Non-Newt. Fluid Mech.* **50** (1993), 173-191.

2. E.J. Hinch, The flow of an Oldroyd fluid around a sharp corner, *J. Non-Newt. Fluid Mech.* **50** (1993), 161-171.

3. R.C. King, M.R. Apelian, R.C. Armstrong and R.A. Brown, Numerically stable finite element techniques for viscoelastic calculations in smooth and singular geometries, *J. Non-Newt. Fluid Mech.* **29** (1988), 147-216.

4. W.J. Lunsmann, L. Genieser, R.C. Armstrong and R.A. Brown, Finite element analysis of steady viscoelastic flow around a sphere in a tube: calculations with constant viscosity models, *J. Non-Newt. Fluid Mech.* **48** (1993), 63-99.

5. M. Renardy, The stresses of an upper convected Maxwell fluid in a Newtonian velocity field near a reentrant corner, *J. Non-Newt. Fluid Mech.* **50** (1993), 127-134.

6. M. Renardy, How to integrate the upper convected Maxwell (UCM) stresses near a reentrant corner (and maybe elsewhere, too), *J. Non-Newt. Fluid Mech.* **52** (1994), 91-95.

7. M. Renardy, A matched solution for corner flow of the upper convected Maxwell fluid, *J. Non-Newt. Fluid Mech.* **58** (1995), 83-89.

8. M. Renardy, High Weissenberg number boundary layers for the upper convected Maxwell fluid, *J. Non-Newt. Fluid Mech.* **68** (1997), 125-132.

9. M. Renardy, Reentrant corner behavior of the PTT fluid, *J. Non-Newt. Fluid Mech.* **69** (1997), 99-104.

10. M. Renardy, The high Weissenberg number limit of the UCM model and the Euler equations, *J. Non-Newt. Fluid Mech.*, to appear.

11. M. Renardy and T. Hagen, Boundary layer analysis of the Phan-Thien Tanner and Giesekus model in high Weissenberg number flow, *J. Non-Newt. Fluid Mech.*, to appear.

CORNER SINGULARITIES IN THREE-DIMENSIONAL STOKES FLOW

H.K. MOFFATT and V. MAK

Department of Applied Mathematics and Theoretical Physics,
Silver Street, Cambridge CB3 9EW, UK.

Stokes flow in a corner region driven by a weakly three-dimensional stirring mechanism at some distance from the corner is considered. It is shown that, when the stirring is antisymmetric about the bisecting plane $\theta = 0$, the flow near the corner exhibits the same type of eddy structure as is familiar from the two-dimensional theory. When the stirring is symmetric about $\theta = 0$ however, a non-oscillatory component is in general present (driven by conditions far from the corner), and this component dominates over the oscillatory component near $r = 0$.

1. Introduction

It is well known (Moffatt 1964a) that if an incompressible viscous fluid, contained between two plane rigid boundaries $\theta = \pm\alpha$ with $\alpha < \alpha_c \approx 73.5°$, is subjected to an arbitrary two-dimensional stirring at some distance from the corner, then under the Stokes approximation the flow near the corner exhibits an infinite sequence of eddies of alternating sense of rotation. These eddies are geometrically and dynamically self-similar and get rapidly weaker as the corner is approached.

It is natural to enquire whether these eddies survive when the stirring mechanism is no longer two-dimensional. For simplicity, we shall suppose that this stirring mechanism is only weakly dependent on the coordinate z parallel to the intersection of the two planes. In the following sections, we consider two specific stirring mechanisms, the first generating a flow that is antisymmetric about the bisecting plane $\theta = 0$, the second a flow that is symmetric. The conclusions concerning the formation of eddies are very different in these two cases.

2. General formulation

We shall use cylindrical polar coordinates (r, θ, z) where r represents the distance from the intersection of the planes $\theta = \pm\alpha$. The velocity field $u = (u, v, w)$ admits the general 'toroidal/poloidal' decomposition

$$u = \nabla \wedge (e_z \psi) + \nabla \wedge \nabla \wedge (e_z \chi), \tag{2.1}$$

where $\psi(r, \theta, z)$, $\chi(r, \theta, z)$ are scalar fields; thus

$$u = r^{-1}\psi_\theta + \chi_{rz}, \quad v = -\psi_r + r^{-1}\chi_{\theta z}, \quad w = -\nabla_2^2 \chi, \tag{2.2}$$

where suffices indicate differentiation, and

$$\nabla_2^2 = \frac{1}{r}\frac{\partial}{\partial r} r \frac{\partial}{\partial r} + \frac{1}{r^2}\frac{\partial^2}{\partial \theta^2} = \nabla^2 - \frac{\partial^2}{\partial z^2} \tag{2.3}$$

21

D. Durban and J.R.A. Pearson (eds.),
IUTAM Symposium on Non-Linear Singularities in Deformation and Flow, 21-26.
© 1999 *Kluwer Academic Publishers. Printed in the Netherlands.*

is the two-dimensional Laplacian operator. Note that in the two-dimensional limit $(\partial/\partial z \to 0)$, we have $u = r^{-1}\psi_\theta$, $v = -\psi_r$, i.e. $\psi(r,\theta)$ is then the stream-function of the flow.

The vorticity field $\boldsymbol{\omega} = \nabla \wedge \boldsymbol{u}$ is given from (2.1) by

$$\boldsymbol{\omega} = \nabla \wedge \nabla \wedge (\boldsymbol{e}_z\psi) - \nabla \wedge (\boldsymbol{e}_z\nabla^2\chi) \,, \tag{2.4}$$

and it follows that

$$\nabla^2\boldsymbol{\omega} = \nabla \wedge \nabla \wedge (\boldsymbol{e}_z\nabla^2\psi) - \nabla \wedge (\boldsymbol{e}_z\nabla^4\chi)$$
$$= -\nabla_2^2\nabla^2\psi\boldsymbol{e}_z + \nabla_2(\nabla^2\psi)_z + (\boldsymbol{e}_z \wedge \nabla_2)\nabla^4\chi \,. \tag{2.5}$$

The Stokes equation in the form $\nabla^2\boldsymbol{\omega} = 0$ is thus satisfied if and only if

$$\nabla_2^2\nabla^2\psi = 0 \,, \tag{2.6}$$

and

$$\nabla_2\frac{\partial}{\partial z}(\nabla^2\psi) + (\boldsymbol{e}_z \wedge \nabla_2)\nabla^4\chi = 0 \,. \tag{2.7}$$

Thus $\nabla^2\psi$ satisfies the two-dimensional Laplace equation; and (2.7) (the Cauchy-Riemann equations) expresses the fact that

$$\frac{\partial}{\partial z}(\nabla^2\psi) + i\nabla^4\chi \tag{2.8}$$

is an analytic function of $x + iy = re^{i\theta}$.

Under the assumption that all fields are weakly varying in the z-direction, we have

$$\frac{\partial^2}{\partial z^2} \ll \frac{1}{r}\frac{\partial}{\partial r}r\frac{\partial}{\partial r} + \frac{1}{r^2}\frac{\partial^2}{\partial\theta^2} \tag{2.9}$$

provided $r = O(1)$; hence it is legitimate to replace ∇^2 by ∇_2^2 in the above equations, an approximation that clearly improves as $r \to 0$. In particular, asymptotically,

$$\nabla_2^4\psi = 0 \tag{2.10}$$

and (using (2.2))

$$\frac{\partial}{\partial z}(\nabla_2^2\psi) - i\nabla_2^2 w \tag{2.11}$$

is an analytic function of $re^{i\theta}$.

3.　Flow antisymmetric about $\theta = 0$

To be specific, let us suppose that the fluid is confined to the domain

$$|\theta| < \alpha \,, \; r < R \,, \tag{3.1}$$

and that the stirring is provided at the boundaries $\theta = \pm\alpha$ in such a way that

$$\boldsymbol{u} = (\pm U(r,z),0,0) \qquad \text{on} \quad \theta = \pm\alpha \tag{3.2}$$

and $\boldsymbol{u} = 0$ on $r = R$. In order to generate weakly three-dimensional flow, we suppose that the imposed boundary velocity U satisfies

$$|\partial U/\partial z| \ll |\partial U/\partial r| \,. \tag{3.3}$$

We suppose moreover that U is nonzero only in a bounded region

$$r_0 < r < r_1 \quad , \quad |z| < z_0 \tag{3.4}$$

where $r_0 > 0$ and $z_0 \gg R$ (consistent with (3.3)).

In terms of ψ and χ, the boundary conditions (3.2) on $\theta = \pm\alpha$ become

$$
\begin{aligned}
r^{-1}\psi_\theta + \chi_{rz} &= \pm U(r,z)\,, \\
-\psi_r + r^{-1}\chi_{\theta z} &= 0\,, \\
-\nabla_2^2\chi &= 0\,.
\end{aligned}
\tag{3.5}
$$

Under the assumption of weak z-dependence, the z-derivative terms in (3.5a,b) are small, and at leading order, these boundary conditions are simply

$$r^{-1}\psi_\theta = \pm U(r,z)\,, \ \psi_r = 0 \text{ on } \theta = \pm\alpha\,. \tag{3.6}$$

We have also

$$r^{-1}\psi_\theta = \psi_r = 0 \text{ on } r = R\,. \tag{3.7}$$

We must first solve the problem (from (2.9))

$$\nabla_2^4\psi = 0\,, \tag{3.8}$$

subject to the boundary conditions (3.6), (3.7). There is a unique solution (up to an arbitrary additive function of z which does not contribute to the velocity field), $\Psi(r,\theta,z)$ say. We are particularly interested in the asymptotic behaviour of Ψ near $r = 0$. From earlier two-dimensional studies (Moffatt 1964a,b) it is known that (under the assumed conditions on $U(r,z)$) this is given by

$$\Psi(r,\theta,z) \sim \mathrm{Re}A(z)r^\lambda \left[\frac{\cos\lambda\theta}{\cos\lambda\alpha} - \frac{\cos(\lambda-2)\theta}{\cos(\lambda-2)\alpha} \right]\,, \tag{3.9}$$

where λ is the (complex) root of the transcendental equation

$$\sin 2(\lambda-1)\alpha + 2(\lambda-1)\sin 2\alpha = 0 \tag{3.10}$$

having smallest real part satisfying $\mathrm{Re}(\lambda) > 1$. The (complex)coefficient $A(z)$ is determined in principle (at each z) by the boundary function $U(r,z)$.

Now we have to find $w(r,\theta,z)$ at leading order. To do this, note that $\nabla_2^2\Psi$ is a harmonic function, and so can be expressed in the form

$$\nabla_2^2\Psi = \mathrm{Re}\mathcal{F}\left(re^{i\theta},z\right) \tag{3.11}$$

for some analytic function $\mathcal{F}$ of the complex variable $re^{i\theta}$. We may note from (3.9) that

$$\mathcal{F}\left(re^{i\theta},z\right) \sim -2(\lambda-2)A(z)\left(re^{i\theta}\right)^{\lambda-2}\sec(\lambda-2)\alpha \tag{3.12}$$

as $r \to 0$. From (2.11), it follows that

$$\nabla_2^2 w = -\mathrm{Im}\ \partial\mathcal{F}\left(re^{i\theta},z\right)/\partial z = H(r,\theta,z)\,,\text{say}\,, \tag{3.13}$$

and that, taking account of the antisymmetry in θ,

$$\nabla_2^2 w \sim 2\mathrm{Im}\left[(\lambda-2)A'(z)r^{\lambda-2}\sin(\lambda-2)\theta\sec(\lambda-2)\alpha\right] \tag{3.14}$$

as $r \to 0$.

The equation

$$\nabla_2^2 w = H(r,\theta,z) \tag{3.15}$$

must now be solved, subject to the boundary condition

$$w = 0 \text{ on } \theta = \pm\alpha \text{ and on } r = R. \tag{3.16}$$

There is obviously a particular integral $w^{(P)}$ which, from (3.14) and symmetry considerations, has the behaviour

$$w^{(P)} \sim \operatorname{Im} \frac{A'(z)r^\lambda}{\cos(\lambda - 2)\alpha} \left[\frac{\sin \lambda\theta}{\sin \lambda\alpha} - \frac{\sin(\lambda - 2)\theta}{\sin(\lambda - 2)\alpha} \right] \tag{3.17}$$

as $r \to 0$. There is also however a complementary function $w^{(C)}$ satisfying $\nabla_2^2 w^{(C)} = 0$ which is inevitably present in order that the combined solution

$$w = w^{(C)} + w^{(P)} \tag{3.18}$$

satisfy the condition $w = 0$ on $r = R$ (cf. Moffatt & Duffy 1980, where this behaviour is analysed in detail). The asymptotic form of $w^{(C)}$ near $r = 0$ is

$$w^{(C)} \sim B(z)r^\nu \sin \nu\theta, \tag{3.19}$$

where

$$\nu\alpha = \pi, \tag{3.20}$$

and $B(z)$ is a (real, slowly varying) function of z, again determined in principle by $U(r, z)$.

Since $\partial w^{(C)}/\partial z \neq 0$, there is an associated flow $\left(u^{(C)}, v^{(C)}\right)$ in each plane $z = \text{cst}$ driven by the continuity equation

$$\frac{1}{r}\frac{\partial}{\partial r}\left(ru^{(C)}\right) + \frac{1}{r}\frac{\partial v^{(C)}}{\partial \theta} = -\frac{\partial w^{(C)}}{\partial z} = -B'(z)r^\nu \sin \nu\theta. \tag{3.21}$$

The solution satisfying the no-slip condition on $\theta = \pm\alpha$ is

$$u^{(C)} = -\frac{B'(z)}{\nu + 1}r^{\nu+1} \sin \nu\theta, \quad v^{(C)} = 0. \tag{3.22}$$

Thus the flow $\left(u^{(C)}, 0, w^{(C)}\right)$ has streamlines on planes $\theta = \text{cst}$.

We now see that there are in general two contributions to the velocity components (u, v) in the neighbourhood of $r = 0$, the first proportional to $r^{\lambda-1}$ being oscillatory (since λ is complex) and the second, arising from the weak z-dependence, proportional to $r^{\nu+1}$ being non-oscillatory (since ν is real). The oscillatory component dominates provided

$$\operatorname{Re}(\lambda - 1) < \nu + 1 = \frac{\pi}{\alpha} + 1, \tag{3.23}$$

or equivalently, provided

$$\xi_1 = 2\alpha\operatorname{Re}(\lambda - 1) < 2\pi + 2\alpha. \tag{3.24}$$

The value of ξ_1 was calculated by Moffatt (1964a) and is less than 4.51 for all values of $\alpha < \alpha_c$ for which λ is complex; hence, for $\alpha < \alpha_c$, the oscillatory term dominates, and the projection (u, v) of the flow on every plane $z = \text{cst}$ ($|z| < z_0$) exhibits the familiar corner eddy structure.

4. Flow symmetric about $\theta = 0$

The treatment here is very similar, but the conclusion is dramatically different. We now suppose that the boundary condition (3.2) is replaced by

$$\boldsymbol{u} = (U(r, z), 0, 0) \qquad \text{on} \qquad \theta = \pm\alpha, \tag{4.1}$$

all other conditions remaining as before. The streamfunction $\Psi(r, \theta, z)$ is now an odd function of θ, and for $r \to 0$ has the form

$$\Psi(r, \theta, z) \sim \mathrm{Re}\, C(z) r^{\lambda} \left[\frac{\sin \lambda \theta}{\sin \lambda \alpha} - \frac{\sin(\lambda - 2)\theta}{\sin(\lambda - 2)\alpha} \right], \qquad (4.2)$$

where λ is now the (complex) root of

$$\sin 2(\lambda - 1)\alpha - 2(\lambda - 1)\sin 2\alpha = 0 \qquad (4.3)$$

having smallest real part with $\mathrm{Re}(\lambda) > 1$. Note the change of sign between (3.10) and (4.3).

The complementary function corresponding to (3.19) is now

$$w^{(C)} \sim D(z) r^{\nu} \cos \nu \theta, \qquad (4.4)$$

where now

$$\nu \alpha = \pi/2. \qquad (4.5)$$

Again, we have two contributions to the flow components (u, v) near $r = 0$, proportional to $r^{\lambda - 1}$, $r^{\nu + 1}$ respectively. We find now however that

$$\xi_2 = 2\alpha \mathrm{Re}(\lambda - 1) > 2\alpha(\nu + 1) = \pi + 2\alpha \qquad (4.6)$$

for all $\alpha \lesssim 78°$ (actually $\xi_2 \gtrsim 7.50$, for symmetric flow – Moffatt 1964a). Hence, it is now the *non-oscillatory* component proportional to $r^{\nu + 1}$ that dominates near $r = 0$, and there are therefore at most a finite number of eddies near the corner in the symmetric case. Of course the non-oscillatory component $u^{(C)}$ is proportional to $D'(z)$ (cf 3.22) and becomes weaker if the three-dimensionality is reduced (so that $D'(z) \to 0$).

5. Conclusions

Stokes flow in a neighbourhood of a sharp corner may, by linearity, be represented as the sum of ingredients that are symmetric and antisymmetric about the bisecting plane $\theta = 0$. We have supposed that the stirring mechanism is located at some distance from the corner, and that it is weakly dependent on the coordinate z parallel to the corner. We have shown that the antisymmetric ingredient of the flow then exhibits the eddy structure that is familiar from the two-dimensional situation. The symmetric ingredient however is dominated near $r = 0$ by a non-oscillatory contribution $\left(u^{(C)}, 0, w^{(C)}\right)$ in planes $\theta = \mathrm{cst}$ which obliterates the corner eddies.

These conclusions are compatible with earlier results of Sano & Hasimoto (1980) who considered the flow associated with a Stokeslet imbedded in the corner region; they are also compatible with the results of a parallel study (Mak & Moffatt 1997) of the three-dimensional flow between two parallel planes $z = \pm z_0$ under general localised boundary forcing – again in that case, flow antisymmetric about $z = 0$ exhibits eddies outside the region of forcing, whereas for flow symmetric about $z = 0$ a non-oscillatory contribution dominates the far-field.

6. Acknowledgement

We are grateful to Osamu Sano for constructive comments and to Anthony Pearson for his careful reading of the manuscript.

REFERENCES

MAK, V. & MOFFATT, H. K. 1997 Three-dimensional Stokes flow between two parallel planes and near a sharp corner. *J. Fluid Mech.* (to appear).

MOFFATT, H. K. 1964a Viscous and resistive eddies near a sharp corner. *J. Fluid Mech.* **18**, 1–18.

MOFFATT, H. K. 1964b Viscous eddies near a sharp corner. *Archiwum Mechaniki Stosowanej.* **16**, 365–372.

MOFFATT, H.K. & DUFFY B.R. 1980 Local similarity solutions and their limitations. *J. Fluid Mech.* **96**, 299–313.

SANO, O. & HASIMOTO, H. 1980 Three-dimensional Moffatt-type eddies due to a Stokeslet in a corner. *J. Phys. Soc. Japan.* **48**, 1763–1768.

FLUID AND SOLID SINGULARITIES
AT THE TIP OF A FLUID-DRIVEN FRACTURE

E. DETOURNAY
Department of Civil Engineering
University of Minnesota
Minneapolis, MN 55455, USA

1. Introduction

Mathematical modeling of a fluid-driven fracture has attracted numerous contributions since the 1950's, see e.g. Khristianovic and Zheltov (1955), Barenblatt (1962), Perkins and Kern (1961), Nordgren (1972), Geerstma and Haafkens (1979) for some early contributions. These studies have been mainly motivated by hydraulic fracturing (a technique used widely in the oil and gas industry to enhance the recovery of hydrocarbons from underground reservoirs), although other applications exist such as magma-driven fracture (Spence and Turcotte, 1985).

Hydraulic fracturing consists of injecting a fluid into a wellbore to initiate and propagate a fracture in the direction perpendicular to the *in situ* minimum compressive stress. The theoretical research in this area is aimed primarily at establishing models for predicting the evolution of the pressure and the size of the fracture given the injection rate. These models require consideration of both fluid and solid mechanics: on the one hand, the lubrication equation to characterize the flow of fluid in the fracture; on the other hand, the elasticity equations to describe the deformation and propagation of the fracture. These models are notoriously difficult to develop because of the strong non-linear coupling between the lubrication and elasticity equations and the non-local character of the elastic response of the fracture.

Detailled numerical simulations carried out in the last few years (e.g. Desroches and Thiercelin, 1993) have provided evidence that the strong coupling is actually confined to a region, near the tip of the advancing fracture, which is small compared to the overall fracture dimension. In this region, rapid variation of the fluid pressure appears to take place. This paper reviews progress recently made in the understanding of the solution

D. Durban and J.R.A. Pearson (eds.),
IUTAM Symposium on Non-Linear Singularities in Deformation and Flow, 27-42.
© 1999 *Kluwer Academic Publishers. Printed in the Netherlands.*

in the near tip region and its implication to the overall solution. It also attempts to define areas where advances have yet to be made.

The paper is organized as follows. First, we consider the problem of a fluid-driven fracture in an impermeable elastic solid with zero toughness. Under the assumption that the fluid flows up to the tip of the fracture, it is shown that a singularity exists at the fracture tip, which is weaker than predicted by linear elastic fracture mechanics (LEFM). This result is then used to construct a similarity solution for a finite hydraulic fracture. Next, the existence of a lag of *a priori* unknown length between the fluid front and the fracture tip is taken into account. The presence of a tip cavity filled with inviscid vapors from the fracturing fluid removes the singularity in the fluid pressure and enables the construction of a solution with the "proper" LEFM square root singularity at the fracture tip. For the stationary problem of a semi-infinite hydraulic fracture propagating in an elastic solid with finite toughness, it is shown that the solution at infinity corresponds actually to the singular solution derived under the conditions of zero toughness. The problem of a finite hydraulic fracture with consideration of a fluid lag at the tip has so far been solved only numerically. The numerical results suggest that the overall solution (i.e. the fracture length, the pressure and crack aperture at the inlet) is hardly influenced by details of the solution at the tip, under conditions where the energy dissipation in the fluid is much larger than the energy expended in fracturing the solid. In these circumstances, the zero-toughness solution provides an excellent approximation for the problem of a hydraulic fracture propagating in a solid with finite toughness.

2. Impermeable Elastic Solid with Zero Toughness

2.1. SINGULAR SOLUTION

The proper singular behaviour at the tip of a hydraulic fracture, completely filled by an incompressible Newtonian fluid and propagating in a zero toughness elastic solid, was first recognized by Spence and Sharp (1985) and also by Lister (1990), from considerations involving the asymptotic behaviour of the singular elastic integral equation and the lubrication equation. The strong coupling between the moving fluid and the zero toughness elastic solid in the tip region actually corresponds to an exact matching singularity between the lubrication and the elasticity equations. This singularity can be deduced using the following simple arguments (Desroches *et al.*, 1994). Consider the stationary propagation of a semi-infinite fluid-driven fracture in an impermeable elastic solid with a far-field compressive stress of magnitude σ_o acting perpendicularly to the fracture. Noting that any singular behaviour in the fluid pressure p must be compatible with the elastic behaviour near the tip of the fracture (since p represents the boundary conditions for the normal stress along the fracture), the most general singular solution for p is given by (e.g. Rice, 1968)

$$p - \sigma_o = A(\alpha)X^\alpha \tag{1}$$

where X is a coordinate with origin at the fracture tip (see Fig. 1), and A and α are real-valued constants. The corresponding fracture width w is given by

$$w = B(\alpha)X^{1+\alpha} \tag{2}$$

On the other hand, the simplifying assumptions stated above lead to a particular form of the lubrication equation

$$\frac{dp}{dX} = \frac{12\mu V}{w^2} \tag{3}$$

where μ is the viscosity of the fluid, and V the crack tip propagation speed. By substituting p and w by their expressions given by (1) and (2) into (3), it can readily be seen that necessarily $\alpha = -1/3$ in apparent contradiction with the classical result $\alpha = -1/2$ of linear fracture mechanics. Using the known elastic expressions for $A(\alpha)$ and $B(\alpha)$, explicit singular expressions for p and w can be found (Desroches et $al.$, 1994). This singular solution can be expressed in a dimensionless form as

$$\hat{\Omega}_* \left(\hat{\xi} \right) = 2^{1/3}3^{5/6}\hat{\xi}^{2/3} \qquad \hat{\Pi}_* \left(\hat{\xi} \right) = -\frac{1}{2^{2/3}3^{2/3}\hat{\xi}^{1/3}} \tag{4}$$

where $\hat{\Omega}_* = w/L_h$, $\hat{\Pi}_* = (p - \sigma_o)/E'$, and $\hat{\xi} = X/L_h$. The lengthscale L_h, which is associated with viscous dissipation, is defined as

$$L_h = \frac{12\mu V}{E'} \tag{5}$$

and E' is the so-called plane strain elastic modulus of the solid, which is related to Young's modulus E and Poisson's ratio ν by $E' = E/(1 - \nu^2)$.

It is worthwhile pointing out that the far-field stress σ_o does not enter in this solution, except as a reference level for the fluid pressure p.

2.2. SIMILARITY SOLUTION FOR A FINITE HYDRAULIC FRACTURE

The above result can be directly used for solving the problem of a two-dimensional fracture propagating in an infinite impermeable elastic solid under plane strain conditions and driven by an incompressible Newtonian fluid injected from a source located at the center of the fracture (see Fig. 1). A self-similar solution exists (Spence and Sharp, 1985; Carbonell and

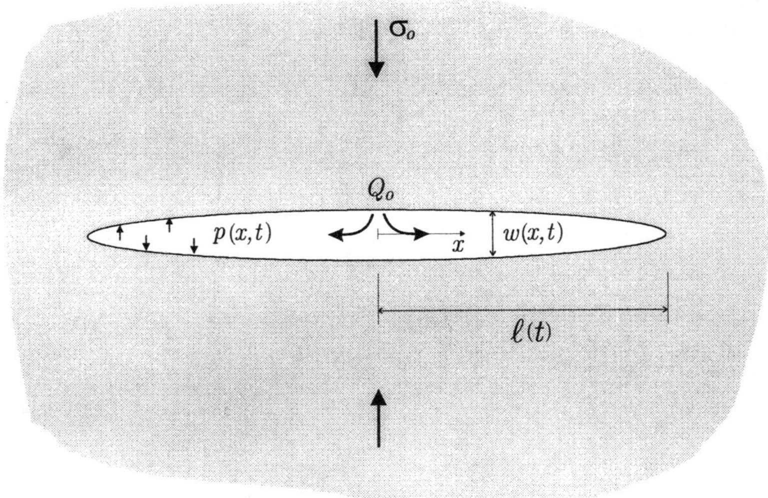

Figure 1. Two-dimensional fluid-driven fracture

Detournay, 1998) under the following assumptions: (1) the fracture is completely filled by the injected fluid; (2) the fracture toughness $K_{Ic} = 0$; (3) the injection rate varies either as a power law or as an exponential function of time. In the following, we outline the mathematical basis of the solution obtained by Carbonell and Detournay (1998), but restrict consideration to a constant injection rate Q_o.

We first define a time scale T and a length scale L, the latter one being introduced through the volume of fluid $V = Q_o T$ injected during time T

$$T = \frac{12\mu}{E'} \qquad L = \sqrt{Q_o T} \tag{6}$$

as well as a dimensionless moving coordinate ξ and a dimensionless time τ

$$\xi = \frac{x}{\ell} \qquad \tau = \frac{t}{T} \tag{7}$$

where ℓ is the current length of one fracture wing, see Fig. 1. We then introduce the scaled fracture length $\chi(\tau)$ and the following dimensionless field quantities: the net pressure $\Pi(\xi,\tau)$, the flow rate $\Psi(\xi,\tau)$, and the crack aperture $\Omega(\xi,\tau)$

$$\chi = \frac{\ell}{L} \qquad \Pi = \frac{p}{E'} \qquad \Psi = \frac{qT}{L^2} \qquad \Omega = \frac{w}{L} \tag{8}$$

Following Spence and Sharp (1985), we search for a self-similar solution for pressure and crack opening of the form

$$\Omega(\xi, \tau) = \kappa \tau^{1/3} \Omega_\xi(\xi) \qquad \Pi(\xi, \tau) = \tau^{-1/3} \Pi_\xi(\xi) \tag{9}$$

as well as a power law expression for the scaled length $\chi(\tau)$

$$\chi(\tau) = \gamma \tau^{2/3} \tag{10}$$

where the *a priori* unknown constant γ is defined in terms of the fracture volume

$$\gamma^{-2} = \int_{-1}^{1} \Omega_\xi(\rho) d\rho \tag{11}$$

The quantities $\Omega_\xi(\xi)$ and $\Pi_\xi(\xi)$ are governed by a system of integro-differential equations

$$\Pi_\xi = -\frac{1}{4\pi} \int_{-1}^{1} \frac{\Omega_\xi' d\rho}{\rho - \xi} \tag{12}$$

$$\frac{1}{3}\Omega_\xi - \frac{2}{3}\xi\Omega_\xi' = \frac{d}{d\xi}\left(\Omega_\xi^3 \frac{d\Pi_\xi}{d\xi}\right) \tag{13}$$

where the prime denotes differentiation with respect to the argument. The system of equations is subject to the following tip conditions

$$(1 \mp \xi)^{-1/2}\Omega_\xi = 0 \qquad \Omega_\xi = 0 \qquad \Omega_\xi^3 \frac{d\Pi_\xi}{d\xi} = 0 \qquad \text{at } \xi = \pm 1 \tag{14}$$

By using the scaling (7) and (8) as well as as the power law (10) for the fracture length, the tip solution (4) can be recast into an asymptotic solution of the two functions $\Omega_\xi(\xi)$ and $\Pi_\xi(\xi)$ near the end points, $\xi = \pm 1$

$$\Omega_\xi = 2^{2/3}3^{1/2}(1\mp\xi)^{2/3}, \qquad \Pi_\xi = -\frac{1}{6}\left[\frac{4}{(1 \mp \xi)}\right]^{1/3} \qquad \text{for } |\xi\pm 1| \ll 1 \tag{15}$$

Determination of the net pressure $\Pi_\xi(\xi)$ and crack opening $\Omega_\xi(\xi)$ is carried out using a numerical method, inspired by the procedure described by Spence and Sharp (1985). This method consists in first finding a representation for Π_ξ and Ω_ξ, such that the elasticity equation (12) and boundary conditions (14) are automatically satisfied. Furthermore, the solution F (i.e., either Π_ξ or Ω_ξ) is expressed as a sum of a "general" and a "particular" source solution, denoted as F^* and F^{**}, respectively

$$F(\xi) = F^*(\xi) + BF^{**}(\xi) \tag{16}$$

where B is a coefficient. On the one hand, the general solution F^* has the expected behavior at the fracture tips but is characterized by a zero pressure gradient at the source $\xi = 0$; on the other hand, the particular solution F^{**} is characterized by a jump of the pressure gradient at $\xi = 0$ and by a non-singular behavior at the fracture tips. Finally, the general solution F^* is represented as an infinite series of base functions $F_i^*(\xi)$, i.e. $F^*(\xi) = \sum_{i=1}^{\infty} A_i F_i^*(\xi)$ where the A_i's are initially unknown coefficients, like B.

In the solution obtained by Carbonell and Detournay (1998), the base functions F_i^* are expressed in terms of Jacobi polynomials, rather than Chebyshev polynomials as done by Spence and Sharp (1985) who searched for a solution based on the LEFM square root singularity. Then, by truncating the infinite series F^*, an n-order approximation $F^{(n)}(\xi)$ for pressure and crack opening is obtained. Substituting $F^{(n)}(\xi)$ in the lubrication equation leads to the formulation of a non-linear algebraic equation in ξ, containing the $(n+1)$ unknown coefficients. These $(n+1)$ free parameters are finally computed by enforcing satisfaction of the non-linear algebraic equation at a set of $(n+1)$ collocation points ξ_j, $j = 1...n + 1$.

The net pressure Π_ξ and fracture opening Ω_ξ, computed for both $n = 3$ and $n = 10$ are plotted in Fig. 2. It is evident that hardly any differences exist between the two approximations, and that the solution can be captured with only a limited number of terms n. The quantities γ, $\Pi_\xi(0)$, $\Omega_\xi(0)$ characterizing the solution are estimated to be $\gamma \simeq 0.629$, $\Pi_\xi(0) \simeq 0.566$, $\Omega_\xi(0) \simeq 1.715$, according to the approximation $n = 10$.

3. Impermeable Elastic Solid with Finite Toughness

3.1. PREAMBLE

The cubic root singularity of the stress field and the negative cubic root singularity of the fluid pressure at the fracture tip, discussed above, are a direct consequence of the assumption that the fluid reaches the tip of the advancing fracture. At first glance, it would appear that this solution is only consistent with the assumption of a zero toughness, since the cubic root singularity of the stress implies a zero energy release rate. However, as shown below, the singular solution (4) has the more general meaning of an intermediate asymptotic solution.

For a hydraulic fracture propagating in a solid with toughness $K_{Ic} > 0$, it can actually be shown that a lag λ is required between the fluid front and the crack tip (see Fig. 1). This lag defines a tip cavity which is filled by inviscid vapor from the fracturing fluid (for impermeable rocks). Such a lag is needed to ensure coherence of the mathematical solution. Indeed, under the condition $\lambda = 0$, combination of the lubrication equation (3) with the LEFM asymptotic fracture opening $w \sim X^{1/2}$ implies that the fluid pressure has a logarithmic singularity, $p \sim \ln X$, which is mathematically

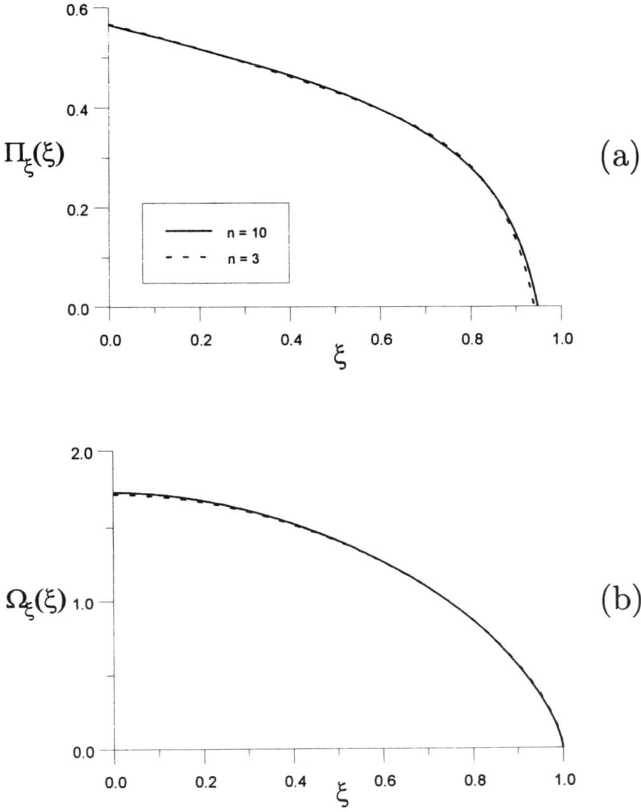

Figure 2. Variation of the net pressure Π_ξ and the crack opening Ω_ξ computed for $n = 3$ and $n = 10$. Note that only the positive part of the net pressure has been shown on the plot (a)

inconsistent with the assumed eigensolution. Hence, the presence of the lag removes the singularity in the fluid pressure, and at the same time enables the classical $X^{-1/2}$ singularity of LEFM to take place.

It could also be argued that a lag would also necessarily form even if $K_{Ic} = 0$, since the fluid cannot sustain a negative pressure (capillary effects being neglected). Although correct, such an argument introduces an extraneous consideration ("a minimum fluid pressure") to the problem as originally defined. Contrary to the case $K_{Ic} > 0$, the assumption of zero lag for $K_{Ic} = 0$ does not lead to any mathematical inconsistency. (This issue actually bears similarity to the question of the existence of a process zone or a plastic zone to ensure finiteness of the stress at the crack tip).

There are actually two related issues that need to be addressed.

- $K_{Ic} = 0$: does the singular solution (4) capture the dominant behaviour of the solution at some distance from the tip where a lag is allowed to develop by imposing the fluid pressure to be positive. In other words, is (4) indeed an intermediate asymptotic solution (Barenblatt, 1996)?

 — $K_{Ic} > 0$: could the solution of a hydraulic fracture be approximated by the zero toughness solution under conditions where the rock toughness is "small", i.e. under conditions where the energy dissipated in the solid by fracture propagation is small compared to the energy dissipated in the fluid through viscous flow?

In the following, we provide arguments that the rock toughness as well as the fluid lag are inconsequential outside of the tip region if (4) emerges as an intermediate asymptotic solution. A criterion for the development of this regime of solution, referred to as viscosity-dominated, is also proposed.

3.2. SEMI-INFINITE FRACTURE

The stationary problem of a semi-infinite hydraulic crack (see Fig. 1) propagating at constant velocity V in an impermeable linear elastic medium, perpendicular to a uniform far-field confining stress σ_o, has recently been solved by Garagash and Detournay (1998). As can be seen in Fig. 3, the existence of a lag of *a priori* unknown length λ between the crack tip and the fluid front is taken into account. Note that the related problem of a semi-infinite buoyancy-driven hydraulic fracture has been solved by Lister (1990), see also Spence *et al.* (1987).

The tip cavity is assumed to be filled by fluid vapors under constant pressure p_λ, assumed to be negligible compared to the reference far-field stress σ_o; hence, $p_\lambda \approx 0$. The fracture internal fluid pressure $p(X)$, $X \geq \lambda$, the crack opening $w(X)$, and the length λ of this cavity are unknown and are part of the solution. Since the crack propagation is stationary, the fracturing fluid front propagates with the same velocity as the crack tip V.

Another lengthscale L_k is introduced to characterize the dissipation due to fracturing of the solid, as well as a small parameter ε

$$L_k = \frac{8}{\pi}\left(\frac{K_{Ic}}{E'}\right)^2, \qquad \varepsilon = \frac{\sigma_o}{E'} \tag{17}$$

The following scaled quantities are then defined: the crack opening $\tilde{\Omega} = \varepsilon^2 w/L_h$, the moving coordinate $\tilde{\xi} = \varepsilon^3 X/L_h$, and the net pressure $\tilde{\Pi} = (p - \sigma_o)/\sigma_o$. After performing the transformation to the moving coordinate ξ and making use of the stationarity condition, the governing system of equations takes the dimensionless form

$$\tilde{\Omega}^2\left(\tilde{\xi}\right)\tilde{\Pi}'\left(\tilde{\xi}\right) = 1 \qquad \text{for} \quad \tilde{\xi} \in (\Lambda, \infty) \tag{18}$$

$$\tilde{\Pi} = -1 \qquad \text{for} \quad \tilde{\xi} \in (0, \Lambda) \tag{19}$$

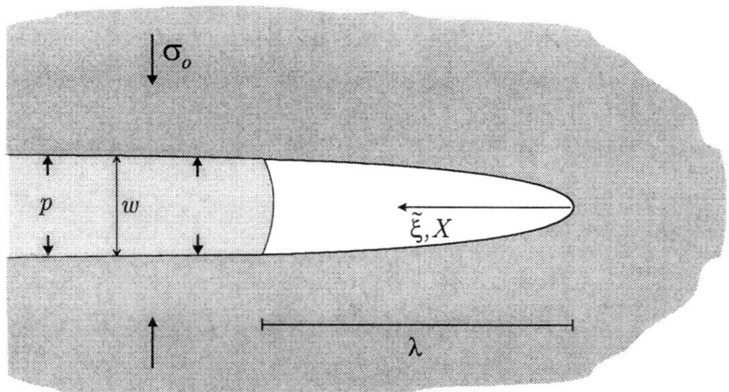

Figure 3. Semi-infinite fluid driven crack with the lag zone adjacent to the tip.

$$\tilde{\Pi}\left(\tilde{\xi}\right) = \frac{1}{4\pi} \int_0^\infty \tilde{\Omega}'\left(\eta\right) \frac{d\eta}{\xi - \eta} \tag{20}$$

$$\tilde{\Omega} = \kappa\sqrt{\tilde{\xi}} + O\left(\tilde{\xi}^{3/2}\right) \tag{21}$$

where (19) expresses the pressure condition in the lag region. In the above, Λ denotes the dimensionless lag, and κ the dimensionless toughness, respectively defined as

$$\Lambda = \frac{\varepsilon^3 \lambda}{L_h}, \qquad \kappa = 2\sqrt{\frac{\varepsilon L_k}{L_h}} \tag{22}$$

Equation (21) is the LEFM asymptotic expression for the opening near the crack tip (e.g. Rice, 1968), with the propagation criterion $K_I = K_{Ic}$ taken into account. Note that only one parameter κ is present in the normalized system of equations and boundary conditions (18)-(21). This system completely defines the crack opening $\tilde{\Omega}(\tilde{\xi}; \kappa)$ and the net pressure $\tilde{\Pi}(\tilde{\xi}; \kappa)$ for the semi-infinite fracture ($0 \leq \tilde{\xi} < \infty$), as well as the position of the fluid front $\Lambda(\kappa)$.

The singular solution (4) can simply be recast in terms of the new quantities $\tilde{\Omega}$, $\tilde{\Pi}$, and the variable $\tilde{\xi}$ as

$$\tilde{\Omega}_*\left(\tilde{\xi}\right) = 2^{1/3}3^{5/6}\tilde{\xi}^{2/3} \qquad \tilde{\Pi}_*\left(\tilde{\xi}\right) = -\frac{1}{2^{2/3}3^{2/3}\tilde{\xi}^{1/3}} \tag{23}$$

Although this singular solution (23) cannot be the solution of the system (18)-(21) as the fluid lag goes to zero $\Lambda \to 0$ (since the net pressure $\tilde{\Pi}$ is singular at the tip according to (4), whereas the solution $\tilde{\Pi}$ of the system

(18)-(21) is finite at the tip in view of the boundary condition (19)), it can be proven that (4) gives the asymptotic behavior of the solution of (18)-(21) at infinity (Garagash and Detournay, 1998)

$$\tilde{\Pi}\left(\tilde{\xi}\right) \;=\; \tilde{\Pi}_*\left(\tilde{\xi}\right) + O\left(\tilde{\xi}^{-1/3-\alpha}\right), \quad \alpha > 0, \quad \text{as } \tilde{\xi} \to \infty \tag{24}$$

$$\tilde{\Omega}\left(\tilde{\xi}\right) \;=\; \tilde{\Omega}_*\left(\tilde{\xi}\right) + O\left(\tilde{\xi}^{1/2}\right), \quad \text{as } \tilde{\xi} \to \infty \tag{25}$$

This asymptote does not depend on the toughness κ.

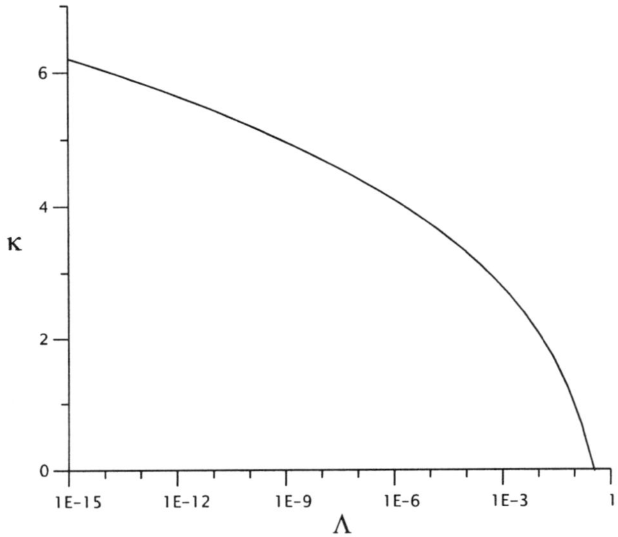

Figure 4.   Plot of the universal relationship between dimensionless toughness κ and dimensionless lag length Λ in semi-log scale.

From the above considerations, the unknown solution behaves according to LEFM in the near tip region, see (19) and (21), and asymptotically as the singular solution (23) at large enough distance from the tip. Hence, there exists a transition between these two asymptotes, which can be computed numerically.

The computed variation of the stress intensity factor κ with the fluid lag Λ is plotted in Fig. 4 in semi-log axes. The fluid lag can be seen to increase with decreasing toughness, to reach a maximum value $\Lambda_o \simeq 0.3574$ for $\kappa = 0$. Note that this value of Λ_o is very close to the value computed by Lister (1990) using a perturbation technique, for the problem of the semi-infinite buoyancy-driven hydraulic fracture.

It is of interest to compute the maximum dimension of the lag, $\lambda_o = \Lambda_o \varepsilon^{-3} L_h$, for some typical values of the physical parameters. Consider the following set: $E' = 3 \cdot 10^4$ MPa, $\mu = 10^{-7}$ MPa·s (100 cp), $\sigma_o = 10$ MPa, and $V = 1$ m/s. Then, the characteristic length $\varepsilon^{-3}L_h = 1.08$ m and $\lambda_o \simeq 0.39$ m. The fluid lag reduces to $\lambda \simeq 0.27$ m for a toughness $K_{Ic} = 1$ MPa·m$^{1/2}$, according to Fig. 4 ($\Lambda \simeq 0.25$ for $\kappa = 0.31$).

The log-log plot of the crack opening $\tilde{\Omega}$ along the crack for various values of the toughness κ (and thus of Λ), shown in Fig. 5, provides transparent evidence that the solution behaves as $\tilde{\xi}^{1/2}$ (but as $\tilde{\xi}^{3/2}$ for $\kappa = 0$) in the region immediately adjacent to the tip (in accordance to LEFM), and that it converges towards the singular solution $\tilde{\Omega}_*(\tilde{\xi})$, shown in dashed line, further away from the tip. There is a transition zone between these two behaviors.

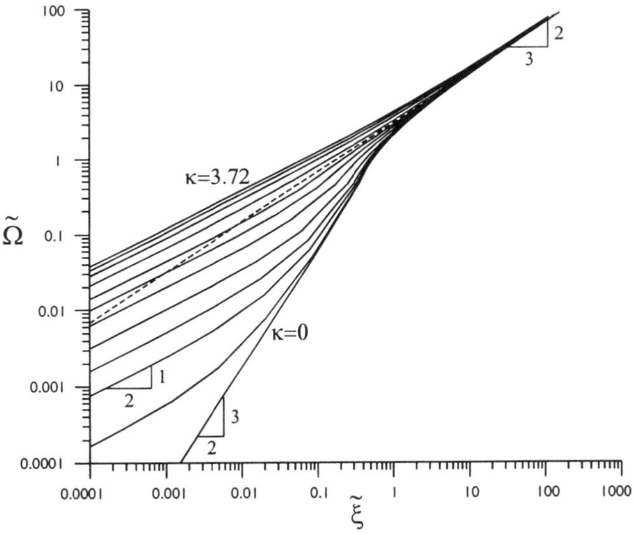

Figure 5.   The dimensionless crack opening Ω along the crack in log-log scale for dimensionless toughness varying from $\kappa = 0$ to $\kappa = 3.7$. The dashed line corresponds to the asymptotic solution at infinity.

The numerical solution indicates that with increasing toughness κ, the fluid lag decreases while the region dominated by the LEFM $\tilde{\xi}^{1/2}$ behavior increases. Also, the transition to the asymptotic behavior at infinity occurs further away from the crack tip, as κ increases.

The characteristic length of the near tip processes, $\varepsilon^{-3}L_h$, is typically several orders of magnitude smaller than the length of hydraulic fractures (10 - 10^3 m). This difference in length suggests that the solution of a semi-infinite fracture could actually describe the near tip asymptotic solution of a finite hydraulic fracture, under conditions where the intermediate asymptotics (Ω_*, Π_*) can develop.

3.3. FINITE FRACTURE

We now turn to the problem of a two-dimensional finite hydraulic fracture, similar to the one described in Section 2.2 except that the solid is characterized by a toughness $K_{Ic} > 0$. This problem has not yet been fully solved, and thus some of the statements below should be treated as conjectures.

A self-similar solution for this problem is obtained by Spence and Sharp (1985) by assuming $\lambda = 0$, while a numerical solution which accounts for the existence of a fluid lag at the tip is described by Desroches and Thiercelin (1993) and Carbonell *et al.* (1998), for the case of a constant injection rate.

The solution with a lag cannot be *strictly* self-similar. Indeed, self-similarity would imply that λ/ℓ is a constant during fracture propagation; this requirement is in contradiction, however, with the expected decrease of λ with time t (in contrast to the increase of ℓ with t) according to the results of Section 3.2. Although the numerical simulations confirm that indeed λ decreases with t, they also suggest "global" self-similarity of the solution. This can be seen in Fig. 6 showing the evolution of the crack opening $\Omega(0, \tau)$ and net pressure $\Pi(0, \tau)$ (at the center of the fracture), and the length $\chi(\tau)$ (refer to Section 2.2 for the notation), for various values of the dimensionless toughness $\mathcal{K} = K_{Ic}/E'L^{1/2}$ ranging from $3.5 \cdot 10^{-5}$ to $3.5 \cdot 10^{-1}$; not only do these quantities follow the power law behaviour expected from a self-similar solution, but also the computed quantities are remarkably close to the zero toughness solution shown by the solid line in Fig. 6. (It is worth mentioning that it can be deduced from the propagation criterion that global self-similarity can only be expected in the particular case of a constant injection rate.)

In summary, the numerical results suggest the solution is "globally" self-similar and that it follows closely the zero toughness solution presented in Section 2.2, at least for $\mathcal{K} \leq 10^{-1}$. Furthermore, the existence of the intermediate asymptotic solution (Ω_*, Π_*) is clearly confirmed by the numerical simulations for $\mathcal{K} \leq 10^{-1}$ (compare the log-log plot of the crack opening versus distance from the tip computed for $\mathcal{K} = 3.5 \cdot 10^{-2}$ and $3.5 \cdot 10^{-1}$ shown in Fig. 7). To cast these observations into a different perpective, we first observe that the power law dependence on time exhibited by the numerical results imply the same power law dependence on time for both the rate of energy $\mathcal{D}$ dissipated within the fluid ($\mathcal{D}$ is the viscous dissipation scaled by the characteristic value $\Gamma = E'L^2/T$) and the energy rate $2\mathcal{G}\dot{\chi}$ expended in creating new fracture surfaces ($\mathcal{G} = \mathcal{K}^2$ denotes the dimensionless energy release rate). Indeed, $\dot{\chi} = \frac{2}{3}\gamma\tau^{-1/3}$ and $\mathcal{D} = \delta\tau^{-1/3}$ with δ defined as

$$\delta = -2\gamma^2 \int_0^1 \Psi_\xi \Pi'_\xi d\xi \tag{26}$$

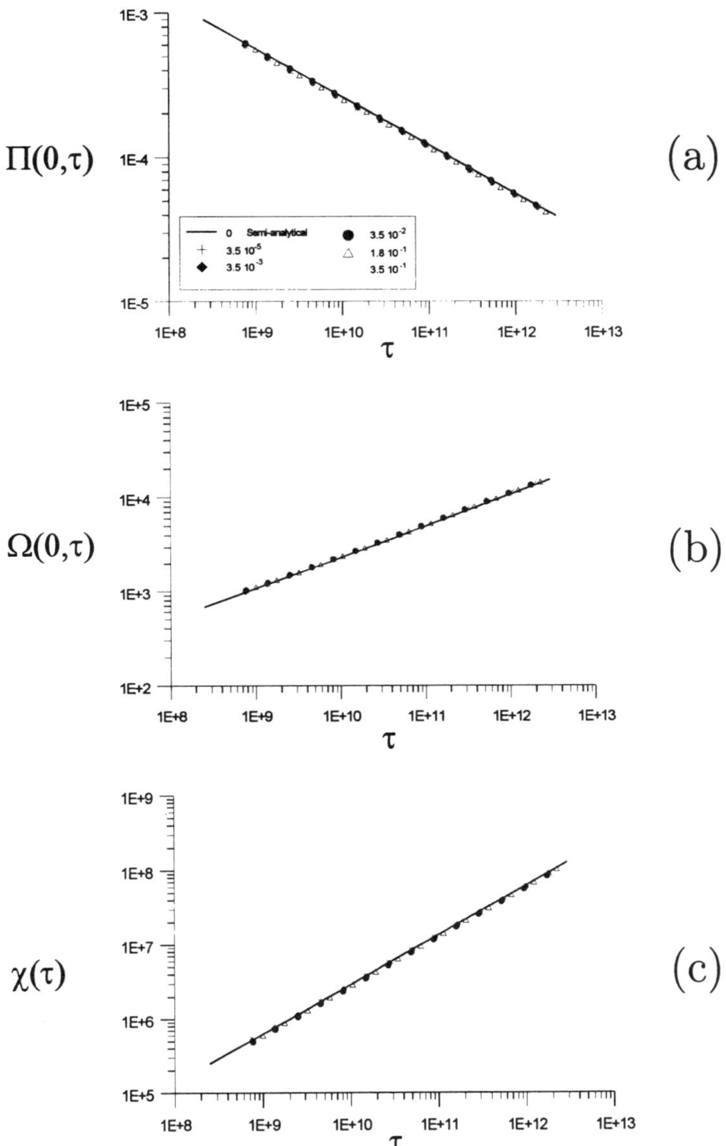

Figure 6. Evolution of the opening and net pressure at the fracture inlet, and fracture length; comparison between numerical simulations (various $\mathcal{K}$) and the $K_{Ic} = 0$ solution.

Hence the dissipation ratio $\vartheta = \mathcal{G}\dot{\chi}/\mathcal{D}$ remains constant during self-similar propagation of the considered hydraulic fracture; it is given by

$$\vartheta = \frac{4\gamma\mathcal{K}^2}{3\delta} \tag{27}$$

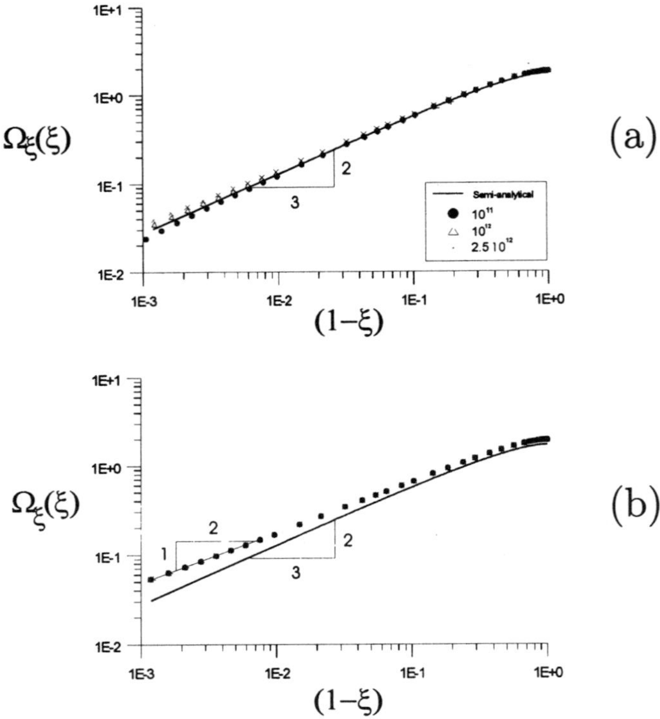

Figure 7. Comparison between zero toughness solution (solid line) and numerical simulations (various symbols for different times): crack opening versus distance from the tip for $\mathcal{K} = 3.5 \cdot 10^{-2}$ (a) and $\mathcal{K} = 3.5 \cdot 10^{-1}$ (b).

Thus, the results of the numerical simulations reported by Carbonell *et al.* (1998) imply the existence of three regimes of solution, which could be discriminated in terms of the dissipation ratio ϑ:

- $\vartheta < 10^{-2}$, viscosity-dominated regime (the $K_{Ic} = 0$ solution applies);
- $10^{-2} < \vartheta < 10^{-1}$, transition regime;
- $\vartheta > 10^{-1}$, toughness-dominated regime.

4. Conclusions and Perspectives

In this paper, we have reviewed some recent advances in the understanding of the structure of the solution near the the tip of a fluid-driven fracture propagating in an impermeable elastic medium. In particular, results reported here suggest that the singular behaviour, derived on the assumptions of zero toughness and no fluid lag, actually characterizes the solution at an intermediate scale under conditions where the toughness of the material and the existence of a tip cavity do not have any influence on the global solution. These conditions appear to correspond to situations where the energy expended in creating new fracture surfaces is small compared to the energy dissipated in the viscous fluid. In this "viscosity-dominated" regime, the global solution behaves as if the material had zero toughness.

The strong non-linear coupling between the fluid and the solid takes place in a region near the tip, which generally has a very small dimension compared to the overall length of an hydraulic fracture. This difference in length scales makes unrealistic the development of purely numerical algorithms which account for the small scale effects near the tip, but yet aim to predict the overall growth of hydraulic fractures. However, it can be expected that by incorporating explicit near tip solutions in the algorithm, robust and efficient numerical models of hydraulic fractures can be developed (Shah *et al.*,1997). Thus, some of the work that has been reported in this paper could somehow be justified by this end use.

An important issue that has not been addressed in this paper, for lack of space is the exchange of fluid between the fracture and the host permeable rock. Although work regarding the tip process in permeable rocks is not as well advanced, some progress has been recently been made, see for example Zazovsky and Panko (1978), Lenoach (1995), and Detournay and Garagash (1998) for particular aspects of this problem.

Finally, note that all the results reviewed in this paper have been obtained within the framework of elasticity, linear fracture mechanics, and the lubrication approximation. These results therefore ignore effects involving other length scales, such as surface tension at the fluid front (Bui, 1996), two-dimensional Stokes flow in the tip region (Bui and Parnes, 1982), and the process zone in the solid (Rubin, 1993), possibly accompagnied by yielding of the rock around the fracture tip (Papanastasiou, 1997). One of the fascinations of the problem of fluid-driven fractures is the multiplicity of length scales that need to be accounted for. Further work in this area will hopefully lead to a classification and rationalization of all these effects.

Acknowledgements

This review is partly based on the outstanding contributions of my graduate students, Roberto Carbonell and Dmitry Garagash. I am indebted to Anthony Pearson with whom I have enjoyed, over the years, numerous stimulating interactions on this topic. I have further benefited from discussions with my former colleagues of Schlumberger Cambridge Research. I would like also to acknowledge the contribution of Jean Desroches who carried out the numerical simulations reported in Section 3.3. This research has been partially funded by a grant from the Graduate School of the U of MN.

References

G.I. Barenblatt, *The mathematical theory of equilibrium cracks in brittle fracture*, Advances in Applied Mechanics **VII** (1962), 55–129.

————, *Scaling, self-similarity, and intermediate asymptotics*, Cambridge Texts in Applied Mathematics, vol. 14, Cambridge University Press, 1996.

H. D. Bui and R. Parnes, *A reexamination of the pressure at the tip of a fluid-filled crack*, Int. J. Engng. Sci. **20** (1982), no. 11, 1215–1220.

H.D. Bui, *Fracture: A topical encyclopedia of current knowledge dedicated to Alan Griffith (ed. G.P. Cherepanov)*, ch. Interaction Between the Griffith Crack and Fluid: Theory of Rehbinder's Effect, Krieger Publishing Company, 1996.

R. Carbonell, J. Desroches, and E. Detournay, *A comparison between a semi-analytical and a numerical solution of a two-dimensional hydraulic fracture*, Int. J. Solids Structures (1998), in press.

R. Carbonell and E. Detournay, *Self-similar solution of a fluid driven fracture in a zero toughness elastic solid*, Proc. Roy. Soc. London, Ser. A (1998), to be submitted.

J. Desroches, E. Detournay, B. Lenoach, P. Papanastasiou, J.R.A. Pearson, M. Thiercelin, and A.H-D. Cheng, *The crack tip region in hydraulic fracturing*, Proc. Roy. Soc. London, Ser. A **A447** (1994), 39–48.

J. Desroches and M. Thiercelin, *Modeling propagation and closure of micro-hydraulic fractures*, Int. J. Rock Mech. Min. Sci. & Geomech. Abstr. **30** (1993), 1231–1234.

E. Detournay and D. Garagash, *The tip region of a fluid-driven fracture in a permeable elastic solid*, J. Fluid Mech. (1998), to be submitted.

D. Garagash and E. Detournay, *Similarity solution of a semi-infinite fluid-driven fracture in a linear elastic solid*, C. R. Acad. Sc. Paris (1998), in press.

J. Geertsma and R. Haafkens, *A comparison of the theories for predicting width and extent of vertical hydraulically induced fractures*, J. En. Res. Tech. **101** (1979), 8-19.

S.A. Khristianovic and Y.P. Zheltov, *Formation of vertical fractures by means of highly viscous fluids*, Proc. 4th World Petroleum Congress, Vol. II, Rome (1955), 579–586.

B. Lenoach, *The crack tip solution for hydraulic fracturing in a permeable solid*, J. Mech. Phys. Solids **43** (1995), no. 7, 1025–1043.

J. R. Lister, *Buoyancy-driven fluid fracture: the effects of material toughness and of low-viscosity precursors*, J. Fluid Mech. **210** (1990), 263–280.

R.P. Nordgren, *Propagation of vertical hydraulic fracture*, Soc. Pet. Eng. J., Trans. AIME **253** (1972), 306–314.

P. Papanastasiou, *The influence of plasticity in hydraulic fracturing*, Int. J. Fracture **84** (1997), 61–97.

T.K. Perkins and L.R. Kern, *Widths of hydraulic fractures*, J. Pet. Tech., Trans. AIME **222** (1961), 937–949.

J. R. Rice, *Mathematical analysis in the mechanics of fracture*, Fracture, an Advanced Treatise, vol. II, ch. 3, pp. 191–311, Academic Press, 1968.

A. M. Rubin, *Tensile fracture of rock at high confining pressure: implications for dike propagation*, J. Geophys. Res. **98** (1993), 15,919–15,935.

K.R. Shah, B.J. Carter, and A.R. Ingraffea, *Hydraulic fracturing simulation in parallel computing environments*, Int. J. Rock Mech. Min. Sci. & Geomech. Abstr. **34** (1997), no. 3/4, 474.

D. A. Spence and P. Sharp, *Self-similar solution for elastohydrodynamic cavity flow*, Proc. Roy. Soc. London, Ser. A **A400** (1985), 289–313.

D. A. Spence and D. L. Turcotte, *Magma-driven propagation crack*, J. Geophys. Res. **90** (1985), 575–580.

D.A. Spence, P.W. Sharp, and D.L. Turcotte, *Buoyancy-driven crack propagation: a mechanism for magma migration*, J. Fluid Mech. **174** (1987), 135–153.

A.F. Zazovskyi and S.V. Panko, *Local structure of the solution of the associated problem of a hydraulic fault crack in a permeable medium*, Izv. AN SSSR. Mekhanika Tverdogo Tela **13** (1978), no. 5, 153–158.

INVERSE PROBLEMS IN HYDRAULIC FRACTURING

O.P. ALEKSEENKO AND A.M. VAISMAN
Institute of Mining, Siberian Branch of Russian Academy of Sciences, 54 Krasnyi Prospekt, Novosibirsk 630091, Russia

AND

A.F. ZAZOVSKY
Schlumberger Cambridge Research, High Cross, Madingley Road, Cambridge CB3 0EL, United Kingdom

Abstract.

This paper discusses the inversion of pressure data collected during hydraulic fracturing operations to yield parameters characteristic of the rock being fractured, its stress state and the geometry of the fracture using a simple approximate fracturing model, the Perkins-Kern-Nordgren (PKN) model. In particular the suitability of the constant pumping rate / shut-in schedule is considered. This example gives a flavour of the mathematical difficulties arising in numerical modelling of hydraulic fracturing and in the processing of fracturing test data. The results might be helpful for development of advanced interpretation techniques required for hydraulic fracturing design and evaluation.

1. Introduction

The PKN model [12, 11] deals with a vertical fracture of constant height which propagates in a horizontal direction when the leakoff mechanism is given by Carter's formula and the spurt volume is negligible. Fluid pressure inside the fracture is considered to be uniform in each transversal cross-section of the fracture and depends only on a horizontal coordinate. The fracture shape is found independently at each distance from the wellbore from the solution of a plane-strain elasticity problem. As a result, the fracture width is proportional to the net pressure, i.e. the difference between the local fluid pressure and the horizontal stress normal to the fracture

43

D. Durban and J.R.A. Pearson (eds.),
IUTAM Symposium on Non-Linear Singularities in Deformation and Flow, 43-54.
© 1999 *Kluwer Academic Publishers. Printed in the Netherlands.*

surface, and the fracture has an elliptical cross-sectional area. The fluid flow inside the fracture is assumed to be laminar. Zero flux and zero net presure conditions are used at the fracture tip during fracture propagation and closure.

The fracturing pressure analysis for the PKN model presented in this paper generalises the original approach by Nolte [7]. The forward problem of PKN fracture propagation and closure is considered in a fully coupled formulation. It is reduced to a non-linear diffusion equation with respect to fracture width having a sink term responsible for leakoff. This equation is solved inside a domain with an unknown moving boundary (bounded by the fracture tip) for pumping at a constant rate followed by shut-in (shut-in test).

Asymptotic analysis of the solution near the fracture tip during fracture propagation has been carried out. Then the forward problem has been solved numerically with a high accuracy using the asymptotically derived singularity in the gradient of fracture width.

The pumping phase has a solution of the forward problem universal in dimensionless coordinates: it does not involve any dimensional parameter because all of them are removed by scaling factors. The shut-in solution depends only on a single parameter which is the dimensionless pumping-in time. A full set of these solutions has been calculated.

The inverse problem is formulated with respect to the unknown parameters: the fracture closure pressure, the fracture height, the elastic modulus and the leakoff coefficient. The fracture closure pressure and the fracture length at the end of pumping-in can be obtained independently from other unknown parameters. This leads to the reduction of the dimensionality of inverse problem.

An interpretation procedure is based on the explicit inversion of the forward problem solution using a history matching method. This new technique has been tested using numerically-generated fracturing test data with different kinds of disturbances including random noise and systematic errors.

2. The PKN Model

We consider one wing of a vertical fracture having constant height H, variable length $x_f(t)$ and elliptical cross-sectional area (Figure 1). The maximum fracture width $w(x,t)$ at each vertical cross section of the fracture $x = const$ is related to the pressure inside the fracture $p_f(x,t)$ by

$$w(x,t) = \frac{H}{G_0} \Delta p(x,t), \quad \text{where} \quad \Delta p(x,t) = p_f(x,t) - p_c, \quad G_0 = \frac{G}{1-\nu}. \quad (1)$$

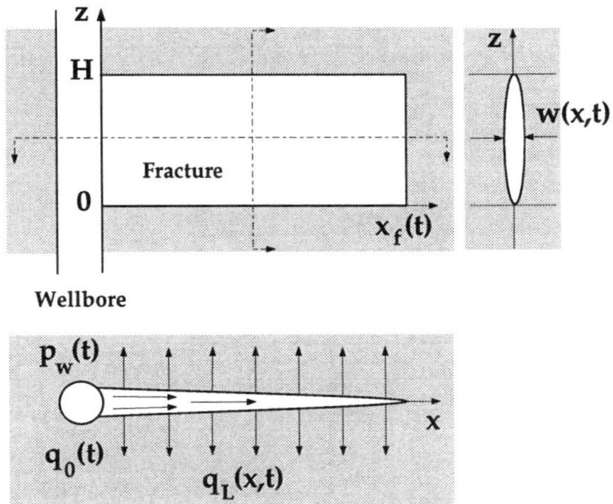

Figure 1. Fracture geometry in the PKN model.

Here Δp is the net pressure, p_c is the fracture closure pressure, G and ν are the shear modulus of the rock supposed homogeneous and its Poisson's ratio respectively.

The fracture closure pressure p_c is of the order of the in-situ horizontal stress in the rock and is about 2.5 times larger than the reservoir pore pressure p_p. For this reason, the net pressure during hydraulic fracturing Δp is usually small compared to the closure pressure p_c and the difference $p_f - p_p$ which determines fluid leakoff from the fracture surface is almost constant along the fracture. This justifies the usage of Carter's formula for the leakoff rate $q_L = 2HC/\sqrt{t}$ into the permeable rock with a constant leakoff coefficient C which is consistent with transient normal flow in a saturated porous medium.

Standard mass and momentum conservation (continuity and laminar viscous flow) equations written for the thin shallow fracture $w(x,t) \ll H \ll L$ lead to the following equation (Nordgren 1972) for the fracture width $w(x,t)$

$$\frac{\partial^2 w^4}{\partial x^2} = \frac{64\mu H}{G_0}\frac{\partial w}{\partial t} + \frac{512\mu HC}{\pi G_0\sqrt{t-\theta(x)}}, \quad 0 \le x \le x_f(t). \tag{2}$$

Here μ is the fluid viscosity and $\theta(x)$ is the time at which fluid loss begins through the fracture surface at a distance x from the wellbore, i.e. the time at which the fracture reaches a length x.

Eqn. (2) is subject to initial conditions

$$w(x,0) = 0, \quad x_f(0) = 0, \tag{3}$$

and boundary conditions

(i) at the fracture tip $x = x_f(t)$

$$w(x,t) = 0, \quad \frac{\partial w^4(x,t)}{\partial x} = 0 \quad \text{at} \quad x = x_f(t) \tag{4}$$

with the physical constraints

$$w(x,t) > 0 \ \text{ for } \ 0 < x < x_f(t), \quad w(x,t) = 0 \ \text{ for } \ x > x_f(t), \tag{5}$$

and (ii) at the wellbore $x = 0$

$$-\frac{\pi G_0}{256\mu} \frac{\partial w^4(x,t)}{\partial x} = q_0(t) \quad \text{at} \quad x = 0 \tag{6}$$

where $q_0(t)$ is the given rate at which fluid is pumped into the fracture

$$q_0(t) = q_i, \quad 0 < t < t_0; \quad q_0(t) = 0, \quad t > t_0. \tag{7}$$

Eqns (3) mean that the fracture is initially closed. The relationships (4) define the fracture tip and the zero flux condition specified at the fracture tip. Eqn. (6) specifies the pumping regime.

Thus the formulation of the forward problem for a shut-in test is given mathematically by Eqn. (2) together with the conditions (3)-(7).

Now let us introduce the dimensionless variables

$$T = \frac{t}{t_*}, \quad T_0 = \frac{t_0}{t_*}, \quad T_c = \frac{t_c}{t_*}, \quad \Theta = \frac{\theta}{t_*}, \quad X = \frac{x}{x_*}, \quad L = \frac{x_f}{x_*}, \tag{8}$$

$$W = \frac{w}{w_*}, \quad Q_0 = \frac{q_0}{q_*}, \quad Q_i = \frac{q_i}{q_*}.$$

Choosing the scaling parameters w_*, x_*, t_* and q_* such that

$$w_* = \left(\frac{16\mu}{HG_0} \frac{q_*^2}{C^2}\right)^{1/3}, \quad x_* = \frac{\pi G_0}{256\mu} \frac{w_*^4}{q_*}, \quad t_* = \left(\frac{\pi w_*}{8C}\right)^2, q_* = q_i, \tag{9}$$

we arrive at the following dimensionless formulation of the forward problem:

$$\frac{\partial^2 W^4}{\partial X^2} = \frac{\partial W}{\partial T} + [T - \Theta(X)]^{-1/2}, \quad \Theta(L(T)) = T, \quad 0 \leq X \leq L(T), \tag{10}$$

$$W(X,0) = 0, \quad L(0) = 0,$$

$$W(X,T) > 0, \ 0 \leq X < L(T); \quad W(X,T) = 0, \ X \geq L(T),$$

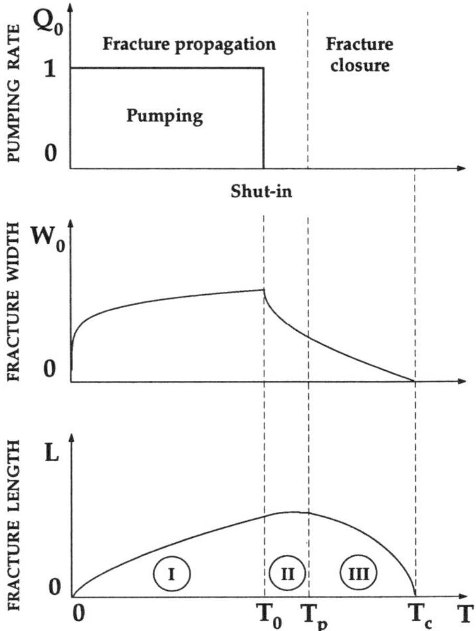

Figure 2. Typical solution for shut-in test.

$$\left.\frac{\partial W^4(X,T)}{\partial X}\right|_{X=0} = -Q_0(T), \quad \left.\frac{\partial W^4(X,T)}{\partial X}\right|_{X=L} = 0,$$

$$Q_0(T) = 1, \quad 0 < T < T_0; \quad Q_0(T) = 0, \quad T > T_0.$$

The problem (10) involves only a single dimensionless parameter, the pumping time T_0, and therefore its solution depends only on T_0.

3. Forward Problem Solutions

A typical solution for a shut-in test looks like the one shown in Figure 2, where T_0 is the pumping time and T_c is the time of fracture closure at the wellbore. The fracture width at the wellbore W_0, which from (1) is also the dimensionless net wellbore pressure $(p - p_c)/\Delta p_*$ where $\Delta p = \mu^{1/3}(4G_0 q_i/H^2 C)^{2/3}$, declines monotonically from $W_0(T_0)$ at the end of pumping to zero at $T = T_c$. The wellbore pressure varies from the peak pressure, p_0, at $t = t_0$ to the fracture closure pressure, p_c, at $t = t_c$. However, in an actual test, the closure pressure p_c is not known so neither is the time of fracture closure at the wellbore t_c.

The fracture length variation during pumping/shut-in test has three phases: (I) fracture growth during pumping $(0 < T < T_0)$, (II) fracture extension after shut-in $(T_0 < T < T_p)$, and (III) fracture recession $(T_p <$

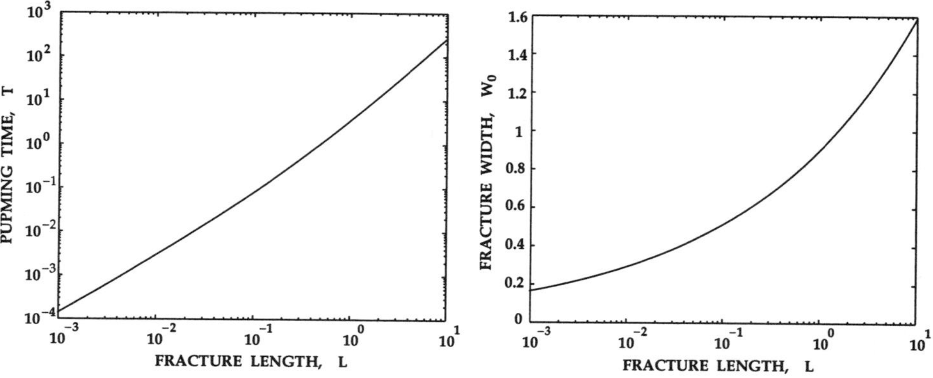

Figure 3.   Pumping solution for $T_0 \to \infty$

$T < T_c$) until the fracture closes at the wellbore (see Figure 2). This feature of the forward problem solution has been neglected in the previous approach [7, 4, 9].

Due to the local relationship (1) between the pressure inside the fracture and the fracture width, only the fracture width at the wellbore $w_0(t) = w(0, t)$ is required for the inversion algorithm, however in order to obtain $w_0(t)$ the full solution for $w(x, t)$ has to be found.

3.1. PUMPING SOLUTION

Although the full solution depends upon the parameter T_0, the pumping phase of the solution, $T < T_0$, does not depend upon T_0 and is therefore part of the universal solution obtained by letting $T_0 \to \infty$. This solution has been found numerically and is shown in Figure 3 where the fracture length L is used as independent variable instead of the time T

$$T = T(L), \quad W_0 = W_0(L), \quad 0 \le L \le L_0 = L(T_0). \tag{11}$$

Asymptotic analysis of the pumping solution has been carried out by Nordgren [11] for small leakoff and large leakoff which are equivalent to small and large times. In terms of our dimensionless variables, the results can be summarized as follows. In both cases there is the same global relationship between the fracture width at the wellbore $W_0(T)$ and the fracture length $L(T)$

$$W_0 \sim L^{1/4}. \tag{12}$$

However, at small time (leakoff), $T \ll 0.06$,

$$L \sim T^{5/4}, \quad W_0 \sim T^{1/5} \tag{13}$$

the fracture propagates faster than at large time (leakoff), $0.06 \ll T < T_0$,

$$L \sim T^{1/2}, \quad W_0 \sim T^{1/8}. \tag{14}$$

Here the dimensionless time $T = 0.06$ corresponds to the phase of pumping when the leakoff volume becomes equal to the fracture volume.

One can find, in particular, that the data $W_0(L)$ in Figure 3 are approximated very well by the global expression

$$W_0 \approx 0.8948 \, L^{1/4} \tag{15}$$

for the whole range of L variation from 10^{-3} to 10.

The relationships (12)-(14) were generalised in [8] for a non-Newtonian fracturing fluid with power-law rheology and other fracture geometries (penny-shaped fracture and vertical fracture with an infinite height known as the KGD model). In current practice, they form a basis for the identification of the fracture geometry using wellbore pressure curves during pumping [10, 4].

Another important feature of the pumping solution is the fracture width behaviour near the fracture tip $X = L(T)$

$$W \approx \left(\frac{3}{4} \, V \, \xi\right)^{1/3}, \quad \frac{\partial W}{\partial \xi} \sim \xi^{-2/3}, \quad V = \frac{dL}{dT}, \quad \xi = L(T) - X \to 0 \tag{16}$$

where ξ is the distance to the fracture tip and V is the velocity of fracture propagation. The exponent $1/3$ in the asymptotic solution for the fracture width found for the PKN model is smaller than the exponent $2/3$ obtained in [5] for the KGD model, i.e. a PKN fracture is open wider near the tip than a KGD fracture.

The asymptotic solution (16) explains the difficulty met in any direct numerical modelling of hydraulic fracturing which is related to the calculation of the fracture tip position or the velocity of fracture propagation. The space derivative of the fracture width is singular near the tip and this usually restricts the local accuracy of finite-difference / finite-element methods.

3.2. SHUT-IN SOLUTIONS

A shut-in test starts after pumping at a constant rate and lasts a finite time up to (and after) fracture closure at the wellbore. The series of shut-in solutions for $L_0 = 0.003, 0.01, 0.03, 0.1, 0.3$ and 1.0 is shown in Figure 4. They are plotted in the coordinates

$$a = \frac{T - T_0}{T_0}, \quad b = \frac{W_0(T)}{W_0(T_0)}. \tag{17}$$

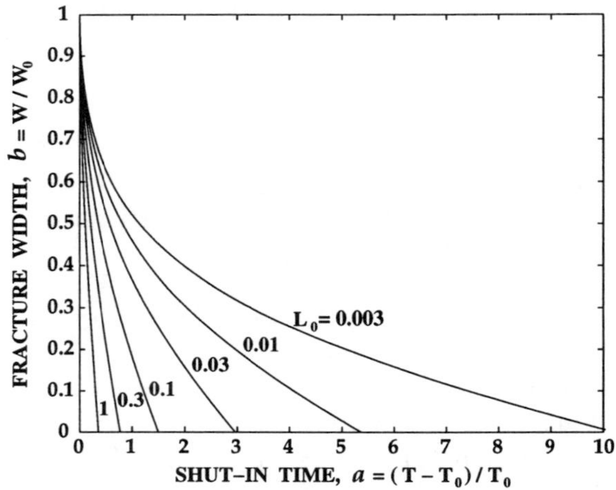

Figure 4. Shut-in solutions.

Here a is the normalised time calculated from the beginning of shut-in test and b is the fracture width normalised with respect to the fracture width at the end of pumping.

It is worth noting that the absolute duration of a shut-in test increases with the pumping time T_0 whereas the relative test duration decreases.

The asymptotic behaviour of the fracture width near the tip during shut-in seems to differ from (16) but has not been investigated so far.

4. Inverse Problem

Analysis shows that the pumping data alone are not sufficient for the determination of unknown fracture and reservoir parameters and therefore shut-in test data have to be involved in any interpretation for them.

Using (1) and (8), the shut-in solutions found above can be represented in the form

$$\frac{p(t) - p_c}{p_0 - p_c} = F\left(\frac{t - t_0}{t_0}, L_0\right), \qquad t_0 < t < t_c. \tag{18}$$

where F is the non-linear function $b = F(a, L_0)$ shown in Figure 4.

The equation (18) contains two unknown parameters: the pressure of fracture closure p_c and the dimensionless fracture length at the end of pumping L_0. The wellbore pressure $p(t)$ and the time t are measured during the shut-in test as well as their values at the end of pumping, p_0 and t_0. This equation provides the basis for the determination of the closure pressure p_c from shut-in test data.

If the parameters p_c and L_0 have been determined somehow the other unknow parameters can be found. Indead, using the plots in Figure 3, one can find T_0 and $W_0(T_0)$. Then, substituting the following expressions for the scaling parameters

$$t_* = \frac{t_0}{T_0}, \qquad w_* = \frac{H\Psi}{G_0}, \qquad \Psi = \frac{p_0 - p_c}{W_0(T_0)} \tag{19}$$

where p_0 is the wellbore pressure at the end of pumping, $t = t_0$, in their definitions (9) and resolving the system of two algebraic equations with respect to three unknown parameters, i.e. the fracture height H, the leakoff coefficient C and the elastic constant G_0, one arrives at the formulae

$$\frac{H}{G_0^{2/3}} = \left(\frac{32q_i}{\pi}\right)^{1/3}\left(\frac{\mu t_*}{\Psi^5}\right)^{1/6}, \qquad CG_0^{1/3} = \left(\frac{\pi^2 q_i}{16\, t_*}\right)^{1/3}(\mu\Psi)^{1/6}. \tag{20}$$

Only the parameters (20) and those that can be obtained from them by algebraic operations can be determined from shut-in test data acquired after pumping at a constant rate. This fact has been confirmed empirically using numerical pressure history inversion for fracturing test interpretation [6]. An attempt to determine combinations of the parameters H, C and G_0, other than those mentioned above, results in a non-uniqueness of solutions. If, on the other hand, G_0 is known, the fracture height H and the leakoff coefficient C can be determined from Eqns. (20). In practice, however, the fracture height might be known from lithology logs; in this case, G_0 and C can be found from (20).

5. Inversion Techniques

Thus, the inverse problem is reduced to determination of p_c and L_0 from Eqn. (18). This problem might be solved by the following two methods:

1. *Reduction to a system of two non-linear algebraic equations.* By substituting in Eqn. (18) the two pressures, $p_1 = p(t_1)$ and $p_2 = p(t_2)$, measured during the test $(t_1,\ t_2 > t_0)$, one obtains a system of two algebraic equations which are linear with respect to p_c and non-linear with respect to L_0. Eliminating the closure pressure p_c between two equations, the inverse problem can be reduced to a single non-linear equation with respect to the fracture length at the end of pumping L_0.
2. *History matching.* This method is based on minimisation of an error function measuring the integral deviation of fracturing test data (the bottomhole pressure [2, 3, 1, 6]) from those predicted by the model (18). Therefore, all of the entire pressure decline data can be involved into the interpretation.

The analysis of these two methods has been conducted using numerically-generated fracturing test data subjected to two different kinds of disturbances (random noise and systematic errors). The main conclusions from this study are listed below:

- The inverse problem for p_c and L_0 is an ill-posed one: small errors in shut-in fracturing data lead to large errors in interpretation results.
- *Method 1* works well only for the perfect data (without disturbances) and fails for noisy data. For example, errors in net pressure measurements of the order of 1% lead to errors in the determination of L_0 of the order of 50%. The errors in the closure pressure determination are even larger.
- *Method 2* works better than *Method 1* for noisy data; however it is sensitive to the form of the error function; for example, conventional pressure history matching does not provide the best results because it does not discriminate long and narrow fractures from short and wide ones if the shut-in fracturing data are noisy.

6. Intelligent History Matching

Progress can be achieved by reducing the inverse problem for p_c and L_0 to two sequential inverse problems, the first one with respect to L_0 and the second one with respect to p_c, and using different error functions at each inversion step.

In order to do this, let us represent Eqn.(18) in the two equivalent forms

$$p = p_c + (p_0 - p_c)\, F(a, L_0), \tag{21}$$

$$\frac{p_0}{p_0 - p_c} = \frac{p_0 - p}{p_0\,[1 - F(a, L_0)]}, \tag{22}$$

where $a = (t - t_0)/t_0$ is the dimensionless shut-in time.

Substituting the measured pressures p_i and times t_i $(i = 1, \ldots, N)$ in Eqn. (22), one arrives at the following error function with respect to L_0

$$\Delta_1(L_0, j) = \sqrt{\frac{1}{N_1} \sum_{i=1,\, i \neq j}^{N_1} \left[\frac{1 - P_i}{1 - F(a_i, L_0)} - \frac{1 - P_j}{1 - F(a_j, L_0)} \right]^2} \tag{23}$$

where $P_i = p_i/p_0$ and the index j can be taken arbitrarily from $1, \ldots, N$.

After determination of the fracture length L_0, the closure pressure $P_c = p_c/p_0$ can be found by the minimisation of the other error function following from (21)

$$\Delta_2(P_c) = \sqrt{\frac{1}{N_2} \sum_{i=1}^{N_2} [P_i - P_c - (1 - P_c)\, F(a_j, L_0)]^2}. \tag{24}$$

Different numbers, N_1 and N_2, are used in (23) and (24) in order to emphasize that the data involved in interpretation at each of these steps can be different.

In our calculations, L_0 varied from 0.003 to 1, N from 2 to 15, the pressures p_i were distributed equidistantly within the interval $[p_0, \; p_c]$, the number j in (23) was always equal to N and the level of noise varied up to 5%. The best results were obtained for $N_1 \approx 10$; further increase in N_1 did not improve the accuracy of the L_0 determination. However increase in N_2 up to 30 led to more accurate p_c predictions.

This approach improves significantly the accuracy of L_0 and p_c determination, especially for data with a moderate level of noise (1%) compared to the pressure history matching method based on minimisation of the error function (21) with respect to two variables simultaneously.

7. Discussion

The analysis conducted here for the PKN model has provided results which are not only instructive in themselves but also might be useful for more comprehensive study of inverse problems in hydraulic fracturing.

Hydraulic fracturing is supposed to be a robust technology which should cope with a great deal of uncertainty about reservoir and fracture properties such as in-situ stresses, elastic moduli of rock, fracture geometry and leakoff parameters. For this reason, fracturing job design normally starts from the collection of information about fracture/reservoir behaviour and is followed by evaluation of fracturing models against field observations.

A shut-in test is a conventional source of such information widely used in practice together with other fracturing tests. During these tests, the wellbore pressure and pumping rate are recorded as functions of time. It is often assumed that a set of field observations on wellbore pressure and pumping rate is enough for determination of unknown parameters necessary for fracturing model calibration and job design. However, this is not obvious and, in each particular case, the possibility of inverting fracturing test data has to be carefully investigated.

The fact that the inverse problem for the PKN model with shut-in test data, considered in this paper, is ill-posed suggests that similar inverse problems for other fracturing models are ill-posed too. Ill-posedness means that the determination of unknown parameters by the inversion of fracturing test data will always be inaccurate. This inaccuracy is partly due to inadequacy of the model, but also to errors in solutions of forward problems and inversion, and to natural errors in pressure measurements. Since for more complicated models one can expect an equivalent or greater ill-posedness, any inversion of fracturing test data has to be treated with extreme care. This conclusion applies, in particular, to pressure history matching, now widely used for fracturing model calibration, because this method is heuristic and it often suffers from non-uniqueness.

8. Summary

The inverse problem for the PKN fracturing model and shut-in test data is ill-posed, i.e. small errors in pressure measurement lead to large errors in calculation of unknown parameters. An intelligent history matching algorithm has been developed which allows us to determine first, the fracture length at the end of pumping, and then the closure pressure, using different error measures for minimization at each step of inversion. After closure pressure determination, other fracture / reservoir parameters can be calculated such as leakoff coefficient and fracture height. The results obtained should be helpful for the improvement of the current approach to fracturing test data interpretation.

References

1. K. Bhala and B.H. Brady. Formation and fracture characterization by inversion of fracture treatment records (Paper SPE 26526). In *68th Annual Technical Conference and Exhibition, Houston, Texas, 3-6 October*, 1993.
2. Ph. Charlez, R. Herail, and D. Despax. Interpretation of hydraulic fracture parametrs by inversion of pressure curves. *Int. J. Rock Mech. Min. Sci. & Geomech. Abstr.*, 26:549–553, 1989.
3. A.R. Crockett, R.M. Willis Jr., and M.P. Cleary. Improvement of hydraulic fracture predictions by real-time history matching on observed pressures. *SPE Production Engineering*, pages 408–416, November 1989.
4. M.J. Economides and K.G. Nolte, editors. *Reservoir Stimulation.* Schlumberger Educational Services, Houston, Texas, 2nd edition, 1989.
5. SCR Geomechanucs Group. The crack tip region in hydraulic fracturing. *Proc. Roy. Soc. A*, 447:39–48, 1994.
6. S.N. Gulrajani, M.G. Mack, and J. Elbel. Pressure history inversion for interpretation of fracture treatments (SPE 36439). In *1996 SPE Annual Technical Conference and Exhibition, Denver, Colorado, 6-9 October*, 1996.
7. K.G. Nolte. Determination of fracture parameteres from fracturing pressure decline (Paper SPE 8341). 1979.
8. K.G. Nolte. Fracture design consideration based on pressure analysis (Paper SPE 10911). 1982.
9. K.G. Nolte, M.G. Mack, and W.L. Lie. A systematic method for applying fracturing pressure decline: Part 1 (Paper SPE 25845). In *SPE Rocky Mountain Regional / Low Permeability Reservoirs Symposium, Denver, CO, 12-14 April*, 1993.
10. K.G. Nolte and M.B. Smith. Interpretation of fracturing pressures. *J. of Petroleum Technology*, 33:1767–1775, 1981.
11. R.P. Nordgren. Propagation of a vertical hydraulic fracture. *SPE Journal*, 12:306–314, 1972.
12. T.K. Perkins and L.R. Kern. Width of hydraulic fractures. *J. of Petroleum Technology*, 13:937–949, 1961.

THE BIMATERIAL NOTCH PROBLEM

LESLIE BANKS-SILLS

The Dreszer Fracture Mechanics Laboratory
Department of Solid Mechanics, Materials and Structures
The Fleischman Faculty of Engineering
Tel Aviv University
69978 Ramat Aviv, Israel

1. Introduction

The bimaterial notch problem has been receiving much attention recently. It is important in many industrial applications in which bonded parts are employed. Consider the geometry in Fig. 1 in which two linear elastic, homogeneous wedges are bonded to one another.

For this geometry, the stresses at the intersection between the interface and free surface are generally singular. That is, the stresses in this neighborhood behave as

$$\sigma_{ij} \sim \frac{K_1}{\left(\frac{r}{L}\right)^{\omega_1}} \, f_{ij}^{(1)}(\theta) \; + \; \frac{K_2}{\left(\frac{r}{L}\right)^{\omega_2}} \, f_{ij}^{(2)}(\theta) \tag{1}$$

where r and θ are the polar coordinates illustrated in Fig. 1, L is a characteristic length and the functions for each stress component $f_{ij}^{(k)}(\theta)$ are known universal functions of geometry and material constants. The power ω_i may be real or complex; frequently, $\omega_1 = \omega_2$.. This singularity depends upon material properties and the notch angles θ_1 and θ_2. The stress intensity factor K depends upon material properties, loading and geometry. It may be noted that there are two stress intensity factors, K_1 and K_2. Each is the amplitude of the singularity of strength ω_i.

As an example, a particular geometry is chosen to elucidate the behavior of ω (see Fig. 2). This figure was taken from Hein and Erdogan [1]. With Poisson's ratio ν_i held constant, the dependence of the singularity upon the ratio of the Young's moduli E_1/E_2 is exhibited. For small values of this

55

D. Durban and J.R.A. Pearson (eds.),
IUTAM Symposium on Non-Linear Singularities in Deformation and Flow, 55-61.
© 1999 *Kluwer Academic Publishers. Printed in the Netherlands.*

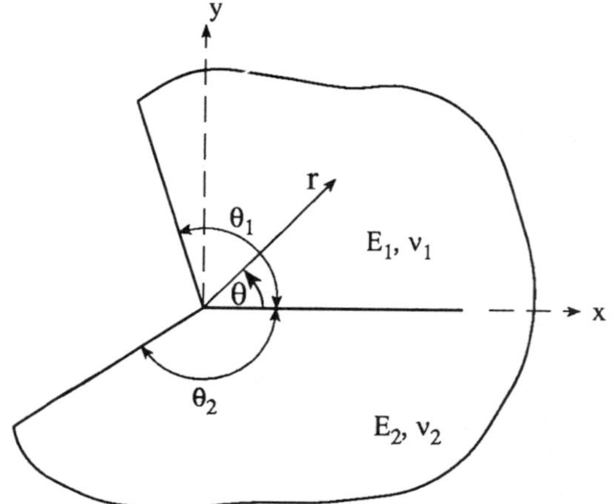

Figure 1.   Bimaterial notch geometry.

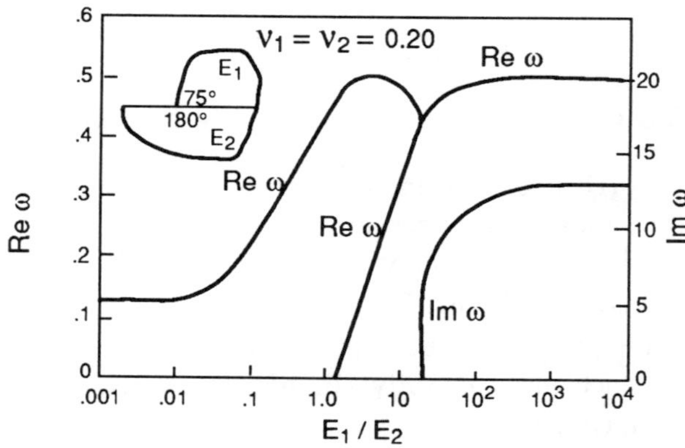

Figure 2.   Behavior of the stress singularity ω in a particular case. This figure is reproduced with kind permission from Kluwer Academic Publishers and Professor F. Erdogan.

ratio, there is one real value. As E_1/E_2 increases, two real roots exist; for higher values, there are two complex conjugate roots.

The singularity is an eigenvalue which satisfies

$$\Delta(\omega) = 0 \tag{2}$$

where

$$\Delta(\omega) = \begin{vmatrix} a_1 & -b_1 & -a_2 & b_2 \\ a_1 & c_1 & -\mu_1 a_2/\mu_2 & -\mu_1 c_2/\mu_2 \\ -\overline{b}_1 & \overline{a}_1 & \overline{b}_2 & -\overline{a}_2 \\ \overline{c}_1 & \overline{a}_1 & -\mu_1 \overline{c}_2/\mu_2 & -\mu_1 \overline{a}_2/\mu_2 \end{vmatrix}. \tag{3}$$

In (3)

$$a_i \;=\; (\omega - 1)(1 - e^{2i\theta_i}) \tag{4}$$

$$b_i \;=\; 1 - e^{-2i(\omega-1)\theta_i} \tag{5}$$

$$c_i \;=\; \kappa_i + e^{-2i(\omega-1)\theta_i} \tag{6}$$

The overbar represents complex conjugate, μ_i, $i = 1, 2$ are the shear moduli of the upper and lower materials, respectively, $\kappa_i = 3 - 4\nu_i$ for plane strain and $(3 - \nu_i)/(1 + \nu_i)$ for generalized plane stress.

For these problems, further reference may be made to [2, 3]. In particular, Pageau $et\ al.$ [4] present standard eigensolutions for bonded wedges.

In this study, two methods are considered for extracting stress intensity factors. These include an influence function method which yields an analytic solution for infinite geometries and conservative integrals to be employed with numerical methods for finite geometries. Each of these is presented in Sections 2 and 3, respectively.

2. Influence Function Method

The influence function method was developed in [5]. For the geometry and loading exhibited in Fig. 3a, the complex stress intensity factor $K = K_1 + iK_2$ is determined from

$$K \;=\; \int_0^{\infty} K^*(r_o)\, dr_o \tag{7}$$

where $K^*(r_o)$ is the influence function and $r = r_o$ is a point along the interface.

The influence function is obtained from the stress intensity factor of the problem of a dislocation along the interface at $r = r_o$ with intensities in the r and θ-directions of

$$G \;=\; \frac{\partial}{\partial r_o}\left(u_r^{(2)} - u_r^{(1)} \right)\bigg|_{\theta=0} \tag{8}$$

$$H \;=\; \frac{\partial}{\partial r_o}\left(u_\theta^{(2)} - u_\theta^{(1)} \right)\bigg|_{\theta=0} \tag{9}$$

To obtain G and H, the two auxiliary homogeneous wedge problems exhibited in Fig. 3b require solution. Each of the upper and lower wedges is solved individually yielding the displacement field along the ray $\theta = 0$. These expressions are substituted into (8) and (9).

The influence function is given by

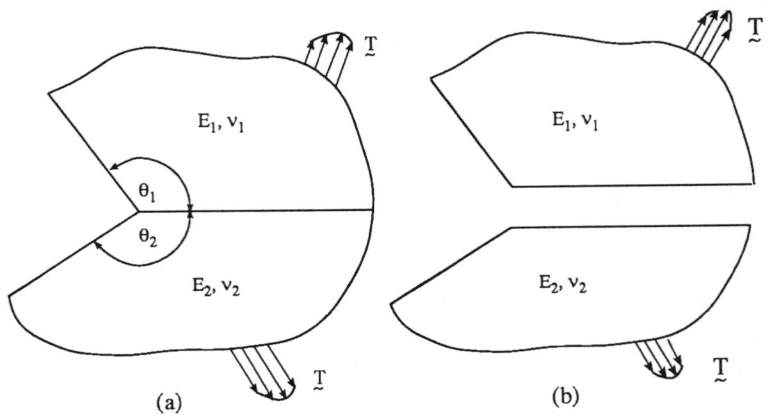

Figure 3.   (a) Bimaterial wedge problem. (b) Corresponding auxiliary homogeneous wedge problems.

$$K^*(r_o) = \lim_{r \to 0} \frac{2\mu_1(\omega - 1)}{\Delta'(\omega)} \left\{ \left[1 - e^{2i(\omega-1)\theta_1} \right] NB_1 - (\omega - 1)\left(1 - e^{-2i\theta_1} \right) N\overline{B}_1 \right\} r^{-}$$

$$(10)$$

where $\Delta'(\omega) = d\Delta/d\omega$, NB_1 is the determinant $\Delta(\omega)$ with its first column replaced by

$$\begin{bmatrix} 0 \\ \frac{(G+iH)r_o^{\omega-1}}{\omega-1} \\ 0 \\ \frac{(G-iH)r_o^{\overline{\omega}-1}}{\overline{\omega}-1} \end{bmatrix} \qquad (11)$$

and $N\overline{B}_1$ is the determinant obtained by replacing the second column of $\Delta(\omega)$ by (11).

Details for the bonded strip are given in [5]. In that investigation, solutions were obtained for a bonded strip subjected to remote tension perpendicular to the bond line, a temperature decrease and pure shear. The stress intensity factors for the first two cases were compared to a finite element solution of a long bimaterial strip [3]. Very good agreement was obtained. The latter solution for the stress intensity factor was seen to be exactly zero as expected. Other new solutions were presented.

3. Conservative Integral

Both conservative line and area integrals were developed in [6] to determine the stress intensity factor K_1 for a bonded strip. These expressions are generalized here for obtaining the complex stress intensity factor K when the eigenvalue is real.

There have been two major studies in which conservative integrals have been presented for obtaining stress intensity factors of notched homogeneous bodies. One approach was taken by Sinclair and co-workers [7, 8] and the second approach was taken by Carpenter [9]–[11]. Both approaches are based upon the Betti reciprocal principle. In the first case, one integral is calculated numerically; in the second case, there are two integrals calculated numerically, one of which is in the neighborhood of the singular stress field. Therefore, the former approach has definite advantages. The studies of Carpenter were extended to bimaterial notched bodies [12, 13]. Indeed, the method presented here has its antecedents in the works of both groups. In this implementation of the Betti reciprocal principle for deriving a conservative integral, it is necessary to perform only one numerical integral outside the singular field.

The conservative line integral is developed from the Betti's reciprocal principle as

$$K = \int_{\Gamma} \left(\tilde{T}_i' u_i - T_i \tilde{u}_i' \right) ds \qquad (12)$$

where Γ is an arbitrary path beginning on the lower notch flank and ending on the upper one (see Fig. 4a). The tractions $T_i = \sigma_{ij} n_j$ and the displacement field u_i are the solution to the problem at hand (to be found numerically). The primed traction and displacement field are complex. The real part is the eigensolution corresponding to the eigenvalue $2 - \omega$ for K_1 and the imaginary part is the eigensolution corresponding to the same eigenvalue but for K_2. It may be shown from the analysis by Bogy [2], that if ω is an eigenvalue, $2 - \omega$ is also an eigenvalue. It is assumed that there are no body forces and ds is differential arc length.

Following Li $et\ al.$ [14] for cracks in homogeneous bodies, the integral in (12) may be converted into an area integral given by

$$K = \int_{A} \left(\sigma_{ij} \tilde{u}_i' - \tilde{\sigma}_{ij}' u_i \right) q_{1,j} \, dA \qquad (13)$$

where σ_{ij} and u_i are the stress and displacement field in the region A obtained from a numerical analysis of the problem, and $\tilde{\sigma}_{ij}'$ and $\tilde{u}_i'$ are the eigensolution for the eigenvalue $2 - \omega$. The region A is chosen here to be a ring as illustrated in Fig. 4b. The function q_1 is taken to be zero on C_1 and

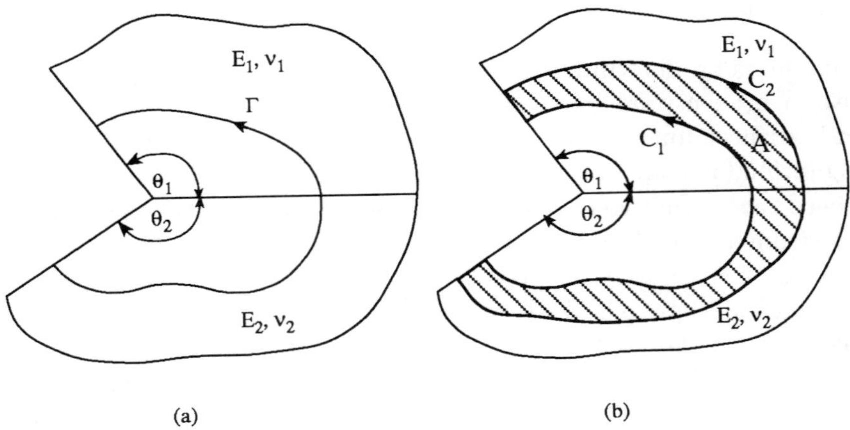

Figure 4.   (a) Path for conservative line integral and (b) ring for conservative area integral.

unity on C_2; it must be continuously differentiable within A. The comma represents differentiation.

For a finite element analysis, A is chosen as a ring of elements and

$$q_1 \; = \; \sum_{i=1}^{8} N_i(\xi, \eta)\, q_{1i} \tag{14}$$

where an eight noded, isoparametric element is employed. In (14), $N_i(\xi, \eta)$ are the shape functions and ξ and η are the coordinates of the parent element. The vector q_{1i} is chosen to fulfill the requirements already defined for q_1 with the additional restriction that the nodal points of the element are mapped in such a way that midpoint nodes remain in that position in the mapped configuration (see [15] for details when applied to cracked bodies). The last restriction increases accuracy. It may be noted that q_1 is actually a mapping of the physical element.

In [6], the integrals in (12) and (13) were employed to calculate K_1 for a finite height bonded strip subjected to remote stress normal to the bond line. Explicit expressions for the primed stresses and displacements were given in an Appendix of that study. All results agreed to within 3.2% of the analytic solution obtained with the influence function method described in Section 2 for an infinite strip. It was seen that results obtained by the area integral were the same for all paths.

4. Conclusions

Two methods for determining stress intensity factors of bimaterial notched bodies have been presented. These include the influence function method and conservative integrals. The influence function method leads to an analytic solution for bodies of infinite extent. The conservative integrals may be employed with numerical calculations such as finite elements for finite bodies. Both methods when applied to the bimaterial strip produce accurate results.

References

1. Hein, V.L., and Erdogan, F.: Stress singularities in a two-material wedge, *International Journal of Fracture Mechanics* **7** (1971), 317–330.
2. Bogy, D.B.: Two edge-bonded elastic wedges of different materials and wedge angles under surface tractions, *Journal of Applied Mechanics* **38** (1971), 377–386.
3. Munz, D., and Yang, Y.Y.: Stress singularities at the interface in bonded dissimilar materials under mechanical and thermal loading, *Journal of Applied Mechanics* **59** (1992), 857–861.
4. Pageau, S.S., Gadi, K.S., Biggers, Jr., S.B., and Joseph, P.F.: Standardized complex and logarithmic eigensolutions for n-material wedges and junctions, *International Journal of Fracture* **77** (1996), 51–76.
5. Banks-Sills, L., Yang, Y.Y., and Munz, D.: An influence function for stress intensity factors of bimaterial notched bodies, accepted for publication: *International Journal of Fracture*.
6. Banks-Sills, L.: A conservative integral for determining stress intensity factors of a bimaterial strip, submitted for publication.
7. Sinclair, G.B., Okajima, M., and Griffin, J.H.: Path independent integrals for computing stress intensity factors at sharp notches in elastic plates, *International Journal for Numerical Methods in Engineering* **20** (1984), 999–1008.
8. Sinclair, G.B.: A remark on the determination of mode I and mode II stress intensity factors for sharp re-entrant corners, *International Journal of Fracture* **27** (1985), R81–R85.
9. Carpenter, W.C.: Calculation of fracture mechanics parameters for a general corner, *International Journal of Fracture* **24** (1984), 45–58.
10. Carpenter, W.C.: Mode I and II stress intensities for plates with cracks of finite opening, *International Journal of Fracture* **26** (1984), 201–214.
11. Carpenter, W.C.: The eigenvector solution for a general corner of finite opening crack with further studies on the collocation procedure, *International Journal of Fracture* **27** (1985), 63–74.
12. Carpenter W.C., and Byers, C.: A path independent integral for computing stress intensities for v-notched cracks in a bi-material, *International Journal of Fracture* **35** (1987), 245–268.
13. Carpenter, W.C.: Insensitivity of the reciprocal work contour integral method to higher order eigenvectors, *International Journal of Fracture* **73** (1995), 93–108.
14. Li, F.Z., Shih, C.F., and Needleman, A.: A comparison of methods for calculating energy release rates, *Engineering Fracture Mechanics* **21** (1985), 405–421.
15. Banks-Sills, L., and Sherman, D.: On the computation of stress intensity factors for three-dimensional geometries by means of the stiffness derivative and J-integral methods, *International Journal of Fracture* **53** (1992), 1–20.

CRACK DEVELOPMENT IN SPATIALLY RANDOM STRESS FIELDS GENERATED BY POINT DEFECTS. FRACTURE IN COMPRESSION

A.V. DYSKIN
Department of Civil Engineering
The University of Western Australia
Nedlands, WA 6907, AUSTRALIA

Abstract. Cracks in random stress fields are assumed to be originated in regions with high local tension. As a legacy of this special location, additional local tractions opening the crack in its centre are developed even in self-equilibrating stress fields. As the crack becomes a mesocrack it will deviate its path to meet the regions with higher possible local tension. In fracture of brittle materials under uniaxial compression wing cracks are developed which, in real 3-D situations, cannot grow extensively and therefore cannot themselves cause failure. Instead, they induce stress fluctuations which generate mesocracks growing towards compression such a way to avoid the wing cracks. Hence, only stresses outside excluded volumes around the wing cracks will affect the mesocrack growth. These stresses have positive mean even if the full stress field is self-equilibrating. This results in a background tension acting perpendicular to the compression axis, amplifying the mesocrack growth and eventually causing failure.

1.   Introduction

Static (quasistatic) crack growth is usually considered as being caused by the action of an externally applied load which is supposed to be known. There are some cases, however, when in addition to this load there exist self equilibrating stress fluctuations acting on the crack. The role of these additional stresses can be played by residual or thermally induced stresses (eg, Bower and Ortiz, 1993; Ortiz and Suresh, 1993) or by stresses induced by the multitudes of defects and heterogeneities (eg, Kunin, 1983). Due to the obvious randomness in the distribution of the residual stresses and/or defects and heterogeneities, this additional stress field may be considered as spatially random (Mori and Tanaka, 1973). A special case is the random stress field produced by defects or heterogeneities in uniaxial or biaxial compression, where the stress components acting on the directions perpendicular to the compression axes have zero average. Then there is a possibility that the experimentally observed initiation and growth of planar cracks parallel to the compression direction (Dyskin *et al.*, 1994a) may be caused by the stress fluctuations. This will be investigated in the present paper.

63

D. Durban and J.R.A. Pearson (eds.),
IUTAM Symposium on Non-Linear Singularities in Deformation and Flow, 63-74.
© 1999 *Kluwer Academic Publishers. Printed in the Netherlands.*

When crack growth is considered in a spatially random stress field there are two cases to be distinguished. The first case is a pre-existing crack immersed in an independent stress field, Figure 1a. If the crack size is much larger than the correlation radius of the field, the stress fluctuations will be averaged out, so the main effect will be delivered by the mean stress, since the total force associated with the fluctuations will be zero. More interesting is the second case, Figure 1b, in which the crack itself is created by the random stress fluctuation. In this case the crack will obviously be created at the place with the maximum stress. This will be shown to result in a non-zero total force associated with the fluctuations, which constitutes a special mechanism of crack growth considered in the present paper.

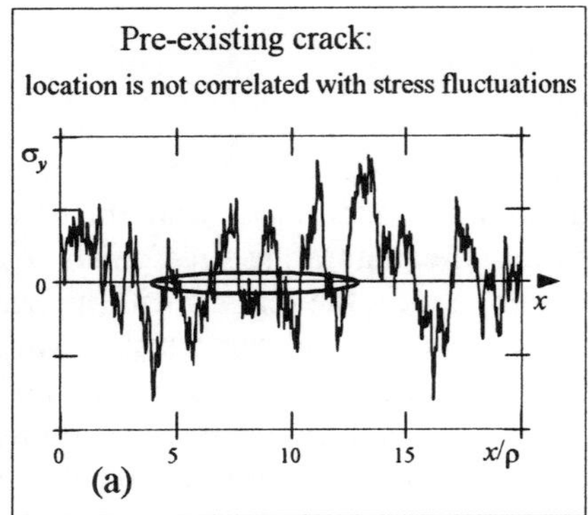

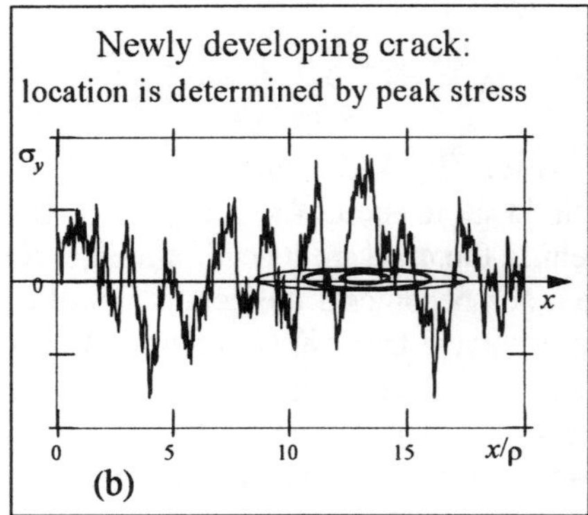

Figure 1. Two cases: (a) crack independent of the stress fluctuations; (b) crack created by the stress fluctuations.

2.   Crack growth in random (Gaussian) stress field

2.1.   THE EFFECT OF FLUCTUATIONS

Suppose that a spatially random stress field $\sigma(\mathbf{x})$ exists within a loaded sample. It is assumed that this stress field is statistically homogeneous (ie. invariant with respect to translations) with the mean $<\sigma>=\text{const}$ and the correlation function $B(\mathbf{x})=<\sigma(0)\sigma(\mathbf{x})>-<\sigma>^2$. It will also be assumed that the random stress field can be approximated by the Gaussian one (numerical experiments by Ortiz and Suresh, 1993, indeed demonstrate the Gaussian nature of residual stresses), ie. it is completely determined by the tensors $<\sigma>$ and $B(\mathbf{x})$.

Suppose the stress magnitude is sufficient to produce local fractures (cracks) somewhere in the material. Obviously this will first happen at a location where the magnitude of the major principal stress, $\sigma(0)>0$, is a maximum. Let us put the origin of a Cartesian coordinate frame at this place. In a vicinity of the origin the stress distribution will be determined by $\sigma(0)$. The average distribution of the corresponding stress component in the intact material on the plane (x_1, x_3) where the crack will be located has the form (eg, Feller, 1971):

$$\langle \sigma(\mathbf{x}) | \sigma(0) \rangle = \langle \sigma \rangle + \langle \Delta\sigma(\mathbf{x}) \rangle, \quad \langle \Delta\sigma(\mathbf{x}) \rangle = [\sigma(0) - \langle \sigma \rangle] B(\mathbf{x})/B(0) \quad (1)$$

where $\langle \cdot | \cdot \rangle$ stands for the conditional mathematical expectation and $\mathbf{x}=(x_1, x_3)$. Figure 2a illustrates this distribution (the x_2-axis is the major principal direction).

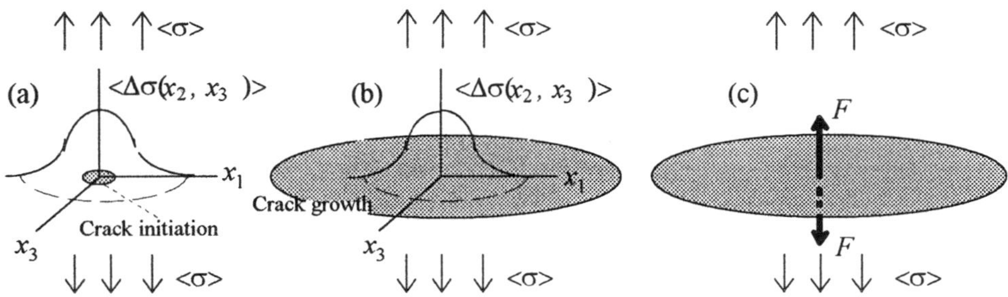

Figure 2. The average load distribution caused by stress fluctuations (a); the load acting on the growing crack (b); the pair of concentrated forces and uniform stress modelling the load (c).

When the crack appears, the problem of determining its opening can formally be solved by considering the crack loaded with tractions equal, with inverse sign, to the corresponding stress components acting in the original material. This means that there is no direct back influence of the crack on the original stress field, unless the crack affects the sources of the stresses (eg. microcracks). Only the case when the crack influence on the stress-generating microcracks can be neglected will be considered here.

As the crack grows, it is subjected to additional tractions with the mean $<\Delta\sigma(x_1, x_3)>$, in excess of the usual mean stress $<\sigma>$, Figure 2b (note, while a 2-D crack can be arrested by regions of high local compression at its tips, in 3-D, it is unlikely that the whole crack contour will be blocked by such regions, so there will always be parts of the contour capable of growing). This additional load is solely due to the special location from which the crack has evolved and is a result of stress fluctuations.

For the sake of simplicity and to make the following analysis possible the correlation function will be presumed to vanish strongly enough as $|\mathbf{x}| \to \infty$, so, in average, the crack can be modelled as a crack loaded by uniform stress $<\sigma>$ and a pair of concentrated forces (Figure 2c) with the magnitude

$$F = \int_{\mathbf{R}^2} \langle \Delta\sigma(\mathbf{x}) \rangle dx_1 dx_2 = \frac{\sigma(0) - \langle \sigma \rangle}{\sqrt{B(0)}} \kappa, \quad \kappa = \frac{1}{\sqrt{B(0)}} \int_{\mathbf{R}^2} B(\mathbf{x}) dx_1 dx_2 \quad (2)$$

2.2.   CRACK GROWTH AT THE SCALE OF THE CORRELATION RADIUS

If the crack is approximated by a disk-like crack of radius R, the propagation criterion can be expressed through the average stress intensity factor (SIF) as follows (eg, Cherepanov, 1979)

$$\langle K_I \rangle = K_{Ic}, \quad \langle K_I \rangle = \frac{F}{(\pi R)^{3/2}} + 2\langle\sigma\rangle\sqrt{R/\pi} \tag{3}$$

where F is given by (2), $<\sigma>$ is the mean stress and K_{Ic} is the fracture toughness of the material. Such a crack will initially grow in a stable manner until it reaches a critical radius, R_{cr}, starting from which the unstable growth will ensue.

$$R_{cr} = \sqrt{3F/2\pi\langle\sigma\rangle} \tag{4}$$

The important conclusion is that the stress fluctuations and the special crack location result in development of a concentrated force, F. This increases the SIF and hence reduces the critical load, ie. the strength, compared to what could be expected if the described mechanism were not taken into account.

2.3.   MESOCRACK GROWTH

When the concentrated force, F, is high, or the mean $<\sigma>$ is very low (as, for example, in uniaxial compression in the direction perpendicular to the load), $R_{cr} >> \rho$. Such a crack will further be called *mesocrack*, indicating that the crack is now much larger than the microscale of the stress fluctuations, but small compared to the external sizes, such that it still can be considered as being in an infinite body. Then, in the course of its stable growth the mesocrack will pass near several regions of high local tensile stresses. It is hypothesised here that the mesocrack will deviate or branch or sprout new cracks in order to pass through the nearest region with the maximum possible tensile stresses (the probability of two or more regions simultaneously happening on the crack path is neglected), Figure 3a. This will produce a tongue of the size of the correlation radius, ρ, of the stress fluctuations, which will then be smoothened by additional growth of other parts of the crack contour, Figure 3b. Thus, a larger crack will be formed of approximately disk shape. This will result in adding a new pair of concentrated forces each time the crack passes the chosen region. Since at each step the crack chooses only one region, the total number of concentrated forces will be proportional to the crack size.

The values of the concentrated forces can be computed from (2) if the stress fluctuation is known. By assumption, each step in the mesocrack propagation involves choosing the region with maximum fluctuation (the region area is $\sim\rho^2$) from the vicinity of the current crack contour (the total area of $2\pi r\rho$). This corresponds to choosing a maximum from $n\sim 2\pi r/\rho$ normally distributed random stress disturbances, $\Delta\sigma_1, ..., \Delta\sigma_n$, which can be considered as approximately independent, since they are located more than the correlation radius apart from each other. The random variable

$(2\ln n)^{1/2}(\Delta\sigma_{max}B(0)^{-1/2}-(2\ln n)^{1/2})$ has asymptotically, for large n, the distribution function $\Lambda_3(x)=\exp(-\exp(-x))$ (eg, David, 1970), hence

$$\frac{E(\Delta\sigma_{max})}{\sqrt{B(0)}} \sim \sqrt{2\ln n} + \frac{\gamma}{\sqrt{2\ln n}}, \quad \gamma = 0.577 \tag{5}$$

Keeping only the leading term in (5) and taking into account that $n\sim 2\pi r/\rho$, one has

$$E\left(\sigma(0)_{max}\right) - \langle\sigma\rangle \sim \sqrt{2B(0)\ln 2\pi r/\rho} \tag{6}$$

This value should now be used in (2) instead of $\sigma(0)-\langle\sigma\rangle$.

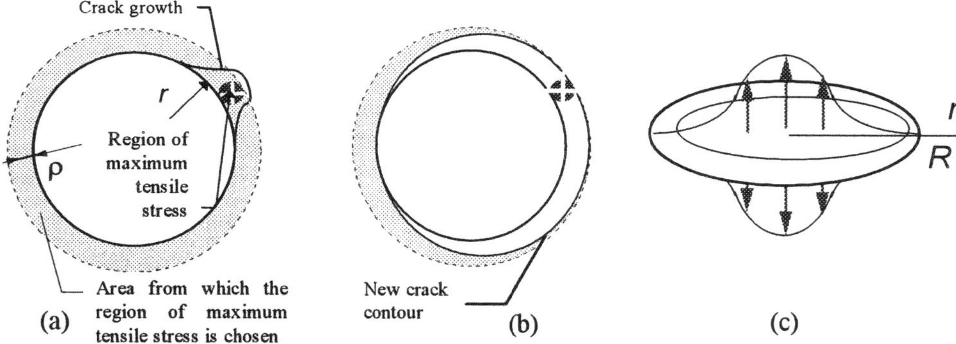

Figure 3. Hypothesised mechanism of mesocrack growth: (a) the crack grows first to the region with maximum tensile stress; (b) other parts of the crack grow to smoothen the shape; (c) a model of mesocrack growth.

For the sake of simplicity the mesocrack will be modelled by a disk-like crack loaded by the uniform load $\langle\sigma\rangle$ and additional tractions modelling the concentrated forces, Figure 3c. Since only one pair of forces is added at each step the average traction acting at the new surface, $2\pi r\cdot(\rho/2)$, will be $F/\pi\rho r$. At each step from all $2\pi r/\rho$ regions located near the crack perimeter only the one with the maximum stress is chosen. Hence the tractions are:

$$q(r) = \frac{F(r)}{\pi\rho r}, \quad F(r) = \kappa\sqrt{2\ln\frac{2\pi r}{\rho}} \tag{7}$$

where κ is given by (2). The average correlation radius can be computed as

$$\rho = \langle\rho(\eta)\rangle, \quad \rho(\eta) = B(0)^{-1}\int_0^\infty B(t\eta)dt \tag{8}$$

Here averaging is performed over all orientations of the unit vector η.

Because the logarithmic dependence (7) is weak, $F(r)$ can be replaced with its value for $r=R$. Using this approximation and the known result for axisymmetrically loaded disk-like crack of radius R (eg, Cherepanov, 1979) one obtains the average SIF, an equation for determining the critical crack radius, R_{cr} (the radius corresponding to the onset of unstable crack growth found from the condition $d\langle K_I \rangle/dR=0$) and the criterion of unstable crack growth:

$$\langle K_I \rangle = \frac{F(R)}{\rho\sqrt{\pi R}} + 2\langle\sigma\rangle\sqrt{R/\pi}, \quad R_{cr} = \frac{F(R_{cr})}{2\rho\langle\sigma\rangle}, \quad \frac{8\langle\sigma\rangle F(R_{cr})}{\pi\rho} = K_{Ic}^{2} \quad (9)$$

This approximation also allows calculating the volume of crack opening, ie the crack contribution to dilatancy, the main precursor of fracture in compression (eg, Brace *et al.*, 1966).

$$V = \frac{1-\nu}{2\mu}\left[\frac{F(R)R^2}{\rho\pi} + \frac{16}{3}\langle\sigma\rangle R^3\right] \quad (10)$$

where μ and ν are, respectively, the shear modulus and Poisson's ratio of the material.

3.   Microcrack - generated random stress fields

3.1.   EXCLUDED VOLUMES AROUND MICROCRACKS ASSOCIATED WITH
       GROWING MESOCRACK

As an example, the statistical properties of the stress field generated by randomly located microcracks will be calculated. In what follows, a special type of stress-generating microcracks will be considered, namely wing cracks developed in brittle materials (eg. rocks), in compression, from approximately disk-like cracks inclined to the compression axis produced, Figure 4a. As shown by Dyskin *et al.* (1994a), in 3-D, the wing cracks have a very limited ability to grow, so further loading can only increase the number of wing cracks. Then, the additional stress fields produced by the wing cracks will initiate new cracks oriented parallel to the compression direction. These new cracks are tensile, capable of growing extensively and thus correspond to mesocracks in the terminology adopted here.

A distinctive feature of wing cracks is that the stress-producing wing opening is mainly controlled by the shearing over the initial cracks which, in its own turn, is fully determined by the high compression. Therefore the change in the opening associated with the influence of additional stresses produced by other cracks can, in the first approximation, be neglected.

To avoid technical difficulties the microcracks will be modelled by point defects producing asymptotically, at large distances, the same stress field as the original microcracks (eg, Gol'dstein and Kaptsov, 1982; Kunin, 1983; Dyskin *et al.*, 1996). This asymptotics has a strong non-integrable singularity at the origin, which requires the introduction of *excluded volumes* surrounding each microcrack (microdefect) whose size is an additional parameter (eg, Kunin, 1983). In general, this is a trick, since real microcracks do not produce non-integrable singularities. However, there is a case when the excluded volumes appear quite naturally.

First, the microcracks are usually produced at some distance from each other, which is related to distribution of the planes of weakness where the microcracks are generated. This implies the existence of excluded volumes in the microcrack locations.

Second, the excluded volumes can also be treated as volumes which the mesocrack will try to avoid during its growth. Germanovich *et al.* (1994) (see also Dyskin *et al.*, 1994b) have found from testing of transparent samples with internal disk-like cracks in uniaxial compression that if one 3-D crack is located on the path of another, the latter will neither get through, nor get arrested as routinely observed in 2-D tests (eg, Horii and Nemat-Nasser, 1985). Instead, it will merely bypass the crack-obstacle and proceed further. The obvious explanation is in the presence of the third dimension: in 3-D, there is enough room for such manoeuvres. Cut-sections of rocks (Peng and Johnson, 1972) also show traces of large cracks growing in the direction of uniaxial compression avoiding small ones.

These observations prompt a hypothesis that microcracks (wing cracks) are surrounded by *excluded volumes* which are avoided by the mesocrack, Fugure 4b. This means that when the microcrack-generated random stress field is considered from the point of view of its effect on the mesocrack propagation, only the points that are outside the excluded volumes should be taken care of. The shape and size of the excluded volumes are dictated by the mesocrack-microcrack interaction. However in the following simplified consideration the excluded volumes will be assumed spherical with the radius a equal to the characteristic size of the microcrack (the initial crack radius, Figure 4a).

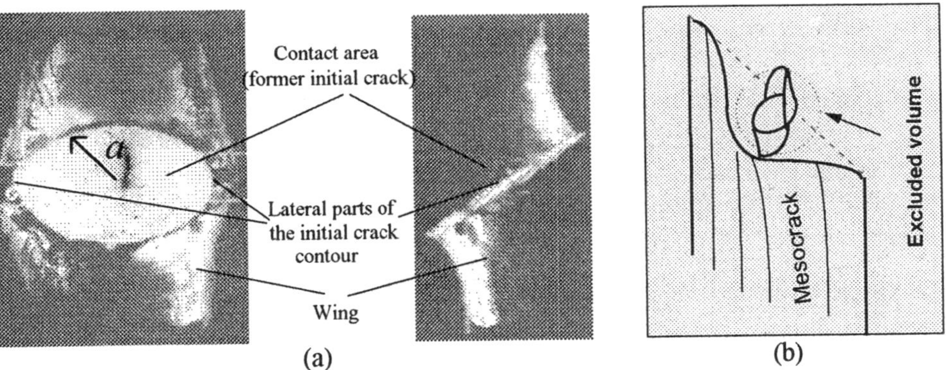

Figure 4. Wing crack and mesocrack growth in uniaxial compression (the compression direction is vertical): (a) 3-D wing crack, after Dyskin *et al.* (1994a); (b) the mesocrack avoids the excluded volume.

3.2.   STATISTICAL PROPERTIES OF MICROCRACK-GENERATED STRESS FIELD

Consider the stress field generated by microcracks located at homogeneously random points ξ_α in a volume V. The (tensor) stress at a point $\mathbf{x}$ is

$$\sigma(\mathbf{x}) = \sigma^0 + \sum_\alpha \sigma_V\left(\mathbf{x}, \xi_\alpha\right) \tag{11}$$

Here σ^0 is the stress which would be created by the external load in the absence of microcracks, and $\sigma_V\left(\mathbf{x}, \xi_\alpha\right)$ is the additional stress generated by a microcrack located at a point ξ_α in the volume V (here $\sigma_V(\mathbf{x}, \xi_\alpha)$ is the exact stress; its dipole asymptotics will be used later).

Suppose the interaction between the microcracks can be neglected. Then for the case of equal excluded volumes, V_0, the expected value is (due to the statistical homogeneity it is sufficient to consider stresses at the origin)

$$\langle\sigma\rangle = \sigma^0 + \frac{M}{V-V_0}\int_{V-V_0}\sigma_V\left(0,\xi\right)dV_\xi$$

$$= \sigma^0 + \frac{M}{V-V_0}\left[\int_V\sigma_V\left(0,\xi\right)dV_\xi - \int_{V_0}\sigma_V\left(0,\xi\right)dV_\xi\right] \tag{12}$$

where M is the number of microcracks in volume V. As $V\to\infty$, since the integral over V can be set equal to zero (eg, Kunin, 1983; Germanovich and Dyskin, 1994), (12) assumes the form

$$\langle\sigma\rangle = \sigma^0 - N\int_{V_0}\sigma\left(\xi\right)dV_\xi \tag{13}$$

For low microcrack concentrations when the influence of the excluded volumes on the microcrack distribution is negligible, the random variables ξ_α can be presumed statistically independent. Then the correlation function assumes the form

$$B(\mathbf{x}) = \langle\sigma(0)\sigma(\mathbf{x})\rangle - \langle\sigma\rangle^2 = N\int_{V-V_0}\sigma\left(\xi\right)\sigma\left(\xi - \mathbf{x}\right)dV_\xi \tag{14}$$

Now let us approximate the microcrack-generated stress field by its far-field asymptotics which can be expressed in a coordinate set (x_1, x_2, x_3) as follows (Gol'dstein and Kaptsov, 1982)

$$\Delta\sigma_{ij}(\xi) = -\frac{(1-2v)\mu}{4\pi(1-v)\xi^3}\left\{\delta_{ij}\left[\frac{1-4v}{1-2v}u_{kk}-3W\right]\right.$$
$$\left.+\xi_{,i}\xi_{,j}\left[\frac{15W}{1-2v}-3u_{kk}\right]-\frac{6v}{1-2v}\left[\xi_{,j}\xi_{,k}u_{ki}+\xi_{,i}\xi_{,k}u_{kj}\right]-2u_{ij}\right\}$$

$$(15)$$

Here $W=u_{ik}\xi_{,i}\xi_{,k}$, $\xi=|\xi|$, u_{ik} is the symmetrical tensor of dipole moments (eg, Dyskin *et al.*, 1996), ",*i*" is derivative with respect to x_i and summation is presumed over repeated indices.

Two cases will be considered: (1) dipole moments corresponding to crack opening in x_2 direction, modelling parallel wing cracks. This situation which will be called *unidirectional opening* corresponds to fracture in biaxial compression where all cracks grow perpendicular to the direction that is free from stress; (2) dipole moments corresponding to equal openings in all directions perpendicular to x_3, which models wing cracks developed in *uniaxial compression* (this case will be called *axisymmetrical opening*). In both cases the load of magnitude $p>0$ is applied along x_3 and σ, B mean σ_{22}, B_{22} (for axisymmetrical opening $\sigma_{11}=\sigma_{22}$, $B_{11}=B_{22}$). The excluded volumes are of radius a.

In the first case the corresponding component of the tensor of dipole moments is assumed equal to the maximum total volume of opening of both wings produced by pre-existing cracks (Dyskin *et al.*, 1996), averaged over all inclinations of initial cracks to the loading directions, in the second case the averaging is performed over all orientations of initial cracks. This gives

$$u_{ik}=\begin{cases}2\delta_{i1}\delta_{k1}V_{wing} & \text{unidirectional}\\ \pi^{-1}\delta_{ik}(1-\delta_{k3})V_{wing} & \text{axisymmetrical}\end{cases}, \qquad V_{wing}\cong pa^3\frac{(2-v)}{(1+v)^2\mu} \qquad (16)$$

After substituting (16) into (15) and then into (7), (8), (13), (14), calculating integrals for $v=0$; 0.25 and 0.5 and fitting a linear dependence for arbitrary v, one has for the variance, the average stress, the average correlation radius and parameter κ:

$$B(0)=\frac{(2-v)^2 I(v)Y^2}{4\pi(1+v)^2(1-v^2)^2}vp^2, \qquad \langle\sigma\rangle=\frac{2}{15\pi}\frac{(2-v)\Phi(v)}{(1+v)(1-v^2)}pv$$

$$(17)$$

$$\rho=J(v)a, \qquad \kappa=\frac{(2-v)K(v)\sqrt{I(v)}}{2\pi(1+v)(1-v^2)}pa^2\sqrt{v}$$

Here $v=Na^3$ is the dimensionless microcrack concentration and the introduced axillary functions are:

Unidirectional opening:
$I(v)=1.83$, $J(v)=0.772$, $Y=2$, $K(v)=7.23$, $\Phi(v)=8(1+2v)$.

Axisymmetrical opening (uniaxial compression):
$I(v)=(4/15)[4v^2+(24/7)v+39/7]$, $J(v)\approx0.863-0.216v$, $Y=\pi^{-1}$, $K(v)\approx\pi^{-1}(5.24-1.20v)$.

For the *unidirectional opening* the correlation function is axisymmetric in the plane (x_1, x_3), the plane of mesocrack growth. Its dependence on distance $|\mathbf{x}|$ in this plane is shown in Figure 5a. For the *axisymmetrical opening* the contour lines for the correlation function in the plane (x_1, x_3) are shown in Figure 5b for $v=0$. In both cases a rapid decay of the correlation function with the increase of $|\mathbf{x}|$ is evident, which justifies the assumption made in the end of section 2.1.

It should be emphasised that due to the presence of excluded volumes the mesocracks avoid during growing, $\langle\sigma\rangle$ is positive, so the cracks are subjected to uniform tensile stress in the directions perpendicular to that of compression. This stress will be called the *background tensile stress*.

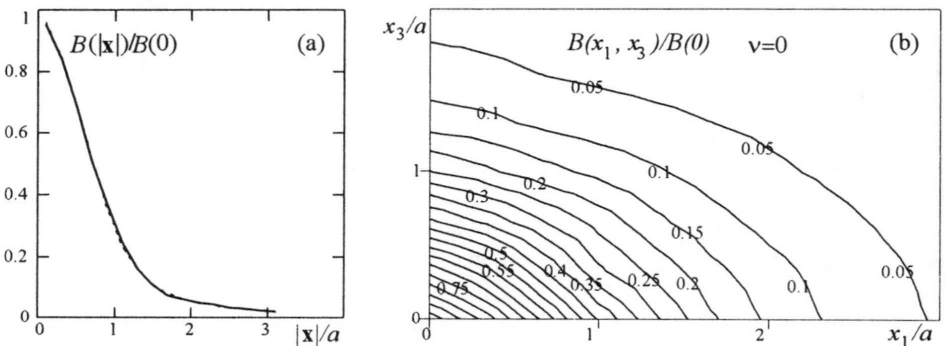

Figure 5. The correlation function in the crack plane for uniaxial tension (a) and compression (b).

4.   Dilatancy and fracture in uniaxial compression

Using (7), (9), (10), expressions (17) for axisymmetrical case and the criterion of crack growth $\langle K_I\rangle = K_{Ic}$, the dilatancy, ie. the volume increment, $\Delta\varepsilon_V$, produced by the opening of the mesocracks, can be expressed as follows

$$\frac{\Delta\varepsilon_V}{\Delta\varepsilon_V^{\max}} = \frac{p\lambda^2(2+\lambda)}{3p_{cr}}, \quad \lambda = \left(\frac{p_{cr}}{p}\right)^2\left[1-\sqrt{1-\left(\frac{p}{p_{cr}}\right)^2}\right]^2$$

$$\Delta\varepsilon_r^{\max} = A\frac{(1-v)}{\sqrt{2}\mu}\left(1+\tfrac{8}{3}\pi\right)Ka^3\sqrt{v}p_{cr} \tag{18}$$

where K is the number of mesocracks per unit volume. This dependence is shown in Figure 6a and compared with the one obtained and verified against all available experimental data by Germanovich *et al.* (1993). In the second dependence the mechanism of critical crack growth is in the interaction between the growing cracks. It is seen that both dependencies are very similar suggesting the importance of both mechanisms, the background tension and crack interaction, for dilatancy and fracture of heterogeneous materials in uniaxial compression.

The uniaxial compressive strength, p_c, can be obtained by solving the last equation in (9) using (7) and (17). Figure 6b shows the dependence of the strength on the concentration, v, of wing cracks. On the plots, the compressive strength is related to the tensile strength, $\sigma_t = 1/2 K_{Ic}(\pi/a)^{1/2}$ (it is assumed that failure in tension is caused by unstable growth of a suitably oriented initial disk-like crack of radius a). It is seen that for low concentration of wing cracks the compressive strength may exceed the tensile one by an order of magnitude and is generally within the range (3-10) typical for brittle materials (eg, Paul, 1968).

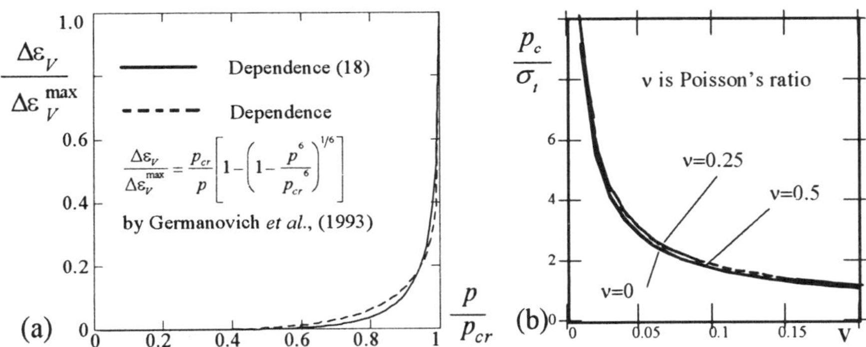

Figure 6. Model of dilatancy and fracture in uniaxial compression: (a) dilatancy vs. compressive load; (b) ratio of compressive to tensile strength vs. wing crack concentration.

5.   Conclusions

Even self-equilibrating spatially random stress fields in a material can cause crack growth provided that the crack has initially been created at a place of high local tension. The additional load induced by the stress fluctuations can be modelled by pairs of concentrated forces reflecting the action of the regions with high tensile stresses. As the crack grows and turns into a mesocrack, it tends to deviate its path to meet the regions with higher possible local tension. As a result, the crack grows further under the action of concentrated forces distributed randomly and uniformly with respect to the crack radius.

In uniaxial and biaxial compression the random stress field is created by wing cracks that are not capable of extensive growth. The mesocrack will tend to avoid them such that the stress field driving the crack is the stress outside the excluded volumes surrounding the microcracks. This outside stress has a positive (tensile) mean, which turns stable mesocrack growth into unstable one and causes failure.

For the case of spherical excluded volumes the correlation radius is found to be less than the microcrack radius, which suggests that the stresses acting on each microcrack can be assumed statistically independent.

6. Literature

Bower, A.F. and Ortiz, M. (1993) An analysis of crack trapping by residual stresses in brittle solids. *J. Appl. Mech.* **60**, 175-182.

Brace, W.F., Paulding, B.M. and Scholz, C. (1966) Dilatancy in the fracture of crystalline rocks. *J. Geophys. Res.* No. 16, 3939-3953.

Cherepanov, G.P. (1979) *Mechanics of Brittle Fracture*, N.Y. McGraw-Hill.

David, H.A. (1970) *Order Statistics.* John Wiley & Sons, Inc. New-York, London.

Dyskin, A.V., Jewell, R.J., Joer, H., Sahouryeh, E. and Ustinov, K.B. (1994a) Experiments on 3-D crack growth in uniaxial compression. *Intern. J. Fracture* **65**, R77-R83.

Dyskin, A.V., Germanovich, L.N., Jewell, R.J., Joer, H., Krasinski, J.S. Lee, K.K., Roegiers, J.-C., Sahouryeh, E. and Ustinov, K.B. (1994b) Study of 3-D mechanisms of crack growth and interaction in uniaxial compression. *ISRM News Journal*, **2**, No. 1, 17-20, 22-24.

Dyskin, A.V., Germanovich, L.N. and Ustinov, K.B. (1996) Modelling of 3-D crack growth and interaction in uniaxial compression. Proc. *1st Australian Congress on Applied Mechanics*, R.H. Grzebieta, A.K. Wang, J. Marco & Y.C. Lam (Eds.) **1**, 139-144.

Feller, W. (1971) *An Introduction to Probability Theory and its Applications.* Vol. II. John Wiley & Sons, Inc. New-York, London, Sydney, Toronto.

Germanovich, L.N., Dyskin, A.V. and Tsyrulnikov, M.N. (1993) A model of brittle rock deformation under compression. *Mechanics of Solids* **28**, 116-128.

Germanovich, L.N. and A.V. Dyskin, (1994) Virial expansions in problems of effective characteristics. Part I. General concepts. *J. Mech. Composite Materials*, 30, No. 2, 222-237.

Germanovich, L.N., Lee, K.K, and Roegiers, J.-C. (1994) Study of three-dimensional crack initiation and propagation in transparent brittle material subject to compression. Part III. Experimental results on uniaxial compression of frozen PMMA samples. *Rock Mechanics Research Center at The University of Oklahoma*: Report RMRC-94-15.

Gol'dstein R.V. and Kaptsov, A.V. (1982) The formation of structures of fracture by low-interacting cracks. *Mechanics of Solids* **17**, 157-166.

Horii H. and Nemat-Nasser, S. (1985) Compression-induced microcrack growth in brittle solids: axial splitting and shear failure, *J. Geophys. Res.* **90**, 3105-3125.

Kunin, I..A. (1983) *Elastic Media with Microstructure. Vol. 2. Three-dimensional models.* Springer Series in Solid-State Sciences, No. 44.

Mori, T. and Tanaka, K. (1973) Average stress in matrix and average elastic energy of materials with misfitting inclusions. *Acta Met.* **21**, 571-574.

Ortiz, M. and Suresh, S. (1993) Statistical properties of residual stresses and intergranular fracture in ceramic materials. *J. Appl. Mech.* **60**, 77-84.

Paul, B. (1968) Macroscopic criteria for plastic flow and brittle fracture. In: Fracture. An Advanced Treatise (H. Liebowitz, ed.) **II**, 313-496.

Peng, S. and Johnson, A.M. (1972) Crack growth and faulting in cylindrical specimens of Chelmsford granite. *Int. J. Rock Mech. Min. Sci. & Geomech. Abstr.* **9**, 37-86.

ELASTIC INTERACTION OF DISLOCATIONS WITH A CRACK IN A DISK

SHMUEL VIGDERGAUZ
R&D Division, The Israel Electric Corp., Ltd.,
P.O.Box 10, Haifa 31000, Israel

Abstract

Under the assumption of linear elasticity numerical solutions of stress field for a circular disk containing a dislocation dipole and a traction-free straight crack are obtained in this paper. The disk may be subjected to self-balanced external forces. The analysis based on solving a system of linear algebraic equations establishes a semi-analytic relation between the disk geometry and the stress intensity factor (SIF) at the crack tip. Some typical results are exemplified graphically.

1. Introduction

Dislocations are inherent in solids. In particular, dislocations which have the net Burgers vector [4] are often used to ideally represent the observed fracture phenomena in an elastic continuum. See, for instance, [2] and references therein. Although such phenomena are evidently important for understanding the mechanical behavior of many engineering materials and they have received proper attention, only a few and rather simplified situations have been studied numerically so far. Mainly, this is due to the fact that every individual case requires the solution of a full-scaled boundary value problem of elasticity involving predefined singularities. Further difficulties frequently encountered when using elastic models are associated with stringers, cracks and other types of discontinuity. So it is important to know, with fair accuracy, the stress distribution in the body undergone to dislocation strains, even if this information only approximates the real defects in a crystal lattice.

In this paper, the numerical analysis is conducted for problems that involve the direct interaction of discrete dislocations with cracks in a finite body subjected to given tractions.

We consider these problems in the continuum linear approximation based on theory of elasticity. The stress field induced by an isolated edge dislocation in an unbounded elastic medium is known to be not self-balanced [5]. In order to avoid the difficulties arising from this circumstance we consider a *dislocation dipole* formed by two oppositely "charged" edge dislocations. For computational convenience we also assume

75

D. Durban and J.R.A. Pearson (eds.),
IUTAM Symposium on Non-Linear Singularities in Deformation and Flow, 75-86.
© 1999 *Kluwer Academic Publishers. Printed in the Netherlands.*

the dipole is located centrally in the disk. This model leads us to the traditional boundary-value problem of solving the equilibrium equations through the use of the Muskhelishvili analytic potentials [5] with given singularities.

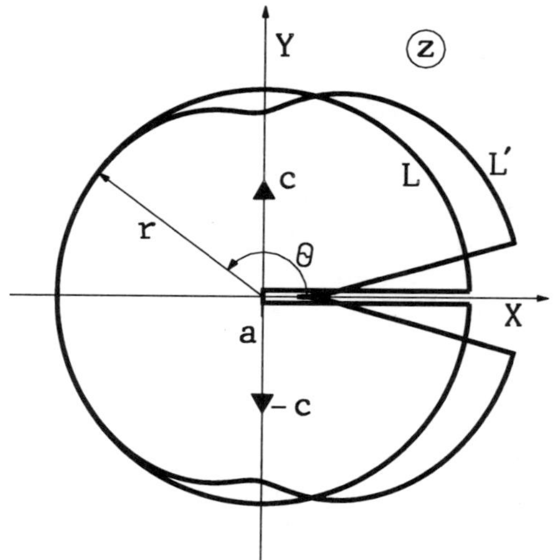

Figure 1. A dislocation dipole in a radially cracked disk. Resulting deformation (not in scale) of its traction-free boundary is also depicted. The clearance between the crack edges shown here in the initial state is exaggerated for the purpose of illustration.

The resultant stress state in the region can be thought of as a superposition of
  (i) the dipole field;
  (ii) the compensating field that eliminates the initial stresses caused by the dipole on the crack faces, and, lastly,
  (iii) the field adjusting the stresses to enforce prescribed loading conditions.
The potentials associated with the last field can not be completely found in a closed form. It is thus necessary to employ effective computational schemes for obtaining accurate results. Solving the problem numerically is done as follows.

By traction-free conditions we first analytically continue the functions sought across the crack boundary, whereupon their Laurent development is used to modify the loading relations on the disk circumference. A number of transformations next reduce these relations to an infinite system of linear algebraic equations. Finally, a series solution of the properly truncated system is computed in terms of given parameters.

We believe that this straightforward approach offers a significant computation saving as against numerically intensive procedures like finite differences. It has already proved to be efficient in the more simple case of an externally loaded cracked disk having no dislocations [7]. Another interesting property of the method used is that no additional computations are required to obtain the SIF values. They are produced immediately from solving the above-mentioned algebraic system.

2. Problem Presentation

Consider an edge dislocation of the Burgers vector $b = (b,0,0)$ in an elastic infinite cylinder parallel to the Z-axis of the Cartesian coordinates x,y,z. Then the deformation will be plane strain [4].

It is convenient to exhibit the 2D linear elasticity solution, including the edge dislocations, cracks and other types of discontinuities by giving the Muskhelishvili potentials $\Phi(z)$, $\Omega(z)$ of a complex variable z in the x,y-plane E [5]. These functions are regular everywhere in the elastic region except, possibly, at a finite number of interior isolated points whose type of singularity is predefined by physical considerations. Thus a dislocation gives rise to a pole at its location z_0 [5]:

$$2\Phi(z) = -i\frac{bB}{z-z_0}; \quad 2\Omega(z) = ibB\frac{z-\overline{z_0}}{(z-z_0)^2}; \quad B = \frac{\mu}{2\pi(1-\nu)} \tag{2.1}$$

whereas the stress field under biaxial load has an inverse square root singularity $r^{-\frac{1}{2}}$ at the crack tip $(a+i0)$ [5]:

$$\Phi(z) = \frac{C_1}{\sqrt{z-a}} + \Phi_1(z); \quad \Omega(z) = \frac{C_2}{\sqrt{z-a}} + \Omega_1(z); \quad r = |z-a| \tag{2.2}$$

Here, r is a distance from the corresponding singular point, μ, ν are given elastic moduli of the medium. The constants C_1, C_2 and regular terms $\Phi_1(z)$, $\Omega_1(z)$ of the potentials can be found by solving the boundary-value problem that involves prescribed tractions or displacements. Up to the proportionality the first coefficient C_1 coincides with the SIF K defined commonly [5] as

$$K = (2\pi)^{1/2} C_1 \tag{2.3}$$

When it comes to analytic manipulations, the second potential $\Omega(z)$ is preferred over its counterpart $\Psi(z) \equiv \Omega(z) - z\Phi'(z)$ commonly used in the literature.

Taking (r,θ) as polar coordinates in E, we have the following relations for determining the normal $\sigma_{rr}(z)$, $\sigma_{\theta\theta}(z)$ and shear $\sigma_{r\theta}(z)$ stress components at any regular point [5].

$$\sigma_{rr}(z) + \sigma_{\theta\theta}(z) = 4\operatorname{Re}\Phi'(z) \tag{2.4}$$

$$\sigma_{\theta\theta}(z) - \sigma_{rr}(z) + 2i\sigma_{r\theta}(z) = 2\frac{z}{\overline{z}}\left[(\overline{z}-z)\Phi'(z) + \Omega(z) - \Phi(z)\right] \tag{2.5}$$

where $\varphi'(z) \equiv \Phi(z)$, $\omega'(z) \equiv \Omega(z)$. A prime indicates differentiation with respect to z and a over-bar stands for complex conjugation.

Up to the rigid body motion the polar displacements $u_r(z)$, $u_\theta(z)$ are defined by

$$2\mu|z|^{-1}z\big[u_r(z) + 2iu_\theta(z)\big] = \kappa\varphi(z) - (\bar{z} - z)\overline{\Phi(z)} - \overline{\Omega(z)}; \quad \kappa = 3 - 4\nu \qquad (2.6)$$

At given tractions, we combine eqn. (2.4-2.5) to obtain the load condition

$$f(t) = \sigma_{rr}(t) + i\sigma_{r\theta}(t) = \Phi(t) + \overline{\Phi(t)} - \frac{t}{\bar{t}}\big[(\bar{t} - t)\Phi'(t) + \Omega(t) - \Phi(t)\big]; \quad t \in L \qquad (2.7)$$

on the boundary L of the cylinder cross-section. It is well known [5], that two equilibrium conditions are necessary for the problem (2.7) to be solvable uniquely. For a unit circle where $t = \exp(i\theta)$ they take the form [5]

$$\int_L f(t)\, dt = 0; \quad \mathrm{Im}\int_L f(\theta)\, d\theta = 0 \qquad (2.8)$$

These conditions express that the resultant vector and moment of the applied forces vanish. They also prevent the potentials $\varphi(z)$, $\omega(z)$ and hence the displacements from being multi-valued. Mathematically, this implies that the Fourier series of the load function $f(t)$ contains neither term proportional to t^{-1} nor pure imaginary constant term.

From (2.1), (2.4) and (2.5) it is immediate that the stresses related to an edge dislocation in the unbounded isotropic material are [4]

$$\sigma_{rr}(z) = \sigma_{\theta\theta}(z) = -bB\,\mathrm{Im}\frac{1}{z - z_0} = -bB\frac{\sin\theta}{r}; \quad \sigma_{r\theta}(z) = bB\,\mathrm{Re}\frac{1}{z - z_0} = bB\frac{\cos\theta}{r}$$

$$z - z_0 = re^{iq}; \qquad f(t) = \frac{ibB}{t - t_0} \qquad (2.9)$$

Here, the relations (2.8) are not met, because the corresponding potentials $\varphi(z)$, $\omega(z)$ contain logarithmic terms and thus are multi-valued in going around the point z_0 [1]. Indeed, by the well-known identities [6]

$$\bar{t} = t^{-1}; \quad \frac{d\bar{t}}{dt} = -\frac{1}{t^2} \qquad (2.10)$$

$$\int_L t^k \, dt = \begin{cases} 2\pi i; & k = -1 \\ 0; & k = 0, \ \pm 1, \ \pm 2,... \end{cases} \qquad (2.11)$$

substituting (2.9) into the integrals (2.8) gives the non-zero real value $2\pi b_x B$ for the first of them.

One way to avoid the multivalueness is to combine two edge dislocations which are equal in magnitude but opposite in sign (a dislocation dipole). Hence, the boundary-value problem we are interested in is that of finding the elastic stress field of a circular disk with a dislocation dipole centrally positioned on the Y-axis and a single traction-free straight crack $\Gamma: a \le x \le 1; \ y = 0$ (Fig. 1, the downwards facing triangle denoted the positive charge). The curvilinear boundary L is taken to be either traction-free or to have prescribed polar stresses $f(t) = \sigma_{rr}(t) + \sigma_{r\theta}(t)$ with the condition (2.8) met in the latter case.

For the dislocation piercing E at the points $z_0 = (0 + ic)$ and $-z_0$, $0 < c < 1$ the potentials related to the dipole are defined from (2.1) as

$$2\Phi_d(z) = i\frac{bB}{z + ic} - i\frac{bB}{z - ic} = \frac{2cbB}{z^2 + c^2}; \qquad \Omega_d(z) = \frac{2cbB(z^2 - c^2)}{(z^2 + c^2)^2} - \Phi_d(z) \qquad (2.12)$$

They clearly satisfy the reflection principle: [1]

$$\Phi(z) = \overline{\Phi(\bar{z})}; \qquad \Omega(z) = \overline{\Omega(\bar{z})} \qquad (2.13)$$

When integrated over z, the functions (2.12) give (again, up to arbitrary constants)

$$\varphi(z) = ibB \ln\frac{z + ic}{z - ic}; \qquad \omega(z) = -2cbB\frac{z}{z^2 + c^2} - ibB \ln\frac{z - ic}{z + ic} \qquad (2.14)$$

As z makes a complete circuit around either of two finite branch points: $\pm ic$, these potentials and hence the corresponding displacements do not return back to their values. By contrast, a complete circuit about both points produces no changes in the branches. In other words, the interval $[-ic; \ ic]$ is a line of discontinuity for the displacements as will be seen in Fig. 3.

By (2.7) and (2.10) the potentials (2.12) yield the following load function:

$$f_d(t) = \sigma_{rr}^d(t) + i\sigma_{r\theta}^d(t) = cbB\left[\frac{1 - 2t^2}{t^2 + c^2} + \frac{t^2}{1 + c^2 t^2} - \frac{2t^2(1 - c^2)}{(t^2 + c^2)^2}\right]; \qquad t \in L \qquad (2.15)$$

Using (2.10, 2.11) we may also derive its Fourier expansion

$$f_d(t) = \sum_{j=-\infty}^{\infty} f_j^d t^j = -2cbB - \frac{bB}{c} \sum_{k=1}^{\infty} (-1)^k c^{2k} \left[t^{2k} + \left(1 - t^2\right) t^{-2k} - 2k\left(1 - c^2\right) t^{-2k-2} \right] \quad (2.16)$$

which consists of only even powers of t with real coefficients. Thus, the traction at the circumference L of the disk due to the dipole, satisfy the equilibrium conditions (2.8). By this means the potentials (2.12) provide a complete solution to a solid disk with the dislocation dipole. However, in the crack case they correspond to the problem of an internally loaded crack, again with given tractions on the external boundary L.

3. Equilibrium And Boundary Conditions. Analytic Considerations

We are now in a position to find the tractions induced by the dipole stress field (2.12) on the crack sides $\Gamma_\pm$: $a \leq x \leq 1$; $y_\pm = 0$. Here, the identities (2.10) are evidently replaced by

$$\bar{\tau} = \tau; \quad \frac{d\bar{\tau}}{d\tau} = 1; \quad \tau \in \Gamma_\pm \tag{3.1}$$

Keeping this in mind, we use (2.4), (2.5), (2.12) and (2.13) to arrive at

$$\sigma_{\theta\theta}^\pm(x) \equiv \sigma_{yy}^\pm(x) = p(x) = \Omega_d(x) + \overline{\Phi_d(x)} = 2cbB \frac{x^2 - c^2}{\left(x^2 + c^2\right)^2}; \quad \sigma_{r\theta}^\pm(x) = \sigma_{xy}^\pm(x) = 0 \tag{3.2}$$

Along with (r,θ), Cartesian coordinates (x,y) will thereafter be used, when required.

In order for the crack sides to be cleared from the tractions (3.2), we must impose the compensating stress field (ii) specified in E by the boundary condition

$$\Phi_2^\pm(x) + \overline{\Omega_2^\pm(x)} = -p(x); x \geq a; y = 0 \tag{3.3}$$

Mechanically, the above potentials may have a singular behavior of the type (2.2) at the crack tip $x = a$.

Introducing the new functions

$$G(z) \equiv \Phi_2(z) - \Omega_2(z) \tag{3.4}$$

$$H(z) \equiv \Phi_2(z) + \Omega_2(z); \quad z \in E \tag{3.5}$$

we next separate the real and imaginary parts of (3.3) to arrive at the following two boundary problems

$$\operatorname{Im} G^{\pm}(x) \equiv 0; \quad x \geq a \tag{3.6}$$

$$\operatorname{Re} H^{\pm}(x) \equiv -p(x); \quad x \geq a \tag{3.7}$$

From (3.6) and (2.13) it follows that the function $G(z)$ is real on the x-axis with no singularities in the whole plane E. Hence, it can be developed in a Taylor series with real coefficients

$$G(z) = \sum_{j=1}^{\infty} \alpha_j z^j \tag{3.8}$$

which converges in any finite circle centered at the origin [1].
The particular solution to the non-homogeneous Hilbert problem (3.7), (2.13) is expressed [6] by the following Cauchy principal-value integral with a square-root singularity

$$J(z) = \frac{-1}{\pi i R_1(z)} \int_a^{\infty} \frac{p(x) R_1(x)}{(x-z)} dx = \frac{2cbB}{\pi i R_1(z)} \int_a^{\infty} \frac{(x^2 - c^2) R_1(x)}{(x^2 + c^2)^2 (z-x)} dx; \quad \overline{J(z)} = J(\bar{z}); \quad z \in E \tag{3.9}$$

We suppose $R_1(z) \equiv \sqrt{(z-a)}$ is the analytic branch of the function which takes real and positive boundary values on the top face of the crack: $x > a, y = 0^+$. One can see that (3.9) does solve the problem because of the different signs of the square root on the two sides of real line from the point $x = a$ to positive infinity.
By routine mathematics the integral (3.9) may be taken in a closed form [3]

$$\frac{J(z)}{2cbB} = \frac{(z^2 - c^2)}{(z+ic)^2 (z-ic)^2} - \frac{i}{2R(z)} \left[\frac{\sqrt{a+ic}}{(z+ic)^2} + \frac{\sqrt{a-ic}}{(z-ic)^2} \right] + \frac{i}{4R(z)} \left[\frac{1}{(z+ic)\sqrt{a+ic}} + \frac{1}{(z-ic)\sqrt{a-ic}} \right] \tag{3.10}$$

which explicitly involves power type singularities at the dislocation points $z = \pm ic$.
By adding the general solution $H_0(z)$ [6] of the homogeneous problem (3.7), (2.13)

$$H_0(z) = i R_1^{-1}(z) \sum_{j=0}^{\infty} \beta_j z^j; \quad \operatorname{Im} \beta_j = 0; \quad j = 0, 1, 2, \dots \tag{3.11}$$

we use (3.3)-(3.5) and (3.8), (3.11) to obtain

$$2\Phi_2(z) = J(z) + \sum_{j=0}^{\infty} \alpha_j z^j + iR_1^{-1}\sum_{j=0}^{\infty}\beta_j z^j$$

$$2\Omega_2(z) = J(z) - \sum_{j=0}^{\infty} \alpha_j z^j + iR_1^{-1}\sum_{j=0}^{\infty}\beta_j z^j \qquad (3.12)$$

Taken together, the above infinite sums correspond to the last field (**iii**) mentioned in the Introduction while the integral term $J(z)$ generates the compensating field (**ii**). Thus all the requirements of the problem are indeed met.

The real coefficients $c_j, d_j, j = 0,1,2,\dots$ serve to satisfy the boundary condition (2.7) which can be written, in view of (3.12), (2.15) and (2.11), as

$$\sum_{j=0}^{\infty}\alpha_j t^j + \overline{\alpha_j}\left[(1-j)t^{-j} + (j+2)t^{-j-2}\right] + i\sum_{j=0}^{\infty}\beta_j R_1^{-1}(t)t^j - \overline{\beta_j}R_1^{-1}(t^{-1})\left[(1-j)t^{-j} + j^{-j-2}\right]$$

$$+i\sum_{j=0}^{\infty}\overline{\beta_j}\,\frac{\partial \overline{R_1^{-1}(t)}}{\partial t}t^{-j} = f(t) - f_d(t) - q(t) \qquad (3.13)$$

where the function $q(t)$ is defined on L by substituting $J(z)$ through (3.12) to the right-hand side of (2.7)

$$2q(t) = 2\operatorname{Re}J(t) + \left(t^{-3} - t^{-1}\right)J'(t) \qquad (3.14)$$

We also expand the functions $f(t)$ and $q(t)$ into the Fourier series on L

$$f(t) = \sum_{j=-\infty}^{\infty} f_j t^j; \quad q(t) = \sum_{j=-\infty}^{\infty} q_j t^j \qquad (3.15)$$

where

$$q_j = \frac{1}{2\pi i}\int_L q(t)t^{j-1}dt \qquad (3.16)$$

and f_j are expressed similarly in terms of the function $f(t)$. Then, comparing the coefficients of t^j; $j = 0,\pm1,\pm2,\dots$ on both sides of (3.13) gives, with the expansion (2.16), an infinite set of linear algebraic equations in the unknowns α_j, β_j. The system thus derived is nearly the same as that we obtained earlier [7] in the more simple case of a cracked disk with no dislocations. The only difference between them is the vector $\{f_j^d + q_j\}, j = 0,1,2\dots$ subtracted now from the right side of the new system while its

matrix remains unaffected by the dislocations. Evidently, this is due to the fact that the potentials of the dipole-induced stress fields **(i)** and **(ii)** have been found explicitly.
One can find the integrals in (3.16), (3.14) numerically or analytically though the resultant expressions seem too cumbersome to be written here. Also, it is not difficult to check that the corresponding coefficients q_j vanish as required by the equilibrity conditions (2.8).
With (2.6), integrating the potentials obtained gives the displacement components $u_r(z)$, $u_\theta(z)$ that involve the terms coming from the integral of the function $J(z)$

$$\int J(z)dz = \frac{-z}{z^2 + c^2} + \frac{iR_1(z)}{2}\left[\frac{\sqrt{a - ic}}{z - ic} + \frac{\sqrt{a + ic}}{z + ic}\right]$$

In practice an infinite system of equations is replaced by the corresponding finite system involving the first $2(N + 1)$ unknowns α_j, β_j; $j = 0,1,\ldots,N$. As in [7], the summation limits M and N for computations are so chosen that the approximate equality $N \approx 3M$ is met.

4. Numerical Examples

To be more specific, we limit ourselves in this section to the case when a radially cracked disk is free of applied tractions $\left(f_j = 0;\quad j = 0,\pm1,\pm2,\ldots\right)$ and the elastics stresses are caused solely by the presence of the dislocation dipole (3.1). By calculating the displacements we must first ascertain that the dipole actually forces the crack edges directly apart thus forming a gap between them, as assumed. The straightforward calculations performed for $c = 0.0001, 0.0005$, and $c = 0.001, 0.002, \ldots, 0.960$ show that this is always the case, as in Fig. 1 where the both contours are made coincident at the left-most point. Therefore, a measure of the crack-dipole interaction effect is given by the pure mode I SIF computed in parallel for the same values of c. We denote this function by $K_I(c)$.
It was deduced numerically that $K_I(c)$ tends to infinity as $c^{-1/2}$ when $c \to 0$. Using least squares the normalized function $K_I(c)\sqrt{c}$ was then found to fall perfectly on the following polynomial curve of the third order

$$K_I(c)\sqrt{c} = 0.704834 - 1.67374c + 5.75823c^2 - 2.69565c^3$$

The error sum of squares taken over the interval $c \in [0.0001;\ 0.960]$ turned out to be smaller than 0.0006.

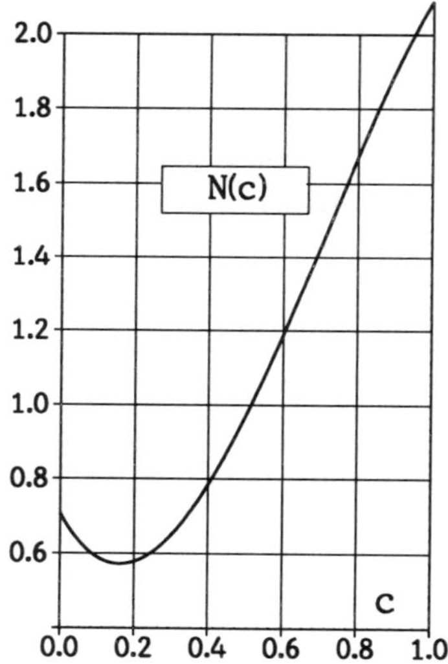

Figure 2. Normalized stress intensity factor $N(c) = K_I(c)\sqrt{c}$ as a function of the dipole half-length c for a load-free disk with a radial crack.

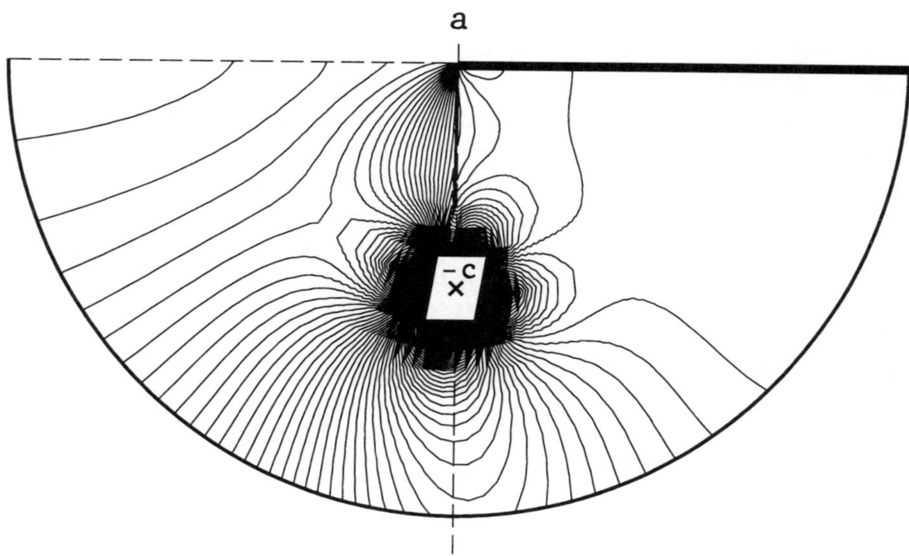

Fig. 3 Radial displacements in a load-free disk with the dislocation dipole ($2c=1$). Thickening to the line ac the contour levels make it visible. The vicinity of the dislocation point is removed lest the figure be overly detailed.

Fig.4 (*a-b*) exemplifies the resultant elastic field computed in a radially cracked disk with $2c = 1$. It is seen that large stress gradients are of a local nature while the rest of the region is much less influenced by the dipole. In the immediate vicinity of the dislocation each stress component changes its sign four times when turning around the piercing point. In contrast, for the dislocation in an unbounded medium this occurs only twice (see Eqn (2.10)). The dissimilarity is caused by the main singular term in (3.9) that prevails near the dislocations as the squared infinities (2.7). Hence we have

$$\sigma_{rr}(z),\ \sigma_{\theta\theta}(z) \sim \frac{\sin 2\theta}{r^2};\quad \sigma_{r\theta}(z) \sim \frac{\cos 2\theta}{r^2}$$

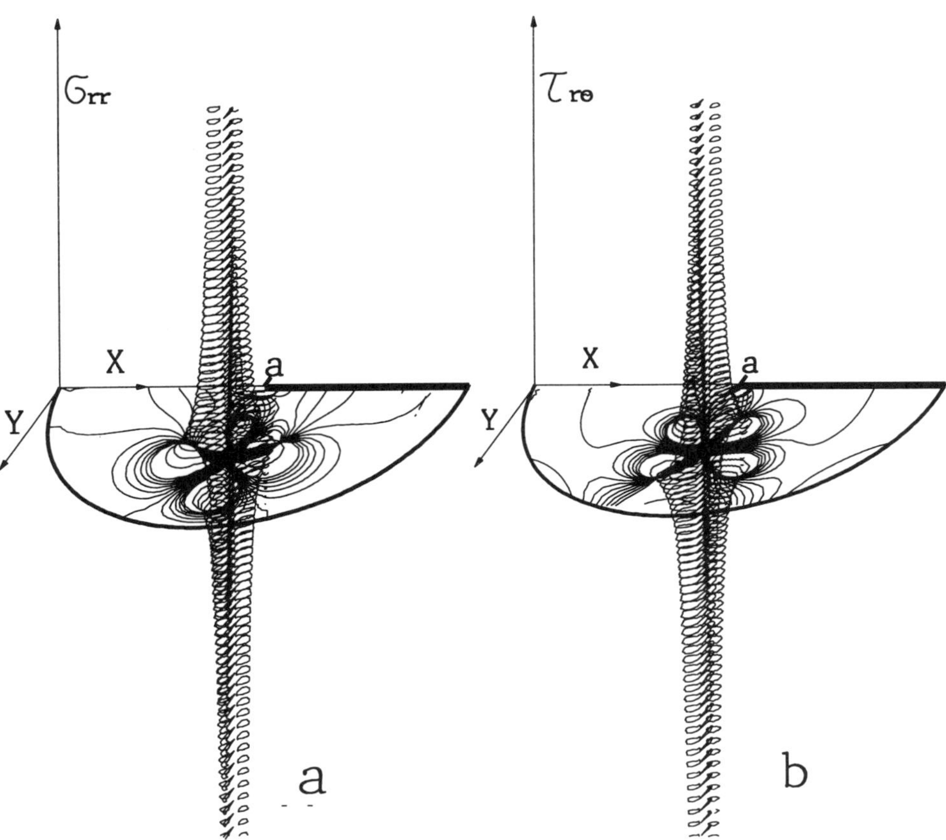

Fig. 4 A load-free disk. Stress levels of σ_{rr} (*a*) and $\tau_{r\theta}$ (*b*). The peaking values are truncated drastically to save room.

5. Concluding Remarks

The paper presented a numerical approach to solving 2D stress fields with different types of singularities via the Muskhelishvili analytical representation translated into the working computer program. A significant advantage of the method is that local singularities are isolated from the computations, making numerical solutions in complex geometries easier to obtain. The computational efficiency and accuracy of the approach place it as a strong contender against today's most popular finite and boundary elements methods. It is especially true for dislocations nearing the disk boundary or crack tips. Another representative geometry where such computations may be effective includes a dipole placed in line with the crack. Combined together, these basic cases cover a wide variety of both of an academic and industrial nature. Further, the method can be extended to circular regions containing multiple cracks, externally loaded cracks or ,possibly, cracks with bonds linking the faces. The results of such computations naturally must be interpreted with the simplifications in mind.

Acknowledgment

I wish to express my heartfelt thanks to Professor Gennady P. Cherepanov, of the Florida International University, who led me to investigate the problem.

References

1. Alhfors, L.V. (1966) *Complex Analysis*, 2nd Ed., McGraw Hill, New York

2. Hurtado, J.A., Dundurs, J. and Mura, T. (1996) Lamellar inhomogeneites in a uniform stress field, *J Mech Phys Solids*, **44**, 1-21.

3. Gradsteyn, I.S. and Ryzhik, I.M. (1994) *Table of Integrals, Series and Products*, 5th Ed., Boston, Academic press.

4. Landau, L.D. and Lifshitz, E.M. (1986) *Theory of Elasticity*, Pergamon Press, Oxford.

5. Muskhelishvili, N.I.. (1975) *Some Basic Problems of the Mathematical Theory of Elasticity*, Noordhoff Int. Publ. Leiden, the Netherlands.

6. Muskhelishvili, N.I.. (1977) *Singular Integral Equations*, Noordhoff Int. Publ. Leiden, the Netherlands.

7. Vigdergauz, S.B. (1996) An Effective Method of Computing Stress Field in A Cracked Disk, *Engn Fract Mech*, No 4, 545-556

THE ASYMPTOTIC SOLUTION OF ANISOTROPIC GRADIENT ELASTICITY WITH SURFACE ENERGY FOR A MODE-II CRACK

I. VARDOULAKIS
Department of Engineering Science, National Technical University of Athens, GR-15773, Athens, Greece

G. EXADAKTYLOS
Laboratory of Rock Mechanics, Department of Mineral Resources Engineering, Technical University of Crete, GR-73100, Hania, Greece

1. Introduction

The purpose of this paper is to present a view of how the basic problems in the theory of *equilibrium* cracks are properly formulated in the context of an anisotropic *gradient elasticity theory with surface energy* proposed by Vardoulakis and his co-workers [1-4], and to discuss the results obtained. This theory, which is a generalization of Casal's [5] original work on the 1-D tension bar problem, contains both a volume energy strain-gradient term, ℓ, and a surface energy strain-gradient term, ℓ', both having the dimension of length.

Aifantis and co-workers - see reference [6] for a comprehensive review - on the other hand, studied only the effect of the volume energy strain-gradient term, and, as he states, his theory may be viewed in a sense, as a particular case of Mindlin's [7] original theory, involving only one material constant; also, the concept of higher order self-equilibrating stresses doing work on higher order strain gradients need not be introduced. However, the consideration of the surface energy in the present theory leads to a *constitutive* character of the boundary conditions. This strengthens Aifantis' [8] conjecture of the constitutive character of boundary constraints in materials with microstructure. I. Vardoulakis, noting this fact, states [9]: '*...The problem of constitutive boundary conditions is open and deserves further attention from the theoretical as well as the experimental point of view...*'. Furthermore, A. Pearson in his opening lecture at the IUTAM Symposium on 'Non-Linear Singularities in Deformation and Flow', in Technion, Haifa in 1997 [10], states: '*...The full constitutive nature of boundary conditions is often neglected in many continuum models and so oversimplified forms are imposed in physical models...*'.

The special theory proposed by Aifantis has been applied in [11] for solving asymptotically in an elegant manner the gradient-elasticity mode-III crack problem in

87

D. Durban and J.R.A. Pearson (eds.),
IUTAM Symposium on Non-Linear Singularities in Deformation and Flow, 87-98.
© 1999 *Kluwer Academic Publishers. Printed in the Netherlands.*

terms of the Westergaard stress function of classical elasticity. However, it seems that Unger and Aifantis [11] did not use appropriate extra boundary conditions for solving the governing homogeneous partial differential equation of fourth-order describing the mode-III crack problem.

2. Formulation of the problem

The 3D generalization of Casal's gradient-dependent anisotropic elasticity with surface energy leads to the following expression for the strain energy density function [12]

$$w = \frac{1}{2}\lambda\varepsilon_{ii}\varepsilon_{jj} + \mu\varepsilon_{ij}\varepsilon_{ji} + \frac{1}{2}\lambda\ell^2\partial_k\varepsilon_{ii}\partial_k\varepsilon_{jj} + \mu\ell^2\partial_k\varepsilon_{ij}\partial_k\varepsilon_{ji} +$$
$$+ \frac{1}{2}\lambda\ell_k\partial_k(\varepsilon_{ii}\varepsilon_{jj}) + \mu\ell_k\partial_k(\varepsilon_{ij}\varepsilon_{ji}) \qquad i,j,k = 1,2,3 \tag{1}$$

where λ and μ are Lamé's constants, $\varepsilon_{ij} = (1/2)(\partial_j u_i + \partial_i u_j)$ is the symmetric part of the displacement field u_i, $\partial_k \equiv \partial/\partial x_k$, ℓ,ℓ' are characteristic lengths of the material defined previously, and $\ell_k = \ell' v_k$ ($v_k v_k = 1$) is a director. Accordingly (1) defines a gradient anisotropic elasticity with constant characteristic directors ℓ_k. From virtual work principle follow the constitutive relations for the equilibrium stress and double-stress tensors, respectively

$$\left.\begin{aligned}\sigma_{ij} &= \lambda\delta_{ij}\varepsilon_{kk} + 2\mu\varepsilon_{ij} - \ell^2\nabla^2(\lambda\delta_{ij}\varepsilon_{kk} + 2\mu\varepsilon_{ij}) \\ \mu_{kij} &= \ell_k(\lambda\delta_{ij}\varepsilon_{ll} + 2\mu\varepsilon_{ij}) + \ell^2\partial_k(\lambda\delta_{ij}\varepsilon_{ll} + 2\mu\varepsilon_{ij})\end{aligned}\right\} \tag{2}$$

as well as the boundary conditions

$$n_j\sigma_{jk} = \tilde{P}_k, \qquad n_i n_j \mu_{ijk} = \tilde{R}_k \tag{3}$$

where δ_{ij} is the Kronecker delta, ∇^2 is the Laplacian operator, and n_i is the outward unit normal on the boundary. It has been shown by Ru and Aifantis [13] and by Exadaktylos and Vardoulakis [12], that for the case of traction boundary value problems the gradient dependent elasticity predicts the same stresses σ_{ij} with the classical theory of elasticity. The third order stress tensor μ_{ijk} is dual in energy to the strain-gradient is called the *double-stress*, and is symmetric with respect to the last two indices (i.e. $\mu_{ijk} = \mu_{ikj}$). Double-force systems without moments are stress systems equivalent to two oppositely directed forces at the same point; such systems have direction but not net force and no resulting moment. Notice that singularities of this kind are discussed by Love [14] and Eshelby [15].

The displacement-equation of equilibrium in vector notation of the present anisotropic gradient-dependent theory of elasticity with surface energy reads as follows

$$\mu \overline{D}^2 \{ k' \nabla \nabla \cdot \underset{\sim}{u} + \nabla^2 \underset{\sim}{u} \} + \underset{\sim}{f} = 0 \; ; \qquad k' = \frac{\lambda + \mu}{\mu} \tag{4}$$

where the curly underline of a symbol means that this is a vector, $\underset{\sim}{f}$ is the body force per unit volume, ∇ is the gradient operator, $\nabla \cdot$ is the divergence operator, and the operator $\overline{D}^2$ is defined as follows

$$\overline{D}^2 \equiv 1 - \ell^2 \nabla^2 \tag{5}$$

It is important to observe that in the limit $\ell \to 0$ the highest derivative term in (5) is lost, suggesting the emergence of boundary layer effects. It can be proved that any solution $\underset{\sim}{u}$ of the displacement-equation of equilibrium (4), in a region V bounded by a surface S, can be expressed as

$$\underset{\sim}{u} = \underset{\sim}{B} - \ell^2 \nabla \nabla \cdot \underset{\sim}{B} - \frac{1}{2} \nabla \left(\overline{D}^2 - \frac{1}{k} \right) \left[\underset{\sim}{r} \cdot \overline{D}^2 \underset{\sim}{B} + B_0 \right],$$

$$\mu \nabla^2 \overline{D}^2 B_0 = \underset{\sim}{r} \cdot \overline{D}^2 \underset{\sim}{f} - 4\ell^2 \nabla \cdot \underset{\sim}{f} \; , \qquad \mu \nabla^2 \overline{D}^2 \underset{\sim}{B} = -\underset{\sim}{f}. \tag{6}$$

where $\underset{\sim}{r}$ is the position vector. It is noted that the functions $\underset{\sim}{B}, B_0$ reduce to *Neuber's-Papkovich's* functions when $\ell = 0$. It turns out that the gradient vector and scalar fields can be identified with the respective fields of classical elasticity as follows

$$\overline{D}^2 \underset{\sim}{B} \equiv \underset{\sim}{B}^c \quad , \quad \overline{D}^2 B_0 \equiv B_0^c \tag{7}$$

with $\nabla^2 \{ \underset{\sim}{B}^c, B_0^c \} = 0$ and $\underset{\sim}{B}^c, B_0^c$ to be the vector and scalar potentials, respectively, of the classical elasticity theory satisfying the same traction boundary conditions, i.e.

$$n_j \sigma_{jk}^c = \widetilde{P}_k \quad on \quad \partial V_\sigma \tag{8}$$

where the superscript 'c' denotes the classical elasticity solution. Finally, an immediate consequence of relations (7) is the following

$$\underset{\sim}{B} = \underset{\sim}{B}^c + \underset{\sim}{B}^g \; , \quad B_0 = B_0^c + B_0^g , \quad \overline{D}^2 \{ \underset{\sim}{B}^g, B_0^g \} = 0. \tag{9}$$

By virtue of (6), (9) in a Cartesian coordinate system (x,y,z) for the case of plane strain parallel to the xy-plane with $u = (u(x,y), \upsilon(x,y), 0)$ and in the absence of body forces, there are the following relations of the general complex representation for the elastic displacements u, υ

$$2\mu(u+i\upsilon) = \kappa\phi_c(z) - \overline{\psi}_c(\bar{z}) - z\overline{\phi'_c}(\bar{z}) - 4\ell^2\overline{\phi''_c}(\bar{z}) + (u^+ + i\upsilon^+),$$
$$\nabla^2\{\phi_c(z), \psi_c(z)\} = 0, \quad \overline{D}^2(u^+ + i\upsilon^+) = 0 \tag{10}$$

where $\kappa = 3 - 4v$ or $\kappa = (3-v)/(1+v)$ is *Muskhelishvili's constant* for plane strain and generalized plane stress conditions, respectively, $(\cdot)' = \partial/\partial z$ and we have used the identities $\partial/\partial z \equiv (\partial/\partial x - i\partial/\partial y)/2$, $\partial/\partial\bar{z} \equiv (\partial/\partial x + i\partial/\partial y)/2$ with $\bar{z} = x - iy$ to be the complex conjugate of $z = x + iy$, and $i \equiv \sqrt{-1}$. The strains and stresses referred to the Cartesian coordinates (x, y) can be found from (10) as follows

$$2\mu(\varepsilon_{xx} + \varepsilon_{yy}) = 2(1-2v)\{\phi'_c(z) + \overline{\phi'_c}(\bar{z})\} + (u^+_{,x} + \upsilon^+_{,y}),$$
$$2\mu(\varepsilon_{yy} - \varepsilon_{xx} + 2i\varepsilon_{xy}) = 2\{\psi'_c(z) + \bar{z}\phi''_c(z) + 4\ell^2\phi'''_c(z) + \frac{\partial}{\partial z}(u^+ - i\upsilon^+),$$
$$\sigma_{xx} + \sigma_{yy} = 2\{[\phi'_c(z) + \overline{\phi'_c}(\bar{z})], \quad \sigma_{yy} - \sigma_{xx} + 2i\sigma_{xy} = 2\{\bar{z}\phi''_c(z) + \psi'_c(z)\}. \tag{11}$$

Herein we also record only the following double-stresses

$$(2\mu)^{-1}(\mu_{yyy} + i\mu_{yyx}) = (\ell' + \ell^2\partial_y)\{[\phi'_c(z) + \overline{\phi'_c}(\bar{z})]$$
$$+ [\psi'_c(z) + \bar{z}\phi''_c(z) + 4\ell^2\phi'''_c(z)] + \frac{1}{2(1-2v)}(u^+_{,x} + \upsilon^+_{,y}) + \frac{\partial}{\partial z}(u^+ - i\upsilon^+)\} \tag{12}$$

We now state the *mixed-mixed* plane strain boundary value problem of a finite straight mode-II crack occupying the line segment $-\alpha < x < \alpha$, $y = 0^\pm$ subject to a uniform internal pressure $S = -\sigma_o$ ($\sigma_o > 0$), with no loading at infinity. Let S be the complement of the line segment $-\alpha < x < \alpha$, $y = 0$ extended on the half-plane $y \leq 0$. We seek the solution in S subject to the following mixed-mixed boundary conditions

$$\tau_{xy}(x,0^-) = S, \quad \mu_{yyx}(x,0^-) = 0 \qquad 0 \leq x < \alpha,$$
$$u(x,0^-) = 0 \qquad\qquad\qquad\qquad \alpha < x < \infty, \tag{13}$$
$$\sigma_{yy}(x,0^-) = 0, \quad \mu_{yyy}(x,0^-) = 0 \qquad 0 \leq x < \infty.$$

as well as to the homogeneous regularity conditions at infinity

$$\sigma_{ij} \to 0 \ (i,j=x,y), \ \mu_{ijk} \to 0 \ (i,j,k=x,y) \quad as \ \sqrt{x^2+y^2} \to \infty \qquad (14)$$

The first, third and fourth of conditions (13) are the classical ones, whereas the remaining conditions are extra boundary conditions required as a result of higher order terms in the constitutive equations. The symmetry with respect to the y-axis (x,z - plane) provides additional conditions

$$u(x,0)=u(-x,0), \ \tau_{xy}(x,0)=\tau_{xy}(-x,0), \ \mu_{yyx}(x,0)=\mu_{yyx}(-x,0) \quad -\infty < x < \infty. \qquad (15)$$

The first and fourth stress boundary conditions in (13) are satisfied if we introduce the following *Westergaard* stress function

$$2\phi_c{'}(z) = Z_{II} = -iS\alpha \int_0^\infty J_1(\alpha\xi)e^{-i\xi z}d\xi \qquad (16)$$

where $J_n(\cdot)$ is the usual Bessel function of the first kind and of order n. It is easily shown from (10) that the displacement field is defined by the components

$$2\mu u = 2(1-v)\,Re(\overline{Z}_{II}) - y\,Im(Z_{II}) - 2\ell^2\,Re(Z'_{II}) + u^+(x,y)$$
$$2\mu\upsilon = (1-2v)\,Im(\overline{Z}_{II}) - y\,Re(Z_{II}) + 2\ell^2\,Im(Z'_{II}) + \upsilon^+(x,y) \qquad (17)$$

Herein, we also record only the following double-stress components, that are considered in the boundary conditions

$$\mu_{yyy} = \ell^2\,Im(Z'_{II}) + y\left[\ell'\,Im(Z'_{II}) + \ell^2\,Re(Z''_{II})\right]$$
$$+2\ell^2\left[\ell'\,Re(Z''_{II}) - \ell^2\,Im(Z'''_{II})\right] + \frac{1}{2(1-2v)}(\ell' + \ell^2\partial_y)[2(1-v)\upsilon^+_{,y} + 2vu^+_{,x}],$$
$$\mu_{yyx} = -\ell'\,Im(Z_{II}) - 2\ell^2\,Re(Z'_{II}) + y\left[-\ell'\,Re(Z'_{II}) + \ell^2\,Im(Z''_{II})\right]$$
$$+2\ell^2\left[\ell'\,Im(Z''_{II}) + \ell^2\,Re(Z'''_{II})\right] + \frac{1}{2}(\ell' + \ell^2\partial_y)[u^+_{,y} + \upsilon^+_{,x}] \qquad (18)$$

3. Reduction of the problem to integral equation

Herein, we shall adopt the notations

$$\bar{f}(\xi) \equiv F[f(x); x \to \xi] \equiv \frac{1}{\sqrt{2\pi}} \int\limits_{-\infty}^{\infty} f(x)e^{i\xi x}dx,$$

$$F_c[f(x); x \to \xi] \equiv \sqrt{\frac{2}{\pi}} \int\limits_{0}^{\infty} f(x)\cos(x\xi)dx, \qquad (19)$$

$$F_s[f(x); x \to \xi] \equiv \sqrt{\frac{2}{\pi}} \int\limits_{0}^{\infty} f(x)\sin(x\xi)dx$$

with ξ being the real-valued transform parameter.

The general solution of the functions $u^+(x,y)$, $\upsilon^+(x,y)$ for the half-plane $y \le 0$, considering the regularity conditions at infinity and symmetry relations furnishes

$$\{u^+(x,y), \upsilon^+(x,y)\} = \left\{ F_c[\{\bar{u}(\xi)e^{ya(\xi)}; \xi \to x], F_s[\{\bar{\upsilon}(\xi)e^{ya(\xi)}; \xi \to x]\right\}, y \le 0 \quad (20)$$

$$a(\xi) \equiv \sqrt{\xi^2 + \ell^{-2}} \qquad (21)$$

Our task is now to explicitly determine the functions $u^+(x,y)$, $\upsilon^+(x,y)$ from the double stress boundary conditions. To this end note that the last of boundary conditions (13) can be written in the following form

$$\mu_{yyy}(x,0) = 0 \Leftrightarrow \bar{\mu}_{yyy}(\xi,0) = 0 \qquad (22)$$

which is accordance to the first of eqns (18), and (20) yields

$$\bar{\upsilon}(\xi) = \frac{1}{2(1-v)a(\ell' + \ell^2 a)} \{2v\xi(\ell' + \ell^2 a)\bar{u}(\xi)$$

$$-2(1-2v)\sqrt{\frac{\pi}{2}}Sa\xi[\ell^2 + 2\ell'\ell^2\xi + 2\ell^4\xi^2]J_1(\alpha\xi)\}, \qquad 0 \le \xi < \infty \qquad (23)$$

where we have put the abridged notation $a \equiv a(\xi)$ in order to shorten the results to be recorded presently. It also follows immediately that the remaining boundary conditions, namely the second and third of (13) will be satisfied if the unknown function $\bar{u}(\xi)$ satisfies the following dual integral equations

$$F_c[\gamma(\xi)\bar{u}(\xi);\xi \to x] = \sqrt{\frac{\pi}{2}}S\alpha F_c[\beta(\xi)J_1(\alpha\xi);\xi \to x] \quad 0 \le x < \alpha,$$

$$F_c[\bar{u}(\xi) + \sqrt{\frac{\pi}{2}}2\ell^2 S\alpha\xi J_1(\alpha\xi);\xi \to x] = 0 \qquad \alpha < x < \infty. \tag{24}$$

in which

$$\beta(\xi) = 2(1-2v)a^{-1}[\ell^2\xi^2(1+2\ell^2\xi^2)+2\ell'\ell^2\xi^3]$$
$$-4(1-v)[\ell'(1+2\ell^2\xi^2)+2\ell^4 a^2\xi],$$
$$\gamma(\xi) = (\ell'+\ell^2 a)[2(1-v)a+2va^{-1}\xi^2], \qquad 0 \le \xi < \infty$$

Now, the object is to obtain an integral equation from the pair (24). For this purpose we take the following representation

$$\bar{u}(\xi) = \sqrt{\frac{\pi}{2}}[-2\ell^2 S\alpha\xi J_1(\alpha\xi) + \xi^{-1}\int_0^\alpha t\psi(t)J_1(t\xi)dt] \qquad 0 \le \xi < \infty \tag{25}$$

In view of the above relation (25), it can be easily shown that the second of (24) is identically satisfied (i.e. the displacement $u(x,0^-)$ vanishes outside the crack region). Next, the expression (25) for $\bar{u}(\xi)$ is introduced into the first of (24), x is replaced by x', and an integration with respect to x', over the range $0 \le x' \le x$ ($0 \le x < \alpha$) is performed to yield

$$\int_0^\omega \rho\psi(\ell\rho)K(\chi,\rho)d\rho = F(\chi) \qquad 0 \le \chi < \omega \tag{26}$$

with the kernel $K(\chi,\rho)$ given by

$$K(\chi,\rho) = K_1(\chi,\rho) + K_2(\chi,\rho), \quad K_1(\chi,\rho) = \int_0^\infty J_1(\rho\zeta)\sin\chi\zeta\,d\zeta$$

$$K_2(\chi,\rho) = \int_0^\infty b_2(\zeta)J_1(\rho\zeta)\sin\chi\zeta\,d\zeta, \quad b_2(\zeta) = (1-v)\zeta^{-2} + k[(1-v)\frac{a'}{\zeta^2} + \frac{v}{a'}]$$

$$F(\chi) = S\omega\int_0^\infty \left\{ -2(1-v) + (1-2v)\zeta[(1+2\zeta^2)a'^{-1} - 2\zeta^2] - \right.$$

$$\left. - 4(1-v)k[a'^{-1}\zeta^2 - \zeta^{-1} - 2\zeta + a'\right\} J_1(\omega\zeta)\sin\chi\zeta\,d\zeta \quad 0 \le \chi < \omega \tag{27}$$

and the non-dimensional variables $\zeta, \chi, \rho, \omega$, as well as, the length-ratio k defined by

$$\zeta = \ell\xi, \quad \chi = x/\ell, \quad \rho = t/\ell, \quad \omega = \alpha/\ell, \quad k = \ell'/\ell, \quad a'(\zeta) = \ell a(\xi) = \sqrt{1+\zeta^2} \tag{28}$$

The above integral equation is a linear Fredholm integral equation of the first kind with non-symmetric kernel. It is well-known in the literature that integral equations of the first kind are generally ill-posed in the sense of *Hadamard*. By following the procedure proposed in [1,3] the above integral equation is reduced to a better behaved Fredholm integral equation of the second kind with a simpler and symmetric kernel, i.e.

$$\Psi(\rho) = \lambda_2 \int_0^\omega \sqrt{\rho\tau}\Psi(\tau)d\tau\int_0^\infty \zeta b_2(\zeta)J_1(\rho\zeta)J_1(\tau\zeta)d\zeta$$

$$-\frac{1}{2}S\omega\int_0^\infty \zeta c_2(\zeta)J_1(\rho\zeta)J_1(\omega\zeta)d\zeta - 2(1-v)S\omega^{1/2}\delta(\omega-\rho) \tag{29}$$

$$-2(1-v)kS\frac{\rho\sqrt{\rho}}{\omega}{}_2F_1(\frac{3}{2},\frac{1}{2};2;\frac{\rho^2}{\omega^2}) \quad 0 \le \rho < \omega; \quad \lambda_2 = 1$$

where

$$\Psi(\rho) = \sqrt{\rho\ell}\psi(\ell\rho), \quad c_2(\zeta) = (1-2v)\zeta[a'^{-1}(1+2\zeta^2) - 2\zeta^2]$$

$$+4(1-v)k[a'^{-1}\zeta^2 - 2\zeta + a'], \quad 0 \le \zeta < \omega$$

$\delta(\cdot)$ is the generalized delta function of *Dirac*, and ${}_2F_1(a,b;c;z)$ is *Gauss's* hypergeometric function. Once $\psi(t)$ is known, the problem can be regarded as solved. Fredholm integral equations of the second kind can be solved by the method of successive approximations. Fredholm integral equation of the 2nd kind (27) can be written in a compact manner as follows

$$\psi(t) = -\kappa\sqrt{\frac{\alpha}{t}}\,S\delta(\alpha-t)+\psi^*(t) \quad 0\le t<\alpha; \quad S=-\sigma_0 \tag{30}$$

where we have turn to the state before introduction of the non-dimensional variables defined in (28) and we consider the lower surface of the crack (i.e. $y=0^-$). By using the *shifting property* of the delta function, and recalling eqns it can be easily shown that the zero-order approximation of the function $\psi(t)$, $-\kappa\sqrt{\alpha/t}\,S\delta(\alpha-t)$, cancells out the displacement predicted by the classical LEFM which is responsible for the infinite slope of the crack displacement at the crack tip. The solution of integral eqn (30) with symmetric kernels may be found by recourse to Fredholm's 1st theorem and the method of resolvent

4. Solution near the tip

Let $\overline{S}$ from here on stand for the open half-plane $y\le 0$ together with its bounding edge, i.e. for the region $(-\infty<y\le 0, -\infty<x<\infty)$. Introducing polar coordinates $r_1,\theta_1,r,\theta,r_2,\theta_2$ through the relations

$$z = re^{i\theta}\,, z-\alpha = r_1 e^{i\theta_1}\,, z+\alpha = r_2 e^{i\theta_2} \tag{31}$$

we seek to determine the behavior of the solution at the (singular) endpoints of the crack. By means of familiar Bessel integral-identities, the following elementary expansion

$$\ell a(\xi) = \xi\ell + \frac{1}{2\xi\ell} - \frac{1}{8\xi^3\ell^3} + o(\xi^{-5}) \tag{32}$$

which follows from (21) and is valid for every fixed positive ℓ as $\xi\to\infty$, the *shifting* property of the delta function, as well as the first of (17), and (30) we get the following estimate, which holds true as $r_1\to 0$ for every fixed positive ℓ for the horizontal displacement

$$2\mu u(r_1,\theta_1) = S\sqrt{2\alpha}\,r_1^{1/2}\,\sin\frac{\theta_1}{2}\cos^2\frac{\theta_1}{2}$$
$$-\frac{2}{3}S\sqrt{2\alpha}\,\psi^*(\alpha-\eta)r_1^{3/2}\,\sin\frac{3\theta_1}{2}+o(r_1^{5/2}) \tag{33}$$

where throughout this paper the order-of-magnitude symbols 'O' and 'o' are used in their standard mathematical connotation; in particular, a function is O(1) if it remains bounded in the underlying limit, whereas it is o(1) if it vanishes in the underlying limit.

Furthermore, in (33) η is a small length (i.e. $\eta <<< \alpha$) which removes the weak logarithmic singularity of $\psi^*(t)$ at $t = \alpha$ for $k \neq 0$. From (33) it is evident that the crack lips form a *cusp* of the first kind with zero enclosed angle and zero first derivative of the displacement at the crack tip. This fact indicates that the present gradient elasticity theory with surface energy predicts the same crack shape as Barenblatt's [16] 'cohesive-zone' theory, but without requiring an extra assumption on the existence of interatomic forces at the outset beyond those implied by the gradient terms in the generalized constitutive equation.

The first derivative of u with respect to x near the crack tip (i.e. the slope of the crack profile) can be found from (33) to be

$$2\mu \frac{\partial u}{\partial x}(r_1,\theta_1) = S \frac{\sqrt{2\alpha}}{4} r_1^{-1/2} \sin\theta_1 \cos\frac{\theta_1}{2} \{\cos\theta_1 - 2[\cos^2\frac{\theta_1}{2} - 2\sin^2\frac{\theta_1}{2}]\}$$
$$+ S\sqrt{2\alpha}\,\psi^*(\alpha - \eta)r_1^{1/2} \{\sin\theta_1 \cos\frac{3\theta_1}{2} - \cos\theta_1 \sin\frac{3\theta_1}{2}\} + o(r_1^{3/2}) \tag{34}$$

It is interesting to find the values of the slope in front and behind the crack tip on the crack plane ($y = 0^-$). From the above formula

$$2\mu \frac{\partial u}{\partial x} = \begin{cases} 0 & \theta_1 = 0 \\ -S\sqrt{2\alpha}\,\psi^*(\alpha - \eta)r_1^{1/2} & \theta_1 = \pi \end{cases} \tag{35}$$

Hence, the present gradient-dependent theory predicts that the slope of the crack displacement is *finite* and *continuous*, i.e.

$$\lim_{x \to \alpha^-} \frac{\partial u}{\partial x}(x,0^-) = \lim_{x \to \alpha^+} \frac{\partial u}{\partial x}(x,0^-) = 0 \tag{36}$$

On the other hand, as it was mentioned previously Linear Elastic Fracture Mechanics (LEFM) predicts infinite slope of the crack displacement at the crack tip, that is

$$2\mu \frac{\partial u^c}{\partial x}(r_1,\pi) = -\sqrt{\frac{2}{\pi}} K_I \frac{(1-v)}{\sqrt{r_1}} \neq 2\mu \frac{\partial u^c}{\partial x}(r_1,0) = 0 \tag{37}$$

Barenblatt [16] has introduced a small cohesive zone ahead of the 'physical' crack tip whose size is determined *explicitly* by requiring the cancellation of stress singularity at the tip of the cohesive zone (or tip of 'effective' crack), or equivalently smooth closure of crack lips. However, the slope of the crack opening displacement in mode-I deformation (that is a case which is directly analogous to the present case of mode-II deformation) turns out to be infinite at the crack tip/cohesive zone boundary,

$2\mu\partial\upsilon/\partial x \propto \sigma_0 \alpha^{3/2}(\alpha-x)^{-3/2}$ $\quad x \to \alpha -$, even though a smooth closure condition is ensured. In the present formulation, however, *cohesive forces* are not introduced at the outset but are inherent in the constitutive structure of the theory and, as it was shown above, these forces - which are actually the self-equilibrated double-forces - result in a smooth closure condition for the faces of the 'mathematical' crack tip and lead to a 'proof' of the existence of the cohesive zone. In fact an asymptotic expansion of the tangential double-stress close to the crack tip gives

$$\mu_{yyx}(r_1,\theta_1)/(S\sqrt{\alpha}) = -\ell'\{\frac{\sqrt{2}}{4}r_1^{-1/2}\ sin\ \theta_1\ sin\ \frac{3\theta_1}{2}+\frac{\sqrt{2}}{2(1-v)}$$
$$\times\psi^*(\alpha-\eta)r_1^{1/2}\ cos\ \frac{\theta_1}{2}\}+\frac{1}{4\sqrt{2}}\ell^2 r_1^{-3/2}\ \{2\ sin\ \frac{3\theta_1}{2}+3\ sin\ \theta_1\ cos\ \frac{5\theta_1}{2}\} \tag{38}$$

From the above expression it may be derived

$$\mu_{yyx}(r_1,\pi)=\frac{S\sqrt{\alpha}}{2\sqrt{2}}\ell^2 r_1^{-3/2};\quad \mu_{yyx}(r_1,0)=-\frac{\sqrt{2}S\sqrt{\alpha}}{2(1-v)}\psi^*(\alpha-\eta)\ell' r_1^{1/2} \tag{39}$$

The above formulae (39) elucidate the role of the volume and surface strain-gradient parameters, namely ℓ,ℓ', rerspectively. That is to say, ℓ is responsible for the action of infinite 'cohesive' double-forces behind the crack tip, leading to a cusping crack. On the other hand, the surface energy parameter captures the existence of a 'process' zone in front of the crack tip and further controls the crack compliance to loading.

In the context of the proposed theory the energy release during an *infinitesimal* advancement of the crack tip by a distance $\delta\alpha$ can be found to be

$$\delta U = G_{II}\delta\alpha = \int_0^{\delta\alpha} \tau_{xy}(\delta\alpha-h,0+)u(h,-\pi)dh = \frac{K_{II}^2}{8\mu}\psi^*(\alpha-\eta)(\delta\alpha)^2,\quad \delta\alpha \to 0 \tag{40}$$

where $K_{II}=\sigma_0\sqrt{\pi\alpha}$ and G_{II}, is the mode-II stress intensity factor, and energy release rate, respectively. It should be mentioned that Barenblatt's and Dugdale's stress-finiteness condition results to a similar dependence as that given by (40), i.e. $G_{II} \propto \delta\alpha$, i.e. the energy release rate turns out to approach zero for an infinitesimal crack extension, hence the term 'equilibrium cracks' given by Barenblatt. If it is assumed that the crack must grow continuously from its initial to final length then as it is clear from (40) whatever the external loads the crack will never extend. This is another typical case where the Griffith-Irwin criterion fails (see for example [17]) and is discussed in [3].

Acknowledgements: The authors gratefully acknowledge the financial support of the European Union under the contract SMT4-CT96-2130.

5. References

1. Vardoulakis, I., Exadaktylos, G., Aifantis, E.: Gradient elasticity with surface energy: Mode-III crack problem, *Int. J. Solids Structures* **33** **(30)** (1996), 4531-4559.
2. Exadaktylos, G.E., Vardoulakis, I., Aifantis, E.C.: Cracks in gradient elastic bodies with surface energy, *Int. J. Fracture* **79** (1996), 107-119.
3. Exadaktylos, G.E.: Gradient elasticity with surface energy: Mode-I crack problem, *Int. J. Solids Structures* (in press) (1997).
4. Vardoulakis, I., Sulem, J.: *Bifurcation Analysis in Geomechanics*, Blackie Academic and Professional, 1995.
5. Casal, P.: La capillarite interne, *Cahier du Groupe Francais d'Etudes de Rheologie C.N.R.S.* **VI**(3) (1961), 31-37.
6. Aifantis, E.C.: On the role of gradients in the localization of deformation and fracture, *Int. J. Engng. Sci.* **30** (1992), 1279-1299.
7. Mindlin, R.D.: Second gradient of strain and surface-tension in linear elasticity, *Int. J. Solids Structures* **1** (1965), 417-437.
8. Aifantis, E.C.: A proposal for continuum with microstructure, *Mech. Res. Comm.* **5**, (1978) 139-145.
9. Vardoulakis, I., Shah, K.R., Papanastasiou, P.: Modelling of tool-rock interfaces using gradient-dependent flow theory of plasticity, *Int. J. Rock Mech. Min. Sci. & Geomech. Abstr.* **29**, No. 6 (1992), 573-582.
10. Pearson, J.R.A.: Length scales, asymptotics and non-linear singularities, In: *Non-linear singularities in deformation and flow* (Durban, D., Pearson, J.R.A., eds.), IUTAM Symposium, Haifa, Israel 1997.
11. Unger, D.J., Aifantis, E.C.: The asymptotic solution of gradient elasticity for mode III, *Int. J. Fracture* **71** (1995), R27-R32.
12. Exadaktylos, G.E., Vardoulakis, I.: Displacement potentials formulation of anisotropic gradient elasticity theory with surface energy, In: *Proc. 2nd National Congress on Computational Mechanics* (Ed's: Sotiropoulos D.A. and Beskos D.E.), Vol. II. Chania, Crete, Greece, 26-28 June, 635-644, 1996.
13. Ru, C.Q., Aifantis, E.C.: A simple approach to solve boundary-value problems in gradient elasticity, *Acta Mech.* **101** (1993), 59-68.
14. Love, A.E.H.: *A treatise on the mathematical theory of elasticity*, Cambridge 1927.
15. Eshelby, J.D.: The force on an elastic singularity. *Philos. Trans. Roy. Soc. London Ser. A* **244** (1951), 87-112.
16. Barenblatt, G.I.: Equilibrium cracks formed during brittle fracture - Rectilinear cracks in plane plates. *J. Appl. Math. and Mech.*, **23** (1959), 1009-1029.
17. Leguillon, D.: Asymptotic analysis of a spontaneous crack growth. Application to a blunt crack, In: *Non-linear singularities in deformation and flow* (Durban, D., Pearson, J.R.A., eds.), IUTAM Symposium, Haifa, Israel 1997.

THE UNSTEADY MOTION OF
THREE PHASE CONTACT LINES

J. BILLINGHAM

School of Mathematics and Statistics
The University of Birmingham
Edgbaston
Birmingham B15 2TT
England

1. Introduction

When two fluids meet at a solid surface a three phase contact line is formed.
The motion of contact lines is of crucial importance in many different natural phenomena and industrial processes, for example multiphase flow in
porous media, multiphase flow metering and industrial coating processes.
Most work on moving contact lines has concentrated on steady motion at
low Reynolds number (Cox, 1986). In this paper, we consider the unsteady
motion of a contact line at high Reynolds number. We shall consider only
inviscid fluids, although it is clear that one of the next steps is to consider
the effect of viscosity.

Some recent, related work (Billingham and King, 1995) was concerned
with the penetration of a flat interface between two fluids by a thin, flat
plate, and analysed both the unsteady motion of the contact line formed in
this process and the capillary waves generated on the interface. We specified
the contact angle as a monotonically increasing function of the contact line
velocity. Our aim in the present paper is to show how a recently proposed
model for the motion of contact lines (Shikhmurzaev, 1993) can be applied
to this problem. Using the new theory, which is based on an analysis of the
thermodynamics of the fluid/fluid and fluid/solid interfaces, the contact
angle is a function of quantities that are defined only on these interfaces
and is part of the solution. The new theory has only ever been applied to
steady flows at low Reynolds number, where the agreement between theory
and experiment is impressive (Shikhmurzaev, 1997a). By considering an

D. Durban and J.R.A. Pearson (eds.),
IUTAM Symposium on Non-Linear Singularities in Deformation and Flow, 99-110.
© 1999 *Kluwer Academic Publishers. Printed in the Netherlands.*

unsteady flow at high Reynolds number we hope to test its validity in a new situation.

Asymptotic solutions can be derived, based on expansions in a small parameter that comes from the low compressibility of the fluid. As time increases the problem is similar to that studied by Billingham and King (1995), except that the contact angle is now a function of the history of both the contact line velocity and the length of the interface. We can also determine the leading order solution of the problem that arises when the contact angle remains close to its initial value, and plan to tackle the full, nonlinear problem numerically in the near future.

2. The Model

It is well–known that for a Newtonian fluid that does not slip at solid surfaces, the velocity field in the neighbourhood of a moving contact line is multi–valued and leads to an unphysical singularity in the force exerted on the solid surface (Dussan V. and Davis, 1974). This singularity can be relieved by relaxing the no slip boundary condition in the neighbourhood of the contact line (Dussan V., 1976, Cox, 1986). A new model derived by Shikhmurzaev (1993) allows this to be done in a manner that is based upon an analysis of the thermodynamics of fluid/fluid and fluid/solid interfaces first proposed by Bedeaux *et al.* (1975). After specifying a few phenomenological constants, including the static contact angle, the dynamic contact angle can, in principle, be determined in any given situation. There is no need to specify the contact angle as a function of velocity *a priori*.

The model is based on three key assumptions:

- Within about 10^{-8} to 10^{-9} m of an interphase boundary, fluid molecules are in a different state to those in the bulk fluid.
- For deformations that take place over a longer length scale, we can define a surface density, pressure and velocity on the bounding surface of each fluid.
- As molecules move from a fluid/fluid interface to a fluid/solid interface they take a finite time to relax to their new equilibrium state.

The surface pressure at an interface is defined to be the negative surface tension. Variations in surface pressure due to the relaxation of fluid molecules from one state to another therefore lead to a self–induced surface tension gradient, and hence a Marangoni effect. This affects the value of the surface tension at the contact line, and hence the contact angle.

In the present paper, we shall consider a model problem for a gas/liquid/solid system, with the gas treated as a void. In this case, we only need consider the surface variables in the single liquid phase. The new model modifies the usual boundary conditions in a manner that we now summarize briefly. We assume that surface pressure and density, p^s and ρ^s, satisfy the linear equation of state

$$p^s - p_e^s = \gamma\left(\rho^s - \rho_e^s\right),\tag{1}$$

where γ is a phenomenological constant and p_e^s and ρ_e^s are the equilibrium values of surface pressure and density. Conservation of surface mass leads to

$$\frac{\partial \rho^s}{\partial t} + \nabla^s.\left(\rho^s \mathbf{v}^s\right) = -\frac{\rho^s - \rho_e^s}{\tau},\tag{2}$$

where ∇^s is the gradient operator restricted to the interface, and τ is the timescale over which fluid molecules relax to their equilibrium state. The source term on the right hand side of this equation is due to a flux of fluid from the bulk to the interface and drives the surface density towards its equilibrium value. For a water/air interface, $p_e^s \approx -5 \times 10^{-2}$ kgs^{-2}, $\gamma \approx 10^5$ m^{2}s^{-2}, $\rho_e^s \approx 10^{-6}$ kgm^{-2} and $\tau \approx 10^{-3}$ s. There has been some controversy over the size of τ, which we shall not dwell upon here. Even with $\tau \approx 10^{-6}$s, which is easier to justify on theoretical grounds, the model relieves the stress singularity at the contact line and the analysis below remains relevant.

Consider first the boundary conditions at a solid surface. The surface layer velocity, $\mathbf{v}^s$, can be related to the bulk velocity, $\mathbf{u}$, at the solid surface and the velocity of the solid surface itself, $\mathbf{V}$. By definition, normal velocities, denoted by the subscript n, are continuous so that

$$\mathbf{u}_n = \mathbf{v}_n^s = \mathbf{V}_n, \text{ at the solid surface.}\tag{3}$$

Surface pressure gradients and shear stresses can drive an apparent slip of the bulk fluid, via

$$\mu\mathbf{n}.\left(\nabla\mathbf{u} + \nabla^T\mathbf{u}\right)_t - \frac{1}{2}\nabla^s p^s = \beta\left(\mathbf{u} - \mathbf{V}\right)_t, \text{ at the solid surface,}\tag{4}$$

where μ is the viscosity of the fluid, $\mathbf{n}$ is the outward unit normal at the solid surface, β is a phenomenological constant, and the subscript t indicates that the quantity is resolved in the direction of the tangent to the solid surface. The tangential components of the surface velocities also satisfy

$$\mathbf{v}_t^s = \frac{1}{2}\left(\mathbf{u} + \mathbf{V}\right)_t - \alpha\nabla^s p^s, \text{ at the solid surface,}\tag{5}$$

where α is a phenomenological constant. These equations and boundary conditions are equivalent to those given by Shikhmurzaev (1997b) as (7) to (11).

A schematic of the velocities, $\mathbf{u}$ and $\mathbf{v}^s$, at a solid surface which is moving in its own plane with speed $\mathbf{V}$ is shown as figure 1. Actual slip between the

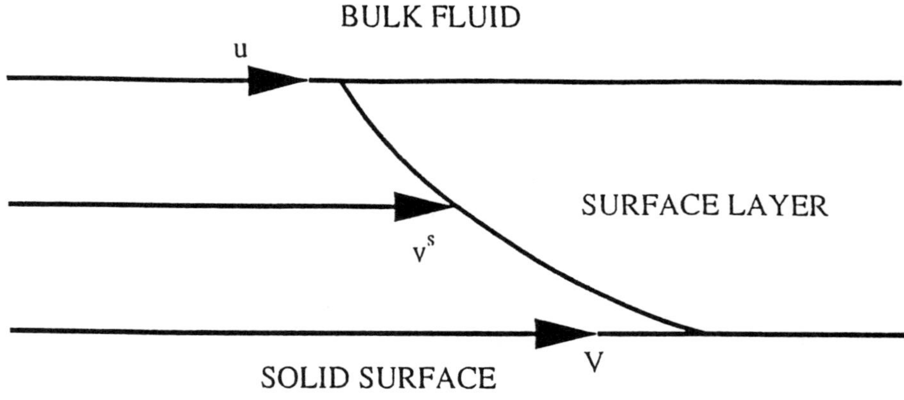

Figure 1. The velocities $\mathbf{u}$, $\mathbf{v}^s$ and $\mathbf{V}$ at a solid surface.

bulk fluid and the solid surface only occurs in the neighbourhood of the contact line, within the dimensions of the interfacial layer (10^{-8} to 10^{-9} m), consistent with molecular dynamics simulations (Koplik *et al.*, 1988). The difference between $\mathbf{u}_t$ and $\mathbf{V}_t$ at a solid surface is only an apparent slip, since the fluid particles on the side of the interfacial layer that meets the solid surface move at the same velocity as that surface.

At liquid/gas interfaces, the appropriate equations and boundary conditions take a similar form, as we shall see below, and include continuity of shear and normal stress, given by Shikhmurzaev (1997b) as (1) to (6). These equations and boundary conditions must be applied in addition to the usual equations of fluid mechanics, along with mass and force balances at the contact line, which we shall describe below.

3. A Simple Initial Value Problem

The series of papers by Shikhmurzaev has already demonstrated that the model described above is in good agreement with experiment for steady flows at low Reynolds number. Our objective is to try to evaluate the predictions of the model for unsteady flows at high Reynolds number. We shall consider a model problem similar to that studied by Billingham and King (1995). Since the introduction of a model for the variation of surface quantities leads to a novel and complicated initial/boundary value problem, we have decided to study the simplest non-trivial problem that we can. We consider a semi-infinite, thin flat plate moving infinitessimally slowly through a liquid towards a gas/liquid interface in the absence of gravity. The leading edge of the plate meets the interface and a contact line is

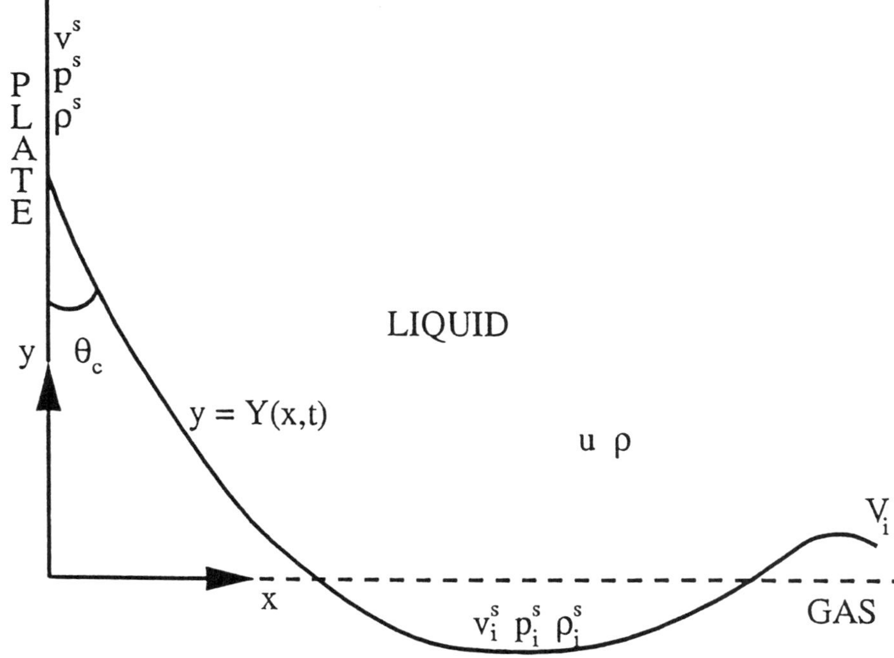

Figure 2. The plate, deforming interface, coordinate system and variable quantities for the model problem.

formed when $t = 0$, at which time the fluid and the plate are at rest and the interface is flat. In other words, there is no external flow, and the motion of the fluid is driven solely by the initial imbalance in the force at the contact line.

There are no length, velocity or time scales imposed by either the geometry of the system or the initial conditions, so the natural quantities with which to non–dimensionalize the variables are $l^* = (\sigma \tau^2 / \rho)^{1/3} \approx 370 \ \mu$m, $u^* = (\sigma / \rho \tau)^{1/3} \approx 0.37 \ \text{ms}^{-1}$ and $t^* = \tau \approx 10^{-3}$ s. Here σ is the equilibrium surface tension and ρ the bulk density of the fluid, here assumed to be oil or water. We also use $\sigma = -p_{ie}^s$ and ρ_{ie}^s to non–dimensionalize the surface densities and pressures. The subscript i denotes surface quantities on the deformable interface, whilst the absence of a subscript indicates a surface quantity on the liquid/solid interface. We treat the gas as a void. The Cartesian coordinate system, (x, y), is as shown in figure 2, along with the deformable interface at $y = Y(x, t)$ and the various bulk and surface variables. Note that we can exploit the symmetry of the system about the y–axis and solve for $x \geq 0$, $y \geq Y(x, t)$ only.

The dimensionless bulk equations are

$$\nabla.\mathbf{u} = 0, \quad \frac{\partial \mathbf{u}}{\partial t} + \mathbf{u}.\nabla \mathbf{u} = -\nabla p + Re^{-1}\nabla^2\mathbf{u}, \tag{6}$$

where $Re = \rho u^* l^* / \mu = (\rho\sigma^2\tau)^{1/3}/\mu \approx 135$ is the Reynolds number. In the present paper we shall neglect viscosity altogether ($Re^{-1} = 0$). We are currently grappling with the difficult problem of including a non–zero viscosity, both in boundary layers and as it affects the singularity that develops when $t = 0^+$. Note that, although no singularity in the stress can arise for an inviscid fluid since it can slip along a solid surface, the new model is not redundant, since it allows us to determine the contact angle dynamically, without specifying it as a function of the contact line speed.

The dimensionless equations for quantities defined at the gas/liquid and liquid/solid interfaces are

$$\frac{\partial \rho^s}{\partial t} + \frac{\partial}{\partial y}\left(\rho^s v^s\right) = -\left(\rho^s - \rho_e^s\right), \tag{7}$$

$$\frac{\partial}{\partial t}\left(\rho_i^s \sqrt{1 + Y_x^2}\right) + \frac{\partial}{\partial x}\left(\rho_i^s v_i^s\right) = -\left(\rho_i^s - 1\right)\sqrt{1 + Y_x^2}, \tag{8}$$

$$p^s - p_e^s = \lambda\left(\rho^s - \rho_e^s\right), \quad p_i^s + 1 = \lambda\left(\rho_i^s - 1\right), \tag{9}$$

where $\lambda = \gamma\rho_{ie}^s/\sigma \approx 20$, is a dimensionless number whose size we shall exploit below when we construct an asymptotic solution of this problem. Note that $v^s(y)$ and $v_i^s(x)$ are the velocities tangential to the liquid/solid and deformable interfaces respectively, and $Y_x = \partial Y/\partial x$.

The boundary conditions at the solid surface are

$$u_x = 0, \quad \frac{1}{\lambda Re}\frac{\partial u_x}{\partial x} - \frac{1}{2}\frac{\partial \rho^s}{\partial y} = \hat{\beta}u_y, \quad v^s = \frac{1}{2}u_y - \hat{\alpha}\frac{\partial \rho^s}{\partial y}, \tag{10}$$

at $x = 0$ for $y > Y(0,t)$, where $\hat{\beta} = \beta/\lambda\rho u^*$ and $\hat{\alpha} = \alpha\lambda\rho u^*$. We expect that $\hat{\alpha}, \hat{\beta} = O(1)$. Note that u_x and u_y are the x- and y–components of the bulk velocity, $\mathbf{u}$.

The boundary conditions on the deformable, gas/liquid interface are

$$\frac{1 + 4\hat{\alpha}\hat{\beta}}{\sqrt{1 + Y_x^2}}\frac{\partial \rho_i^s}{\partial x} = 4\hat{\beta}\left(u_x + Y_x u_y - v_i^s\right), \tag{11}$$

$$\frac{1}{\lambda Re}\,(\text{viscous stresses}) - \frac{1}{\sqrt{1 + Y_x^2}}\frac{\partial \rho_i^s}{\partial x} = 0, \tag{12}$$

$$-\frac{p_i^s Y_{xx}}{\sqrt{1 + Y_x^2}} - p - \frac{1}{Re}\,(\text{viscous stresses}) = 0, \tag{13}$$

$$\frac{\partial Y}{\partial t} = u_y - u_x Y_x, \tag{14}$$

at $y = Y(x, t)$ for $x > 0$. Since we shall assume that $Re^{-1} = 0$ we have not written out the full expressions for the terms that represent the viscous stresses. Equation (14) is the usual kinematic boundary condition.

At the contact line we have a mass balance,

$$\rho^s \left(v^s - \frac{\partial Y}{\partial t} \right) + \rho_i^s v_i^s = 0, \tag{15}$$

and a force balance

$$p^s - p_e^s - p_i^s \cos \theta_c - \cos \theta_s = 0, \tag{16}$$

at $x = 0$, $y = Y(0, t)$. The contact angle, $\theta_c(t) = -Y_x/\sqrt{1 + Y_x^2}$ at $x = 0$, is indicated in figure 2, and θ_s is the static contact angle, the final phenomenological constant. Equation (16) determines how the surface pressures at the contact line, p^s and p_i^s, affect the dynamic contact angle, θ_c.

The initial conditions are

$$\mathbf{u} = 0, \quad p = 0, \quad Y = 0, \quad v^s = 0, \quad \rho^s = \rho_e^s, \quad p^s = p_e^s,$$

$$v_i^s = 0, \quad \rho_i^s = 1, \quad p_i^s = -1, \text{ when } t = 0, \tag{17}$$

with far field boundary conditions

$$\mathbf{u} \to 0, \quad p \to 0, \quad \text{as } x^2 + y^2 \to \infty; \quad Y \to 0, \text{ as } x \to \infty, \tag{18}$$

$$v^s \to 0, \quad \rho^s \to \rho_e^s, \quad p^s \to p_e^s, \text{ as } y \to \infty, \tag{19}$$

$$v_i^s \to 0, \quad \rho_i^s \to 1, \quad p_i^s \to -1, \text{ as } x \to \infty. \tag{20}$$

When $Re^{-1} = 0$, (8) and (11) to (13) along with (17) and (18) show that

$$\rho_i^s = 1, \quad p_i^s = -1, \quad \frac{\partial v_i^s}{\partial x} = -\frac{\partial}{\partial t} \left(\sqrt{1 + Y_x^2} \right). \tag{21}$$

A surface tension gradient cannot form on the deformable interface, which is effectively inextensible, as shown by (12) for an inviscid fluid. Note that the tangential velocity is discontinuous at the liquid/gas interface. For a fluid with a small but finite viscosity, the tangential velocity in the bulk fluid will vary rapidly across a thin boundary layer.

The length of the interface can only increase via a flow of fluid through the contact line. This is most easily expressed by defining the excess length of the interface as

$$L(x, t) = \int_x^\infty \left(\sqrt{1 + Y_x^2} - 1 \right) dx, \tag{22}$$

so that $v_i^s = \partial L/\partial t$. On the solid surface (10) gives

$$v^s = -D\frac{\partial \rho^s}{\partial y}, \quad \text{where } D = \frac{1 + 4\hat{\alpha}\hat{\beta}}{4\hat{\beta}}, \tag{23}$$

and hence from (7)

$$\frac{\partial \rho^s}{\partial t} = D\frac{\partial}{\partial y}\left(\rho^s \frac{\partial \rho^s}{\partial y}\right) - (\rho^s - \rho_e^s), \quad \text{for } t > 0 \text{ and } y > Y(0, t). \tag{24}$$

The surface density on the solid surface therefore satisfies a nonlinear reaction–diffusion equation, with boundary and initial conditions given by (15), (16), (17) and (19) as

$$\rho^s = \rho_e^s, \quad \text{for } y > Y(0, t) \text{ when } t = 0, \tag{25}$$

$$\rho^s \to \rho_e^s, \quad \text{as } y \to \infty, \tag{26}$$

$$\rho^s = \rho_e^s + \lambda^{-1}\left(\cos\theta_s - \cos\theta_c\right), \quad \text{at } y = Y(0, t), \tag{27}$$

$$D\frac{\partial \rho^s}{\partial y} = \frac{1}{\rho^s}\frac{\partial L}{\partial t}(0, t) - \frac{\partial Y}{\partial t}(0, t), \tag{28}$$

$$Y = 0, \quad \text{at } x = 0 \text{ when } t = 0. \tag{29}$$

This surface problem determines how the contact angle θ_c depends upon the deformation of the interface through $\partial Y/\partial t$ and $\partial L/\partial t$, and shows how variations in the surface density at the contact line can affect the surface density along the solid/liquid interface.

Since the bulk flow is inviscid and initially irrotational, it remains irrotational and we can define a velocity potential, $\phi(x, y, t)$ so that

$$\mathbf{u} = \nabla\phi. \tag{30}$$

In terms of ϕ the boundary value problem satisfied by the bulk fluid is

$$\nabla^2\phi = 0, \quad \text{for } x > 0, \, y > Y(x, t), \tag{31}$$

subject to

$$\phi_x = 0, \quad \text{at } x = 0 \text{ for } y > Y(0, t), \tag{32}$$

$$Y_t = \phi_y - \phi_x Y_x, \quad \text{at } y = Y(x, t), \text{ for } x > 0, \tag{33}$$

$$\phi_t + \frac{1}{2}|\nabla\phi|^2 = -\frac{Y_{xx}}{(1 + Y_x^2)^{3/2}}, \quad \text{at } y = Y(x, t), \text{ for } x > 0, \tag{34}$$

$$\phi = 0, \quad Y = 0, \quad \text{when } t = 0, \tag{35}$$

$$\phi \to 0, \quad \text{as } x^2 + y^2 \to \infty, \quad Y \to 0, \quad \text{as } x \to \infty. \tag{36}$$

This problem is driven by the motion of the contact line through its coupling to the one–dimensional initial/boundary value problem on the solid surface defined by (24) to (29). Note, however, that when $\theta_s > \pi/2$ the inital force imbalance at the contact line tends to prevent the motion of the interface along the solid surface, and no flow occurs.

If instead the problem were driven by imposing a constant contact angle, θ_c (Keller and Miksis, 1982), or by specifying θ_c as an increasing function of the contact line velocity (Billingham and King, 1995), the solution would be known. For the case of a constant contact angle the solution is of similarity form with lengths scaled by $t^{2/3}$. The interface carries capillary waves that decay as $x \to \infty$, and propagate away from the flat plate. When the contact angle is a function of the contact line velocity, this similarity solution can arise as the leading order asymptotic solution for $t \ll 1$.

For the present problem we will proceed by considering the asymptotic solution when $\lambda \gg 1$. This corresponds to an almost incompressible liquid at the liquid/solid and liquid/gas interfaces. The boundary condition (27) suggests that we should seek an asymptotic solution of the form

$$\rho^s = \rho_e^s + \lambda^{-1}\bar{\rho}^s + o(\lambda^{-1}), \quad Y(0,t) = \lambda^{-1}\bar{Y}(t) + o(\lambda^{-1}), \quad \text{for } \lambda \gg 1. \quad (37)$$

At leading order the behaviour of $Y(0,t)$, the position of the contact line, is completely determined by the behaviour of the liquid at the liquid/solid interface. This motion of the contact line then drives the bulk flow. We can solve for $\bar{\rho}^s$ and $\bar{Y}$ and find that

$$\bar{\rho}^s \sim \frac{\cos\theta_s}{\sqrt{\pi}} \int_{y/\rho_e^s Dt^{1/2}}^{\infty} e^{-v^2/4}dv, \quad \bar{Y} = O(t^{1/2}), \quad \text{for } t \ll 1, \quad (38)$$

since the behaviour of $\bar{\rho}^s$ is dominated by linear diffusion as $t \to 0$. This behaviour is not compatible with what we know about the behaviour of the bulk fluid, since it suggests that the contact line moves much more rapidly than the fastest possible motion given by the similarity solution with $\bar{Y} = O(t^{2/3})$. We conclude that (37) is incorrect, and that in fact the contact angle changes instantaneously, as is the case for the similarity solution with constant θ_c (Keller and Miksis, 1982).

We can construct a solution valid for $t \ll 1$ using the expansions

$$\cos\theta_c(t) = \cos\theta_s + O(t^{1/6}), \quad \rho^s = \rho_e^s + t^{1/6}\bar{\rho}^s(z,t) + o(t^{1/6}),$$

$$z = t^{-1/6}\left\{y - Y(0,t)\right\}, \quad (39)$$

with $z, \bar{\rho}^s = O(1)$ as $t \to 0$. The flow of the bulk fluid is governed by an $O(t^{2/3})$ similarity solution. The terms of $O(t^{1/6})$ arise naturally as the ratio of $t^{2/3}$ from this similarity solution and $t^{1/2}$ from the diffusive initial value problem on the solid surface. There is no space to discuss this further here, but we find that for $\lambda \gg 1$ this small time solution becomes non–uniform when $t = O(\lambda^{-6})$.

Appropriate scalings for the variables on the solid surface in the next asymptotic region are

$$\rho^s = \rho_e^s + \lambda^{-1}\tilde{\rho}\,(\tilde{z}, t) + o(\lambda^{-1}), \quad y = \lambda^{-4}\tilde{Y}(0, \tilde{t}) + \lambda^{-3}\tilde{z},$$

$$L = \lambda^{-4}\tilde{L}, \quad Y = \lambda^{-4}\tilde{Y}, \quad t = \lambda^{-6}\tilde{t}, \tag{40}$$

with $\tilde{\rho}^s$, $\tilde{Y}$, $\tilde{z}$, $\tilde{L}$, $\tilde{t} = O(1)$ as $\lambda \to \infty$. At leading order we obtain a simple linear diffusion problem that relates the contact angle to the history of the deformation of the interface via a convolution integral as

$$\cos\theta_c(\tilde{t}) = \cos\theta_s - \left(\frac{1}{\pi D}\right)^{1/2}\int_0^{\tilde{t}} (\tilde{t} - s)^{1/2}\left\{\frac{\partial \tilde{Y}}{\partial \tilde{t}}(0, t) - \frac{\partial \tilde{L}}{\partial \tilde{t}}(0, t)\right\} ds. \tag{41}$$

Suitable scaled variables for the bulk fluid flow are

$$x = \lambda^{-4}\tilde{x}, \quad y = \lambda^{-4}\tilde{y}, \quad \phi = \lambda^{-2}\tilde{\phi}, \tag{42}$$

with $\tilde{x}$, $\tilde{y}$, $\tilde{\phi} = O(1)$ as $\lambda \to \infty$. The bulk equations (31) to (36) are invariant under this transformation, so that the new variables satisfy the same boundary value problem, except that the contact angle is given by (41) as a function of the history of the deformation. We know the behaviour of the solution for $\tilde{t} \ll 1$, which is of similarity form, with the behaviour at the solid surface given by expansions (39). In order to determine what happens for $\tilde{t} = O(1)$ we must solve numerically the initial/boundary value problem given by (31) to (36) and (41), and plan to do so in the near future.

At present, we have only studied the special case, $\theta_s = \pi/2 - \epsilon$ with $\epsilon \ll 1$. The deformation of the interface and all other displacements are then small, and the problem can be linearized. The interfacial boundary conditions can be linearized onto the $\tilde{x}$–axis to give a quarter plane problem. This can be solved using a Laplace transform with respect to $\tilde{t}$ and a Mellin transform with respect to $\tilde{r} = \sqrt{\tilde{x}^2 + \tilde{y}^2}$. As $\tilde{t} \to 0$ this agrees with the linearized problem studied by Billingham and King. As $\tilde{t} \to \infty$ we find that

$$\tilde{Y}(\tilde{x}, \tilde{t}) \sim \frac{\epsilon\,(D\tilde{t})^{1/2}}{12\pi}\frac{1}{2\pi i}\int_{c-i\infty}^{c+i\infty} 27^{p/3}(p - 1)\frac{\cos(\pi p/2)}{\sin\{\pi(p + 1)/3\}}$$

$$\times\ \frac{\Gamma\left(\frac{1}{3}p + \frac{1}{3}\right)\Gamma\left(\frac{1}{3}p\right)}{\Gamma\left(\frac{2}{3}p + \frac{3}{2}\right)}\left(\frac{\tilde{x}}{\tilde{t}^{2/3}}\right)^{-p} dp, \tag{43}$$

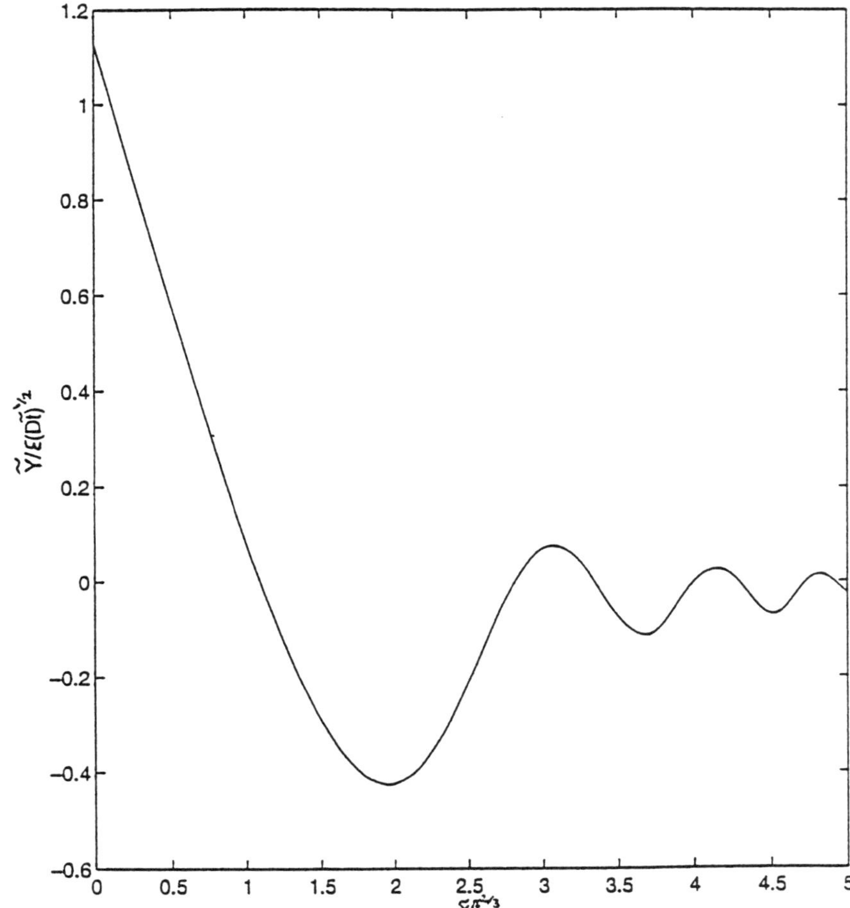

Figure 3. The solution (43), which is valid for $\tilde{t} \gg 1$.

where c is a constant with $0 < c < 2$, and hence

$$\tilde{Y}(0, \tilde{t}) \sim \frac{2\epsilon \left(D\tilde{t}\right)^{1/2}}{\pi^{1/2}} \quad \text{as } \tilde{t} \to \infty. \tag{44}$$

The solution (43) is illustrated in figure 3. This is qualitatively similar to the similarity solution that arises for $\tilde{t} \ll 1$, with decaying capillary waves on the interface. However, in this case the diffusion of surface density on the solid surface leads to $\tilde{Y} = O(\tilde{t}^{1/2})$ as $\tilde{t} \to \infty$, whilst the length scale for the deformation in the $\tilde{x}$–direction is of $O(\tilde{t}^{2/3})$. The slope of the interface therefore tends to zero as $\tilde{t} \to \infty$, in particular, $\theta_c = \pi/2 + O(\tilde{t}^{-1/6})$. In this special case, where the static contact angle is close to $\pi/2$, our analysis shows that, in the absence of both gravity and viscosity, the dynamic contact angle asymptotes to $\pi/2$ as $t \to \infty$. If the physical explanation for this behaviour, that diffusion dominates on the solid surface and leads to $Y = O(t^{1/2})$ whilst surface tension balances inertia in the bulk and leads to a deformation length scale of $O(t^{2/3})$ as $t \to \infty$, remains correct when $\theta_s - \pi/2 \not\ll 1$ then there is a suggestion that the contact angle always asymptotes to $\pi/2$ as $t \to \infty$.

4. Conclusions and Further Work

The work presented above represents a first attempt at examining the predictions of the model derived by Shikhmurzaev in a situation other than that of creeping flow. We have confined our attention to a particularly simple case of inviscid, irrotational flow. We have only been able to obtain a complete asymptotic solution in one particular case, but there is a suggestion that the contact angle asymptotes to $\pi/2$ for all values of θ_s. This remains to be demonstrated by calculating numerical solutions, which we plan to do in the near future. If this suggestion is correct, and indeed even if it is not correct, the next vital step is to include the effect of viscosity. This will manifest itself, not only in boundary layers at the surface of the flat plate and at the free surface, but also in a modification of the singularity when $t \ll 1$. We have seen that the length scale for the solution is of $O(t^{2/3})$, and hence that velocities are singular, of $O(t^{-1/3})$ for $t \ll 1$. This means that the neglected, viscous terms in the governing equations must make their presence felt for t sufficiently small.

References

Bedeaux, D., Albano, A.M. and Mazur, P. (1976) Boundary conditions and non-equilibrium thermodynamics, *Physica A* **82**, 438–462.

Billingham, J. and King, A.C. (1995) The interaction of a moving fluid/fluid interface with a flat plate, *J. Fluid Mech.* **296**, 325-351.

Cox, R.G. (1986) The dynamics of the spreading of liquids on a solid surface, *J. Fluid Mech.* **168**, 169-194.

Dussan V., E.B. and Davis, S.H. (1974) On the motion of a fluid–fluid interface along a solid surface, *J. Fluid Mech.* **65**, 71–95.

Dussan V., E.B. (1976) The moving contact line: the slip boundary condition, *J. Fluid Mech.* **77**, 665–684.

Keller, J.B. and Miksis, M.K. (1983) Surface tension driven flows, *SIAM J. Appl. Maths* **43**, 268–277.

Koplik, J., Banavar, J.R. and Willemsen, J.F. (1988) Molecular dynamics of Poiseuille flow and moving contact lines, *Phys. Rev. Lett.* **60**, 781–794.

Shikhmurzaev, Y.D. (1993a) A two–layer model of an interface between immiscible fluids, *Physica A* **192**, 47–62.

Shikhmurzaev, Y.D. (1993b) The moving contact line on a smooth solid surface, *Int. J. Multiphase Flow* **4**, 589–610.

Shikhmurzaev, Y.D. (1997a) Moving contact lines in liquid/liquid/solid systems, *J. Fluid Mech.* **334**, 211–249.

Shikhmurzaev, Y.D. (1997b) Free-surface cusps and moving contact lines. A common approach to the problems, *this volume.*

SINGULARITIES ON VISCOUS INTERFACES

S. H. DAVIS

Northwestern University

McCormick School of Engineering and Applied Science

Northwestern University

Evanston, IL 60208

1. Introduction

When one thinks of singularities or cusps in fluid mechanics, one usually envisions models of inviscid fluids with interfaces lacking surface tension. Thus, for example the water wave of maximum height, the Stokes wave, has a cusp of angle of $120°$. Further, if one studies idealized flows like line vortices, one will find singular behaviors along the axes.

In the present paper singularities and cusps will be discussed in models of viscous fluids with interfaces possessing surface tension. Attention will be focussed on when to expect such behaviors and how to interpret them. One should recall that when one derives the Navier-Stokes equations, the volume balances give the point equations locally only if the velocity field is sufficiently smooth. However, singular behaviors are not precluded from existing on domain boundaries and it is on boundaries that the singularities do occur.

In the present paper, two types of behaviors will be examined. Firstly, there are singularities that occur due to the splitting of trajectories as illustrated by moving contact lines and by rupture/coalescence mechanics. Secondly, there are cusps and corners on interfaces produced either by external drive or by phase transformation.

2. Trajectory splitting

Consider the kinematics of moving contact lines as illustrated in Figure 1 according to Dussan V. and Davis (1974). Here two immiscible fluids, #1 and #2, contact a smooth rigid plate and #1 moves forward displacing #2. The contact line CL is common to #1, #2 and the solid. In a coordinate system moving with the CL, the

111

D. Durban and J.R.A. Pearson (eds.),
IUTAM Symposium on Non-Linear Singularities in Deformation and Flow, 111-118.
© 1999 *Kluwer Academic Publishers. Printed in the Netherlands.*

plate moves to the left at speed u_{CL}, which is, say, constant.

In the displacing fluid, points on the fluid-fluid interface move forward and arrive at the contact line in a finite time. The motion in #1 is a rolling motion, as shown. In #2, however, the motion is more complex. Points on the #2-solid boundary move leftward and by conservation of mass are mapped into the #2 bulk, as shown in the figure and by experiments in Dussan V. and Davis (1974).

For the present purposes, the most important observation is that the CL is the site of *trajectory splitting*. A point in the #1-#2 interface has dual identity. As it moves toward the CL and then arrives, it splits, part of which "rolls under" #1, and part of which is expelled along the dividing streamline. The splitting gives rise to a strong singularity in the velocity field; the velocity gradients behave like r^{-1}, where r is the distance to the CL. Whenever trajectories split, the velocity vector is multivalued and the gradients are infinite.

The trajectories of fluid points are governed by the system

$$\frac{d\mathbf{x}}{dt} = \mathbf{v}(\mathbf{x}, t) \,,$$

$$\mathbf{x}(0) = \mathbf{x}_0 \quad , \tag{1}$$

where $\mathbf{v}$ is the Eulerian velocity, t is time, and $\mathbf{x}(t)$ is the position of a point. When $\mathbf{v}$ is Lipschitz continuous, the trajectory through each point is unique. When $\mathbf{v}$ is less smooth, uniqueness is not guaranteed. The simple example in one dimension:

$$\frac{dx}{dt} = x^{\frac{1}{2}} \,, \quad x(0) = 0 \,, \tag{2}$$

shows that the non-differentiable velocity $x^{\frac{1}{2}}$ has two trajectories through the origin, $x(t) \equiv 0$, and $x(t) = \frac{1}{4}t^2$.

Another example of trajectory splitting occurs during the spontaneous rupture of a liquid film on a solid substrate when van der Waals attractions are effective i.e. when the local film thickness $z = h(x, t)$ is in the range $100 - 1000\mathring{A}$. Williams and Davis derived an evolution equation for such a film with viscosity μ and surface tension σ, viz.

$$\mu h_t + A \left(h^{-1}h_x\right)_x + \sigma \left(h^3 h_{xxx}\right)_x = 0 \ . \tag{3}$$

Here A is proportional to the Hamaker constant in the vdW model. Burelbach et al. (1988) show that an initial interface perturbation causes $h \to 0$ in a finite time, t_R, and furthermore, find numerically that the minimum film thickness h_m satisfies $h_m \approx (t_R - t)f(x)$ which is an approximate solution of equation (3) with surface tension σ set to zero. Thus, in the rupture process surface tension is negligible! If the rupture point opens to create a dry spot on the substrate, then trajectory splitting occurs and a singular behavior is expected.

When a film with two free surfaces ruptures, the kinematics is identical with the breaking of a mass of liquid into two. Clearly, whether van der Waals attractions are included (see Erneux and Davis 1993) or not (see Eggers 1995), at rupture a single (mathematical) point must split making the instantaneous local velocity field singular as discussed above. The local similarity solution of the breaking of a wedge of inviscid fluid with surface tension (Keller and Miksis 1983) shows how the rupture process can induce inertial effects that produce capillary-wave motion on the interfaces. It is common practice in certain quarters when numerically simulating rupture (or co-alescence) to stop the calculation just before rupture and then restarting just after, without matching together the two processes by jump conditions. This procedure is a dangerous one that omits the possibly important mechanics of trajectory splitting.

3. Cusps and corners

Joseph et al. (1991) showed that by co-rotating two cylinders beneath a liquid-liquid interface, the interface can be distorted into an apparent cusped configuration as sketched in Figure 2a. Jeong and Moffatt (1992) simplified the problem by submerging a vortex dipole of strength α a distance d below the interface as shown in Figure 2b; the fluid satisfies the Stokes equations with viscosity μ and the dynamics of the second fluid is neglected. In two dimensions they find that the radius of curvature $\mathcal{R}$ at the "cusp" satisfies

$$\mathcal{R}/d \sim \frac{256}{3} \exp(-32\pi\mu\alpha/d^2\sigma) \ , \tag{4}$$

an exponentially small effect of surface tension as $\sigma \to 0$. The "cusp" is not actual for nonzero σ. However, when the estimate of $\mathcal{R}$ is made for typical situations with typical fluids, $\mathcal{R} \approx 10\text{Å}$, well beyond the continuum limit. Thus, whenever continuum theory applies, the interface *is* cusped.

Figure 2c, taken from Joseph (1992), sketches the presumed streamlines in the two-fluid problem. Kinematically, this is identical to that of the moving CL with contact angle $180°$. The reflux in the second fluid suggests the presence of the trajectory-splitting singularity. Hence, one suspects that the inclusion of the dynamics of the second fluid will influence the result of analysis.

Another example of interfacial cusps arises in phase transformations. Figures 3 are taken from Anderson et al. (1996). Figure 3a shows a droplet of water on a solid plate; the fluid is static. Figure 3b shows the same droplet after the substrate is undercooled and the drop is frozen.

Despite the fluid-fluid interface having surface tension, the solid drop which is the locus of the contact line (water, ice, gas) as the drop is frozen, is cusped. Anderson et al. (1996) show that it is the contact-line mechanics in the partially frozen drop that gives rise the cusp. When the contact angle θ, is measured in the liquid, is zero, as it should be for *pure* water, then cusps never occur. When θ is a nonzero constant, then cusps may occur but the frozen drop is then a pyramid. When θ is allowed to depend on the speed of the contact line, which means on the rate of freezing, then the cusps appears and the bi-concave shape of the frozen drop is reproduced. This result is evidence that such non-equilibrium conditions at contact lines should be enforced.

4. Conclusions

Trajectory-splitting singularities and cusps/corners can occur on interfaces with surface tension in liquids with viscosity. An understanding of when such structures are present is not merely of intrinsic interest, but is essential to the accuracy of numerical codes that give convergent reliable approximate solutions.

References

[1] Burelbach, J. P., Bankoff, S. G. and Davis, S. H. (1988) Nonlinear stability of evaporating condensing liquid films, *J. Fluid Mech.* **195**, 463-494.

[2] Eggers, J. (1995) Theory of drop formation, *Phys. Fluids A* **7**, 941-953.

[3] Dussan V., E. B. and Davis, S. H. (1974) On the motion of a fluid-fluid interface along a solid surface, *J. Fluid Mech.* **65**, 71-95.

[4] Anderson, D.M., Worster, M. G. and Davis, S. H. (1996) The case for a dynamic contact angle in containerless solidification, *J. Crystal Growth* **163**, 329-338.

[5] Joseph, D. D., Nelson, J., Renardy, M. and Renardy, Y. (1991) Two-dimensional cusped interfaces, *J. Fluid Mech.* **223**, 383-409.

[6] Joseph, D. D. (1992) Understanding cusped interfaces, *J. non-Newtonian Fl. Mech.* **44**, 127-148.

[7] Jeong, J. T. and Moffatt, H. K. (1992) Free-surface cusps associated with flow at low Reynolds number, *J. Fluid Mech.* **241**, 1-22.

[8] Erneux, T. and Davis, S. H. (1993) Nonlinear rupture of free films, *Phys. Fluids A* **5**, 1117-1122.

[9] Williams, M. B. and Davis, S. H. (1981) Nonlinear theory of film rupture, *J. Coll. Interf. Sci.* **90**, 220-228.

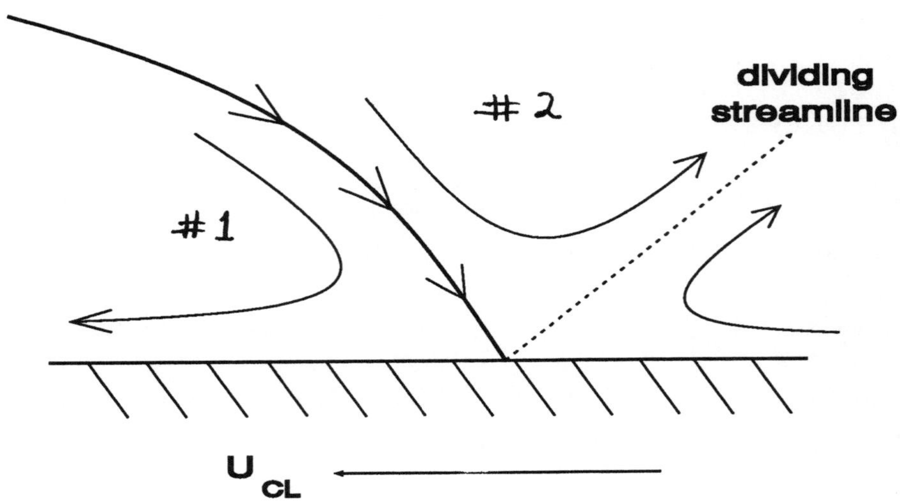

Fig.1 Kinematics of Moving
Contact Lines

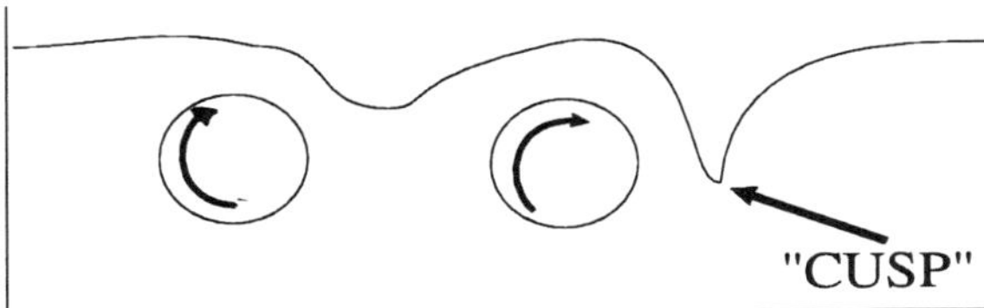

Fig. 2(a)

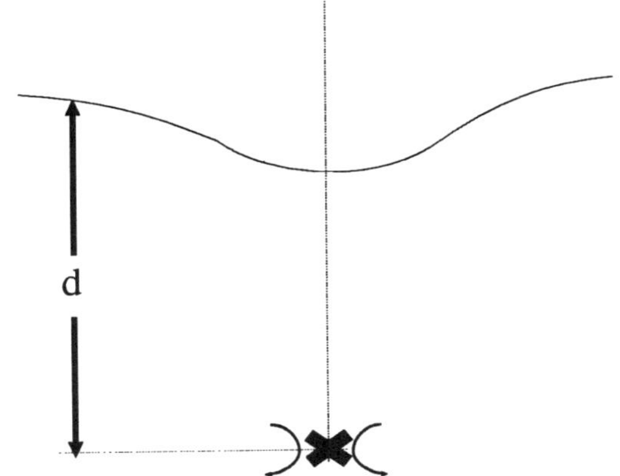

Fig. 2(b)

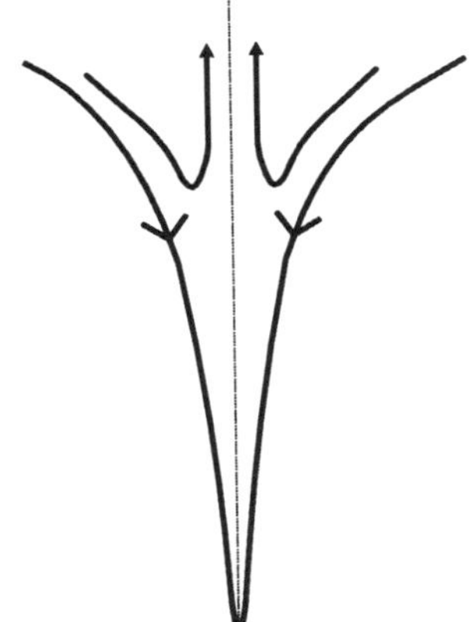

Fig. 2(c)

Fig. 3(a)

Fig. 3(b)

SPIRALS, JETS, AND PINCHES

MICHAEL J. SHELLEY
Courant Institute of Mathematical Sciences
New York University
New York City, NY 10012

1 Introduction

It remains an open question in the mathematical and physical understanding of homogeneous, incompressible, three-dimensional fluids whether a singularity can form dynamically from smooth initial data. However, when a fluid system is multiphase, that is, composed of two immiscible fluids separated by an interface under a surface tension, then our everyday experience tells us the answer; we commonly observe the formation of drops and bubbles through mixing of two fluids, the formation of spray from a crashing wave, or the break-up into droplets of a falling stream of water. The pinching off of a droplet means that the bounding interfaces of the fluid have collided at a finite time. Such events are true fluid singularities, at least in continuum descriptions such as the Navier-Stokes and Euler equations, as the bounding interfaces are also material surfaces, and their collision implies that the fluid velocity gradients must diverge. From another perspective, these singularities occur usually at a single point in the fluid, yet are the mediating events through which the global structure of a fluid can be reorganized. And finally, since such collisions involve arbitrarily small length and time scales, they can bring into play molecular processes, such as viscosity or slight miscibility, and so illuminate the limitations of standard continuum models.

The last several years have seen a large amount of activity directed towards understanding such "pinching" singularities, at least in some idealized circumstances. A sampling of recent studies includes work in Stokes flows [31, 32], lubrication models of thin-film flows and Hele-Shaw flows [7, 11, 2, 13], Hele-Shaw flows [10, 1], and shallow water approximations and experiments of axially symmetric jets [12, 29, 6].

In this paper I will review recent work on understanding the formation of pinching singularities that form through the Kelvin-Helmholtz (K-H) instability of two immiscible fluids shearing past one another (see Thorpe [33] for some early experiments). The K-H instability is a prototypical instability of high-Reynolds number flow, and we consider the simplest case: The fluids are two-dimensional, separated by a sharp interface under surface tension, and are inviscid, irrotational, and density matched. The nonlinear growth and evolution of the separating interface is governed then by only the joint effects of the K-H instability and the dispersion due to surface tension.

119

D. Durban and J.R.A. Pearson (eds.),
IUTAM Symposium on Non-Linear Singularities in Deformation and Flow, 119-128.
© 1999 *Kluwer Academic Publishers. Printed in the Netherlands.*

2   The Kelvin-Helmholtz instability and surface tension

In Hou, Lowengrub, & Shelley (1994) [15] and Hou *et al.* (1997) [16] (referred to henceforth as HLS1 and HLS2, respectively), we considered the dynamics of such an interface, under surface tension, that is 1-periodic in the x-direction. The dynamics of the interface is given elegantly and compactly by the Birkhoff-Rott equation for vortex sheet motion in the so-called "Lagrangian" frame [27, 25]:

$$\frac{\partial \mathbf{X}}{\partial t}(\alpha, t) \;=\; \frac{1}{2\pi} P.V. \int_{-\infty}^{+\infty} \gamma(\alpha', t) \frac{(\mathbf{X}(\alpha, t) - \mathbf{X}(\alpha', t))^{\perp}}{|\mathbf{X}(\alpha, t) - \mathbf{X}(\alpha', t)|^2} d\alpha' \,, \qquad (1)$$

$$\frac{\partial \gamma}{\partial t}(\alpha, t) \;=\; We^{-1} \frac{\partial \kappa}{\partial \alpha} \,. \qquad\qquad (2)$$

Here $(x, y)^{\perp} = (-y, x)$, α is a Lagrangian variable, $\mathbf{X}(\alpha, t)$ is the location of the interface (with κ its curvature), and γ is the "vortex sheet strength", which is proportional to the velocity jump across the interface. Finally, We is the Weber number, which is inversely proportional to the surface tension parameter, and which measures the strength of the K-H instability relative to the dispersive stabilization associated with surface tension.

The linearized motion about the flat equilibrium $x(\alpha, t) = \alpha$, $y(\alpha, t) = 0$, and $\gamma(\alpha, t) = 1$, gives the growth rate of a perturbation with wavenumber k as:

$$\sigma_k^2 \;=\; (2\pi)^2 \left(\frac{1}{4} k^2 - 2\pi \frac{We^{-1}}{2} |k|^3 \right). \qquad (3)$$

This relation gives instability for wavenumbers $0 < |k| < We/4\pi$, and dispersion for wavenumbers $|k| > We/4\pi$. The wavenumber of maximum growth is $|k| = We/6\pi$. The surface tension dispersively controls a (catastrophic) high wavenumber instability. This instability is the Kelvin-Helmholtz instability due the to shearing motion across the interface.

While they are compact, it is difficult to use Eqs. (1) - (2) to simulate the evolution of the interface. Surface tension introduces terms that have a large number of spatial derivatives, and that are embedded both nonlinearly and nonlocally into the equations of motion. If an explicit time integration method is used, these terms can induce strong stability constraints upon the time-step that are generally time dependent, and are made more severe by any differential clustering of points along the interface. For example, if the "Lagrangian" formulation were used (as in [25, 26, 5]) to simulate an evolution of interest here (see below), then the stability bound on the time-step, for a fixed spatial resolution, would decrease by a factor of 10^6 over the course of the simulation.

In HLS1 we presented a different formulation and new methods for computing the motion of a fluid interface with surface tension in a two-dimensional, irrotational and incompressible fluid (see also [3, 5, 14]). This formulation has all the nice properties for time integration methods that are associated with having a linear highest order term. The reader is directed to HLS1 for details, but

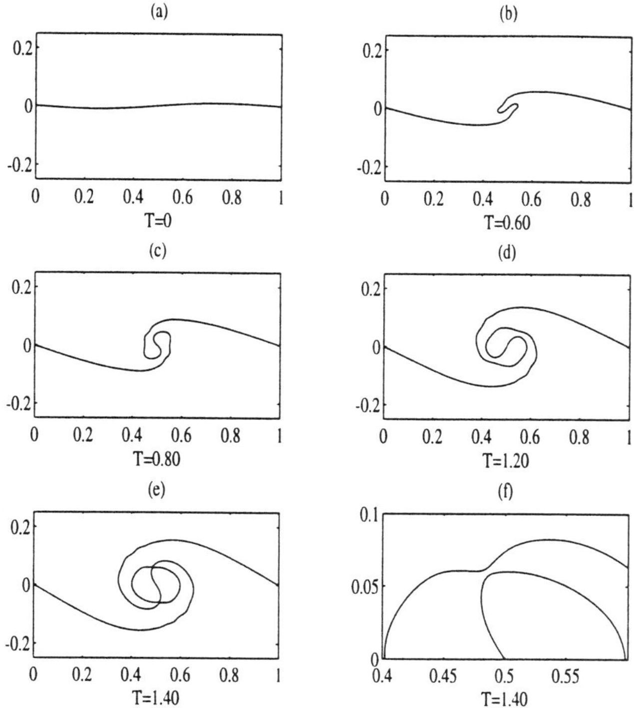

Figure 1: The evolution of the K-H instability for $We = 200$ with a blow-up of the thinning neck.

the resulting numerical methods significantly reduce the time-step stability constraints usually associated with surface tension, and have since been applied to many related problems, even beyond the fluid mechanical context.

Having circumvented the severe numerical stiffness constraints, in HLS1 we computed the long-time behavior of both an expanding gas bubble in a radial Hele-Shaw flow, and the K-H instability in the presence of surface tension. Our study of the latter problem was continued and expanded in HLS2. Our simulations of K-H instability revealed a rich variety of behavior. For small We (dispersively dominated) the interface simply oscillates in time with no apparent development of new structure, while for intermediate We (with a few unstable length-scales), the interface forms elongating fingers of fluid that penetrate each fluid into the other. Most relevant here, we found that at a sufficiently large Weber number (or small surface tension), the interface rolls up into a "K-H spiral". This is seen in Fig. 1, which shows such a simulation for $We = 200$, proceeding from initial data that is nearly flat and γ a positive constant. As is seen in the figure however, its smooth evolution is eventually terminated by the collision of the turns of the

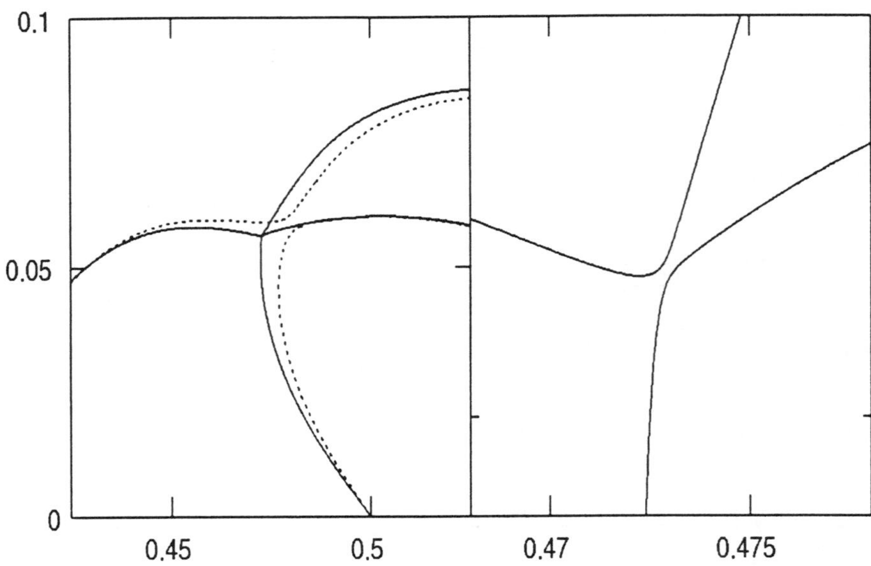

Figure 2: Left box: A blow-up of the thinning neck at $t = 1.4135$ (dashed) and $t = 1.427$ (solid). Right box: A further magnification by a factor of 10 of the neck region. The singularity time is estimated as $t_p \approx 1.4273$.

spiral with one another, forming at that instant a trapped bubble of fluid. The lower right box shows a blow-up of the "pinching" region, where the turns of the spiral are about to collide. Such an event in a real fluid would be the signature of an imminent topology change.

In the absence of surface tension, it is well known that the unregularized K-H instability drives the system into a finite time singularity in which the curvature of the interface diverges [22, 21, 18, 8, 9, 23, 28]. This occurs well before the development of any large-scale structure, such as roll-up. Moore [22] provided the first analytical evidence of this singularity's formation, and so we refer to it as the *Moore singularity*. We do not observe the Moore singularity in the presence of surface tension, though at large We its shadow is seen in the production of dispersive waves along the interface, seen in Fig. 1 moving outwards from the spiral. (See [19, 4, 34] for the effect of other physical regularizations.) This pinching singularity, which occurs at later times, is of a fundamentally different character. The Moore singularity occurs through the rapid *compression* of conserved circulation along the interface. In the presence of surface tension, the topological singularity is associated with the rapid *production* of new, localized circulation (we will return to this point). Siegel [30] has predicted corner formation in this problem, using a version of Moore's analysis, but the singularity there is isolated and not associated with pinching.

In HLS2, we studied this singular event exhaustively, employing high-order

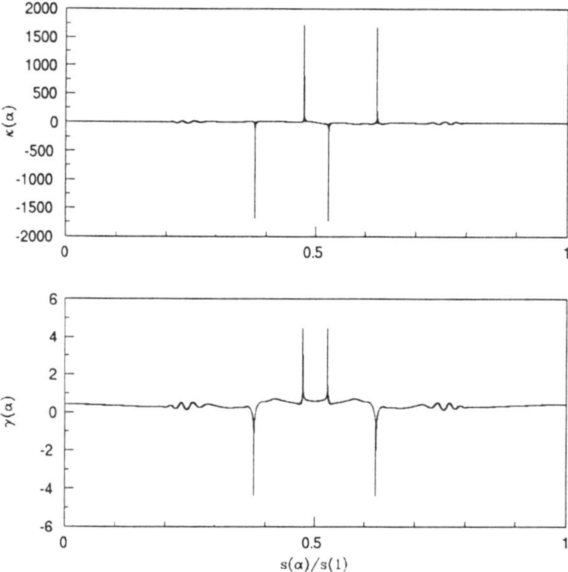

Figure 3: The interfacial curvature and sheet strength immediately before the singularity time, as a function of arclength.

time-stepping methods combined with graded spatial gridding, so as to resolve the oncoming singularity. Expectations of self-similarity would suggest that the colliding interfaces form a corner at the singularity time, and that

$$d^{\star}_{\min} \sim (t_p - t)^{2/3}, \quad |\gamma|_{\max} \sim (t_p - t)^{-1/3}, \quad |\kappa|_{\max} \sim (t_p - t)^{-2/3}, \tag{4}$$

where $d_{\min}$ is the width of the collapsing neck of fluid, and t_p is the time of the pinching singularity (see [17] for a construction of similarity solutions in a related problem). The details appear in HLS2, but here I list a few of the main points we gleaned from our simulations:

1. Our data analysis suggests that the interface does form a corner as the pinching time is approached. Fig. 2 shows two successive blow-ups in the pinching region, as the opposing sides of the spiral approach one another. This simulation uses 8192 points to resolve the interface, with the graded mesh giving a further 8-fold spatial refinement in the neighborhood of the incipient pinch.

2. The pinching singularity seems only described partially by self-similarity. More specifically, $d_{\min}$ behaves very nearly as predicted by self-similarity, but both γ and κ, while diverging, show persistent discrepancies from its predictions. Fig. 3 shows the nearly singular spatial structure of κ and γ very near the singularity time.

3. The local structure of the flow in the neck region is that of an intense jet that pulls fluid from outside the spiral, and pushes it inwards. In the right box of Fig. 2 this jet is fluxing fluid from the upper right through the narrowing neck. As γ is a circulation density, this jet is produced by negatively signed sheet strength on the upper interface of the neck (the first spike from the left in γ in Fig. 3), coupled to positively signed sheet strength on the lower interface (the third spike in γ).

It is important to note that in the absence of surface tension, γ would be conserved along particle paths, and such dynamical production of oppositely signed circulation would not be possible.

3   Jets

In modeling work, joint with M. Pugh (Pugh & Shelley [24]), we abstract what seems the crucial ingredient of the pinching singularity observed by Hou, Lowengrub, & Shelley, and study the role of surface tension on the dynamics of a symmetric, planar jet. In this setting, the "jet" is an inner fluid being fluxed through an outer, density matched, immiscible fluid. That is, there are now two interfaces under surface tension, located at $(x, \pm h(x, t))$ ($2h$ is the jet width), with equal and opposite vortex sheet strengths $\pm\gamma(x, t)$. As a further simplification, we study the dynamics of *thin* jets (i.e. shallow water theory), using a large aspect ratio expansion to derive reduced PDE descriptions for $h(x, t)$ and $\gamma(x, t)$.

A straightforward expansion of the equations of motion to second order in the aspect ratio yields a reduced system that is ill-posed, while the full system is not. Fortunately, the cause of this ill-posedness is clear and can be removed by making the system implicit in h_t, yielding:

$$h_t + (h\mathcal{H}[h_t])_x = -(h\gamma)_x \tag{5}$$

$$\gamma_t + (\gamma\mathcal{H}[h_t])_x = -\gamma\gamma_x + We^{-1}h_{xxx}. \tag{6}$$

The jet width $h(x, t)$ is taken to be initially positive. The nonlocal Hilbert transform, $\mathcal{H}$, arises from the asymptotic expansion of Birkhoff-Rott type integrals. Eq. (5) is a statement of mass conservation, and is in "shallow water form" $h_t + (hU)_x = 0$. This form encodes the statement that the finite-time collapse of the jet width, $h \downarrow 0$, implies a flow singularity: a simple argument (see [11]) shows that if h is smooth and $h(x, t) \downarrow 0$ at a point in finite time then, at the very least, $U_x \uparrow \infty$ at that point.

This system captures the competition between the K-H instability of a jet, and the dispersive effect of surface tension. It also fully determines exponents of self-similarity, agreeing with those for the unapproximated problem (as in expression (4)). Moreover, our numerical simulations show that this system also forms corner pinching singularities in finite time. One such simulation is shown in Fig. 4, where We is chosen so that there is but one unstable mode in the period.

In contrast to HLS2, we find that γ and h_{xx} (the "long-wave" curvature) now behave roughly in accordance with self-similarity in their temporal behavior, while

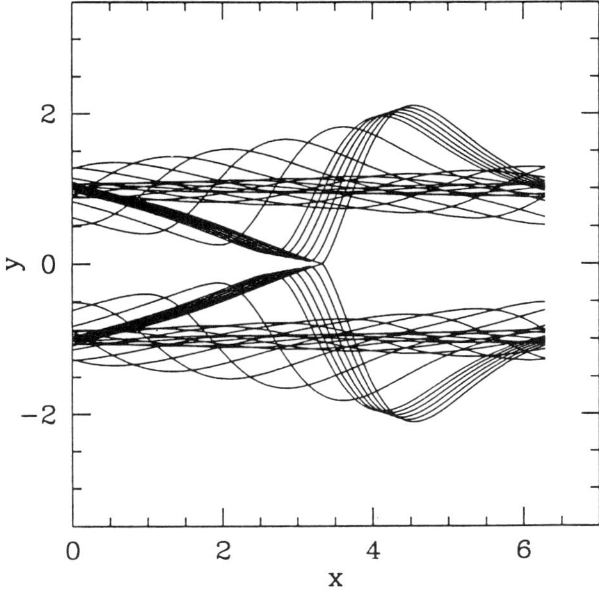

Figure 4: The pinching of a shallow-water jet.

h_{min} shows a persistent discrepancy. These differences may arise from the assumption of symmetry, not found in the well-analyzed full simulations of HLS1. We note too that if the singularities of the reduced system are of a self-similar type, then the asymptotic assumptions made to derive the reduced system are violated, and so these differences may follow from the system straying from the shallow water regime.

In Pugh & Shelley [24], we also study whether the nonlocal terms of (5–6) are needed for a pinching singularity to occur. Retaining only the surface tension term from the higher-order asymptotic contributions, we have the purely local model,

$$
\begin{aligned}
h_t + (h\gamma)_x &= 0 \\
\gamma_t + \gamma\gamma_x &= We^{-1}h_{xxx}.
\end{aligned}
$$

This is a very intriguing system. Here the surface tension contribution appears as a dispersive perturbation to a system that can be solved exactly in its absence. Specifically, for zero surface tension ($We = \infty$) exact solutions have finite-time singularities where $h \uparrow \infty$. We can analytically preclude such finite-time blow-up for the local model in the presence of surface tension. Further, our simulations of this simpler system show the formation of finite-time pinching singularities. Most surprisingly, their structure is very similar to those of the thin jet model, even though this system does not fully determine exponents of self-similarity.

Finally, I report briefly on recent progress in comparing the results of Pugh

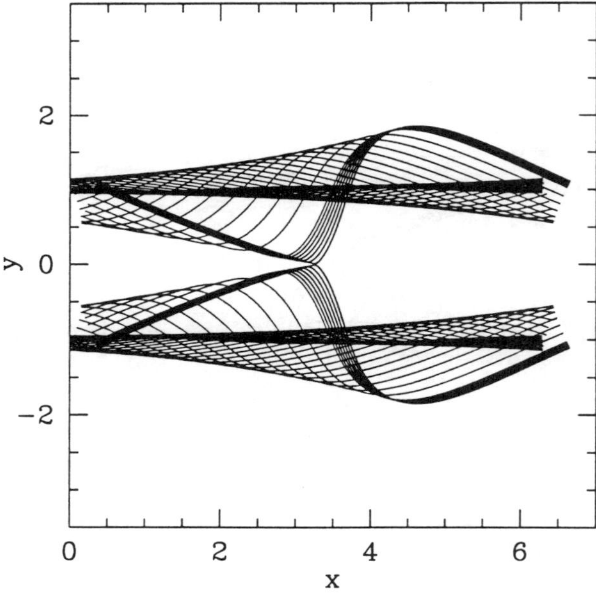

Figure 5: The pinching of a jet for the unapproximated system.

& Shelley on thin jet approximations with simulations of the *unapproximated* dynamics. This is joint work with J. Lowengrub and M. Pugh [20]. Fig. 5 shows the simulation of the full jet, for the same initial data as that used in Fig. 4, as the jet collapses into a corner singularity. The similarity of the reduced system to the full system is striking. Aside from the visual similarity between the two systems, we find too that the temporal behavior of the collapse width h agrees more closely with the thin jet model, Eqs. (5)-(6), than with the predictions of self-similarity.

4   Conclusion

The Kelvin-Helmholtz instability provides a fundamental mechanism through which two immiscible fluids can mix together. We have shown that the simplest model of such a system gives "pinching" singularities, which are the events signalling an imminent change in topology, i.e., droplet formation. We have shown further that these singularities are associated with the formation of an intense jet within the turns of the developed Kelvin-Helmholtz spiral, and have constructed models of isolated jets that reproduce the collapse in a relatively simpler setting.

I wish to thank my collaborators John Lowengrub, Mary Pugh, and Tom Hou, and acknowledge support from Department of Energy grant DE-FG02-88SER25053, National Science Foundation grants DMS-9396403 (PYI) and DMS-9404554, and the Exxon Educational Foundation.

References

[1] R. Almgren, Singularity formation in Hele-Shaw bubbles, *Phys. Fluids* **8**, 344 (1996).

[2] R. Almgren, A. Bertozzi, and M. Brenner, Stable and Unstable Singularities in the Unforced Hele-Shaw Cell, *Phys. Fluids* **8**, 1356 (1996).

[3] G. Baker, and A. Nachbin, Stable methods for vortex sheet motion in the presence of surface tension, *SIAM J. Sci. Comp.*, to appear.

[4] G.R. Baker, and M.J. Shelley, On the connection between thin vortex layers and vortex sheets, *J. Fluid Mech.* **215**, 161 (1990).

[5] J.T. Beale, T.Y. Hou and J.S. Lowengrub, Convergence of boundary integral methods for water waves with and without surface tension, *SIAM J. Num. Anal.*, to appear.

[6] S. Bechtel, C. D. Carlson, and M.G. Forest, Recovery of the Rayleigh capillary instability from slender 1-D inviscid and viscous models, *Phys. Fluids* **12**, 2956 (1995).

[7] A.L. Bertozzi, M.P. Brenner, T.F. Dupont, and L.P. Kadanoff, Singularities and similarities in interface flows, in *Trends and Perspectives in Applied Mathematics*, L. Sirovich, Ed., Springer-Verlag Applied Mathematics Series, New York, 1994.

[8] R. Caflisch & O. Orellana, Long time existence for a slightly perturbed vortex sheet, *Comm. Pure Appl. Maths.* **34**, 807-838 (1986)

[9] R. Caflisch & O. Orellana, Singular solutions and ill-posedness of the evolution of vortex sheets, *SIAM J. Math. Anal.* **20**, 293-307 (1989).

[10] S. Cardoso and A. Wood, The formation of drops through viscous instability, *J. Fluid Mech.* **289**, 351 (1995).

[11] P. Constantin, T.F. Dupont, R.E. Goldstein, L.P. Kadanoff, M. Shelley, and S.-M. Zhou, Droplet breakup in a model of the Hele-Shaw cell, *Phys. Rev. E* **47**, 4169 (1993); T.F. Dupont, R.E. Goldstein, L.P. Kadanoff, and S.-M. Zhou, Finite-time singularity formation in Hele-Shaw systems, *Phys. Rev. E* **47**, 4182 (1993).

[12] J. Eggers, Universal pinching of 3D axisymmetric free surface flow, *Phys. Rev. Lett.* **71**, 3458 (1993); J. Eggers and T.F. Dupont, Drop formation in a one-dimensional approximation of the Navier-Stokes equation, *J. Fluid Mech.* **262**, 205 (1994).

[13] R. Goldstein, A. Pesci and M. Shelley, Topology transitions and singularities in viscous flows, *Phys. Rev. Lett.* **70**, 3043 (1993). R. Goldstein, A. Pesci and M. Shelley, Attracting manifold for a viscous topology transition, *Phys. Rev. Lett.* **75**, 3665 (1995). R. Goldstein, A. Pesci and M. Shelley, Instabilities and Singularities in Hele-Shaw Flow, *Phys. Fluids*, to appear.

[14] T. Hou & H. Ceniceros, Convergence of a non-stiff boundary integral method for interfacial flows with surface tension, *Mathematics of Computation* **67**, 137 (1998).

[15] T. Hou, J. Lowengrub, M. Shelley, Removing the Stiffness from Interfacial Flows with Surface Tension, *J. Comp. Phys* **114**, 312 (1994).

[16] T. Hou, J. Lowengrub, M. Shelley, The long-time motion of vortex sheets with surface tension, *Phys. Fluids* **9**, 1933 (1997).

[17] J. B. Keller & M. Miksis, Surface Tension driven Flows, *SIAM J. App. Math* **43**. 268-277 (1983).

[18] R. Krasny, A study of singularity formation in a vortex sheet by the point vortex approximation, *J. Fluid Mech.* **167**, 65 (1986).

[19] R. Krasny, Desingularization of periodic vortex sheet roll-up, *J. Comp. Phys.* **65**, 292-313 (1986).

[20] J.S. Lowengrub, M. Pugh, & M.J. Shelley, in preparation.

[21] D. I. Meiron, Baker, G. R., & Orszag, S. A., Analytic structure of vortex sheet dynamics. Part 1. Kelvin-Helmholtz instability, *J. Fluid Mech.* **114**, 283-298 (1982).

[22] Moore, D., The spontaneous appearance of a singularity in the shape of an evolving vortex sheet, *Proc. R. Soc. Lond. A* **365**, 105-119 (1979).

[23] D.A. Pugh, Development of vortex sheets in Boussinesq flows- formation of singularities, Ph.D. Thesis, Imperial College, London, 1989.

[24] M. Pugh and M. Shelley, Singularity Formation in Models of Thin Jets with Surface Tension, *Communications in Pure and Applied Mathematics*, to appear.

[25] D.I. Pullin, Numerical studies of surface tension effects in nonlinear Kelvin-Helmholtz and Rayleigh-Taylor instability, *J. Fluid Mech.* **119**, 507 (1982).

[26] R. Rangel and W. Sirignano, Nonlinear growth of the Kelvin-Helmholtz instability: effect of surface tension and density ratio, *Phys. Fluids* **31**, 1845 (1988).

[27] P.G. Saffman and G.R. Baker, Vortex interactions, *Ann. Rev. Fluid Mech.* **11**, 95 (1979).

[28] M. Shelley, A study of singularity formation in vortex sheet motion by a spectrally accurate vortex method, *J. Fluid Mech.* **244**, 493 (1992).

[29] X.D. Shi, M.P. Brenner, and S.R. Nagel, A cascade of structure in a drop falling from a faucet, *Science* **265**, 219 (1994); M.P. Brenner, X.D. Shi, and S.R. Nagel. Iterated instabilities during droplet formation, *Phys. Rev. Lett.* **73**, 3391 (1994).

[30] M. Siegel, A Study of Singularity Formation in the Kelvin-Helmholtz Instability with Surface Tension, *SIAM J. Appl. Math.* **55** (1995).

[31] H.A. Stone and L.G. Leal, Relaxation and breakup of an initially extended drop in an otherwise quiescent fluid,*J. Fluid Mech.* **198**, 399 (1989); M. Tjahjadi, H.A. Stone, and J.M. Ottino, Satellite and subsatellite formation in capillary breakup, *J. Fluid Mech.* **243**, 297 (1992).

[32] S. Tanveer and G.L. Vasconcelos, Bubble breakup in two-dimensional Stokes flow, *Phys. Rev. Lett.* **73**, 2845 (1994).

[33] S. Thorpe, Experiments on the instability of stratified shear flows: immiscible flows, *J. Fluid Mech.* **39**, 25 (1969).

[34] G. Tryggvason, W. J. A. Dahm and K. Sbeih, Fine structure of vortex sheet rollup by viscous and inviscid simulation *J. Fluids Engin.* **113**, 31 (1991).

LOCALIZATION OF STRAIN AND THE MELTING WAVE IN HIGH-SPEED PENETRATION

M. V. AYZENBERG
Institute for Industrial Mathematics
Beer Sheva 84213 Israel

AND

L.I. SLEPYAN
Department of Solid Mechanics, Materials and Structures
Tel Aviv University, Ramat Aviv 69978 Israel

Abstract. Localization of shear under high-speed penetration is shown to be accompanied by initiation of a melting wave. Transient, self-similar and steady-state problems of heating and melting of the material under idealized conditions of penetration are studied, and the role of plastic hardening is examined. A critical value of the discontinuity in the velocity in the shear bend is found: the melting wave arises independently of the hardening modulus if the discontinuity exceeds this point. Resistance to shear in the melting wave is shown to decrease drastically. This ensures the separation of the flow jets from the surrounding material. Thus, the plastic jet model of penetration is justified.

1. Introduction

1.1. HYDRODYNAMIC MODELS OF PENETRATION

Well-known hydrodynamic models of penetration have a half-century history and take their origin from the work of Birkhoff *et al.* (1948), where penetration is considered as a collision of two jets of ideal fluids. In this model, the penetration velocity follows immediately from the Bernoulli equation. Allen and Rogers (1961) and then Alekseevskii (1966) and Tate (1967) introduced into this equation the so-called flow strength parameters to take into account the strength of the projectile and target materials. A comprehensive historical review and analysis of potentialities of these models, their advantages and drawbacks can be found in Zukas (1990).

D. Durban and J.R.A. Pearson (eds.),
IUTAM Symposium on Non-Linear Singularities in Deformation and Flow, 129-140.
© 1999 *Kluwer Academic Publishers. Printed in the Netherlands.*

Such a model permits the penetration depth and the projectile erosion to be determined. At the same time, it provides no way for determining the crater geometry and the mushrooming radius of the projectile. In this connection, a version of the hydrodynamic model has been elaborated by Slepyan (1978) for describing the movement of a rigid projectile or an ideal fluid jet in a deformable medium. The flow of the materials is assumed to exist only within an area of a finite, current radius, which is determined together with the other parameters of the process based on a given strength parameter of the target and the current velocity of the penetration. The surrounding material is assumed to be at rest.

This model can be used for an estimation of the resistance to the movement, the mushrooming radius of the projectile (for the fluid jet), and the size of the crater created. However, in the case of plastic materials such a hydrodynamic model cannot be used immediately. First, this model does not take into account energy loss in plastic deformation of the backward jets. Second, the conditions which allow interaction between the ideal jets and the surrounding immobile material to be neglected have not been established.

In the present paper, the latter problem is considered. It is shown that for high-speed penetration, localization of shear as a discontinuity in the flow of the material leads to initiation of a melting wave. Resistance to shear in the melting wave drastically decreases, and this results in separation of the flow jets and the surrounding, immobile material with negligible shear stresses in the interface. This allows the plastic jet model to be justified.

As to the energy loss in the backward jets, we only note here that the plastic work can be determined based on the scheme of proportional strain of the materials, which allows this work to be defined in terms of the initial and final parameters of the flow present in the hydrodynamic model formulation. This leads to the modified Bernoulli equation valid for plastic jets, and finally, to the closed system of governing equations.

1.2. LOCALIZATION OF SHEAR

As is known, instability of a uniform plastic strain can arise under dynamic shear and this manifests itself in shear bands in thermoplastic solids (Recht, 1964; Anand *et al.*, 1986; Barta, 1987; Molinary and Clifton, 1987; Wright and Walter, 1987, 1996; Meunier *et al.*, 1992; Bai and Dodd, 1992; Gioia and Ortiz, 1996). Scores of works are devoted to measurement and description of initiation and propagation of the localized shear bands under various conditions. In such a band, local temperature is shown to rise by several hundred degrees (Hartley, 1987; Marchard and Duffy, 1988; Zehnder and Kallivayalil, 1991; Bai and Dodd, 1992; Zender and Rosakis, 1992a,b; Zhou *et al.*, 1996a,b). It can increase with the rate and duration of the shear, and under certain conditions this may lead to melting of the material (Marchard and Duffy, 1988; Nicolas and Rajendran, 1990; Zhou *et al.*, 1996a). Our goal is to derive an estimation of (a) the conditions which give rise to the melting, and (b) the resistance to shear in the melting wave.

A theoretical treatment of the process of strain localization with initiation of a melting wave requires knowledge of the dependence of stresses on the high-level strain, strain rate and temperature. As far as we know, there are no sufficient data concerning such a dependence, and the considerations below are necessarily based on an idealized formulation. In particular, a linear hardening is assumed. However, it is shown that a critical value of the jump in velocity in the shear bend does exist: the melting wave arises independently of the hardening modulus if the jump exceeds this point.

2. Temperature in the localized shear band

To justify the acceptability of a corresponding idealized formulation, we begin with the analysis of the role of strain hardening in heating of the material by dynamic plastic shear.

First of all let us introduce parameters of the material used in calculations and estimations:

density $\rho = 8 \cdot 10^3 \ kg/m^3$,

critical shear stress (equal to half the yielding limit) $\tau_0 = 2.5 \cdot 10^8 \ N/m^2$,

hardening modulus k which varies in the calculations,

coefficient of viscosity $\mu = 10^{-2} \ Ns/m^2$,

heat capacity $c = 5 \cdot 10^2 \ Nm/(kg \ ^oK)$,

heat conductivity $\lambda = 50 \ N/(s \ ^oK)$,

melting point $\Theta = 1.8 \cdot 10^3 \ ^oK$ and

latent heat of melting $L = 2.5 \cdot 10^5 \ Nm/kg$.

Using these constants the natural units are introduced as

length-unit $l_0 = 2\sqrt{\lambda\mu/(\rho c \tau_0)} \approx 4 \cdot 10^{-8} \ m$

time-unit $t_0 = \mu/\tau_0 = 4 \cdot 10^{-11} \ s$, the corresponding

speed-unit $d_0 = 10^3 \ m/s$ and

temperature-unit $\theta_0 = \tau_0/(\rho c) = 62.5 \ ^oK$.

2.1. ONE-DIMENSIONAL TRANSIENT PROBLEM OF HEATING

In our problem, heating and melting of the material are induced by a given, tangential particle velocity, v_0 , at the boundary of a half-space of a thermoplastic material and the propagation velocity of this external action, b. These velocities are assumed to correspond to the above-mentioned jump in velocity in the shear bend and its propagation during the penetration.

Consider a half-space, $x > 0$, of a rigid-plastic material under antiplane dynamic shear and under the following conditions. Shear stresses are assumed to correspond to a linear hardening:

$$\tau = \tau_0 \operatorname{sign} \frac{\partial w(x,t)}{\partial x} + k \frac{\partial w(x,t)}{\partial x}, \tag{1}$$

where w denotes displacements directed along a normal to x; τ_0 and k are positive constants. In the considered uncoupled problem, where τ_0 and k are assumed to be independent of temperature, the dynamic equation of motion is the one-dimensional wave equation:

$$\frac{\partial^2 w(x,t)}{\partial t^2} - a^2 \frac{\partial^2 w(x,t)}{\partial x^2} = 0, \quad a^2 = \frac{k}{\rho}, \tag{2}$$

where a is the wave speed.

It is assumed that the work of plastic strain transfers into heat totally, that is, a small part of such work going into energy of micro-strain is neglected. Taking the work of plastic strain into account, the Fourier equation is assumed to govern heat conductivity over the half-space as

$$\rho c \frac{\partial \theta(x,t)}{\partial t} - \lambda \frac{\partial^2 \theta(x,t)}{\partial x^2} = \frac{\partial U(x,t)}{\partial t}, \tag{3}$$

where U is the plastic strain energy density per unit volume.

The initial and boundary conditions are as follows:

$$w = \frac{\partial w}{\partial t} = 0, \quad \theta = \theta_0 = \text{const} \quad (t = 0, x > 0)$$

$$v \equiv \frac{\partial w}{\partial t} = v_0, \quad N = -\lambda \frac{\partial \theta}{\partial x} = 0 \quad (x = 0, t > 0). \tag{4}$$

Here N is the heat flux. This formulation leads to the solution

$$U = \frac{v_0}{a} \left(\tau_0 + \frac{k v_0}{2a} \right)$$

$$\theta = \theta_0 + \left(\frac{\tau_0 v_0}{\rho c a} + \frac{v_0^2}{2c} \right) \left[1 - exp(\alpha t) \operatorname{erfc}(\sqrt{\alpha t}) \right]$$

$$= \theta_0 + \frac{1}{c} \left(\frac{\tau_0 v_0}{\sqrt{k\rho}} + \frac{v_0^2}{2} \right) \left[1 - \frac{2}{\pi} \int_0^\infty \frac{exp(-\alpha t z^2)}{1 + z^2} dz \right], \tag{5}$$

where $\alpha = \rho c a^2 / \lambda = kc/\lambda$ and

$$\operatorname{erfc}(y) = 1 - \operatorname{erf}(y) = \frac{2}{\sqrt{\pi}} \int_x^\infty exp(-z^2) \, dz$$

It can be observed that temperature increases monotonically with time, as fast as α is large, and tends to the limit

$$\theta = \theta_0 + \frac{\tau_0 v_0}{\rho c a} + \frac{v_0^2}{2c} \qquad (6)$$

if $k > 0$. Otherwise, if $k = 0$

$$\theta = \theta_0 + \frac{2\tau_0 v_0}{\sqrt{\pi \lambda \rho c}} \sqrt{t}. \qquad (7)$$

These dependences are valid when the temperature is under the melting point (and if it does not influence the resistance to plastic strain as assumed). The main question is whether the temperature achieves the melting point and how long the corresponding time of heating is. As follows from these results, the limiting temperature increases with decrease in hardening, and the limiting velocity, v_0, required for the temperature to achieve the melting point decreases together with hardening. The critical value v_* of the velocity v_0 is found such that provides temperature of the melting point independently of hardening. Under the above-mentioned parameters, $v_* = 1225\ m/s$ (for the initial temperature $\theta_0 = 300^o K$). If v_0 exceeds v_* only a little, the temperature achieves the melting point very soon, and it does not matter what hardening of the material is. The characteristic time is of the order of $4 \cdot 10^{-9} s$. In the figure below, dependencies of the temperature on time are plotted at $v_0 = 1250\ m/s$.

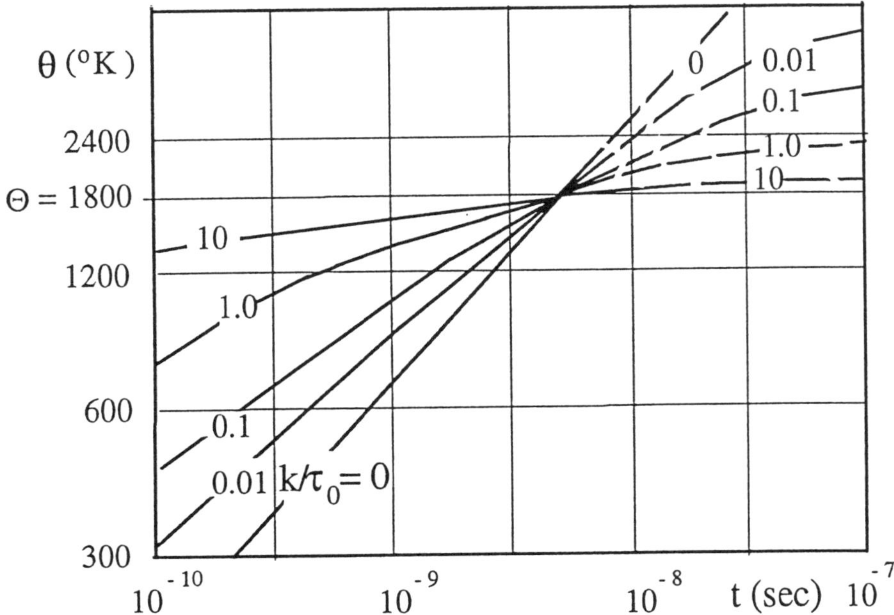

2.2. STEADY-STATE, 2D PROBLEM OF HEATING

Consider a space, x, y, z, filled by the rigid-plastic material (1), the same as above, under the condition at the interface $x = 0$:

$$\frac{\partial w}{\partial t} = \pm v_0 \, H(\eta) \,\, (x = \pm 0), \quad \eta = bt - y. \tag{8}$$

This action is assumed to induce plastic waves propagating with a velocity $a < b$. The formulation leads to the solution

$$\theta = \theta_0 + \frac{2\tau_0 v_0}{\pi \rho c b} \left[1 + \nu \int_0^\eta K_0(\nu \eta) \exp(\nu \eta) \, d\eta \right], \tag{9}$$

where the Bessel function, K_0, and the parameter ν are defined by the relations

$$K_0(x) = \int_1^\infty \frac{\exp(-xt)}{\sqrt{t^2 - 1}} \, dt, \quad \nu = \frac{\rho c b}{2\lambda} = 4 \cdot 10^7 \,\, 1/m. \tag{10}$$

Note that for large $\nu \eta$

$$\theta - \theta_0 \sim \frac{2\tau_0 v_0}{\sqrt{\pi \lambda \rho b c}} \sqrt{\eta}. \tag{11}$$

This result completely corresponds to the above-considered one-dimensional problem (7).

It can be seen that for the high-speed penetration (when $b \approx 1,000 \, m/s$ or higher), the distance η where the asymptote (11) is valid is very small [see (9) and expression (10) for ν]. Thus the 1D formulation of the problem is acceptable here.

3.   Melting wave

3.1. ONE-DIMENSIONAL MELTING WAVE

Thus the 2D temperature field produced by the shear action, propagating with the velocity b of the order of 1,000 m/s, approaches closely the 1D field very soon (at a very small distance from the front of the action, $y = bt$). The same conclusion is valid for the melting wave considered below, because the equations for viscous fluid dynamics (dynamics of the melt-down material) and for heat conductivity are of the same type and, as is shown below, the wave of particle velocities propagates even slower than the temperature wave. This allows restriction by the 1D formulation for the melting wave. Note that it is common for the description of a viscous boundary layer.

Also, as was shown, the temperature achieves the melting point very soon for any hardening if the velocity, v_0, exceeds the critical value v_*. This necessarily results in localization of strain, and this allows, for describing the melting wave, an idealized rigid-plastic material without hardening to be considered.

With the goal to describe the corresponding melting wave and to find the distributions of temperature and shear stresses in such a wave, consider the antiplane problem for a visco-plastic material described below. The same coordinates are used as in the above-considered 1D problem of heating. So, the particle velocities $v = v_z$ and temperature are assumed to depend on the coordinate x and time t.

The material is assumed to be a rigid-plastic solid if temperature $\theta < \Theta$, where Θ is the melting point, and it is a viscous liquid if $\theta > \Theta$. The latent heat of melting, L, is required to melt it down. The limiting shear stresses in the material are τ_0 for both the solid and liquid states. In the solid state, there is no strain rate if the stress $\tau = \tau_{xz} < \tau_0$, and slip can exist in the plane and in the direction of the maximal stress if $\tau = \tau_0$. In the liquid state

$$\tau = \mu \frac{\partial v}{\partial x}, \quad 0 \leq v \leq v_+ \leq v_0, \tag{12}$$

with a continuous velocity: $v_+ = v_0$ (v_+ is the limit of v at the boundary from the right) if $\tau < \tau_0$ or $v_+ \leq v_0$ if $\tau = \tau_0$.

The flow of the material is represented consisting of two regions separated by a moving interface. In the first region the material is melt down, it is considered as a viscous fluid, and the material is rigid in the second region. The temperature at the interface is equal to the melting point, Θ. At the interface as the melting wavefront, there is an energy release equal to the latent heat of melting, L. This energy release is provided by a jump in heat flux, namely,

$$N_- - N_+ = \rho L W \tag{13}$$

where N_+ and N_- are the heat fluxes in front of and behind the moving interface, respectively, and W is its speed (the speed of the wavefront). At the same time, the particle velocity, v (in z-direction) is assumed to be continuous at the moving interface. The temperature at the interface is continuous too.

The process can be divided into three periods. In the first, $0 < t < t_1$, temperature $\theta < \Theta$, and $v = 0$. A growing layer of the melt-down material arises at $t = t_1$. However, in the second period, $t_1 < t < t_2$, the velocity of the material at $x = 0$ does not achieve the applied velocity v_0 ($v_+ < v_0$). At last, in the third period ($t > t_2$), $v_+ = v_0$.

Thus consider the half-space, $x > 0$, filled by a plastic material which is initially at rest. Under a given particle velocity, v_0, at the boundary, $x = 0$, the melting wave is expected to arise in the increasing region, $0 <$

$x < X(t), dX/dt \geq 0$, where the material becomes a viscous liquid. The equation of motion is valid as

$$\rho\frac{\partial v}{\partial t} = \frac{\partial \tau}{\partial x},\tag{14}$$

where $v(0,t) = v_+$. Under the linear viscosity relationship (12), where $\mu = $ const (a dependence of μ on temperature is neglected), the velocity obeys the equation

$$\rho\frac{\partial v}{\partial t} = \mu\frac{\partial^2 v}{\partial x^2}\tag{15}$$

in the region $0 < x < X$, where $\theta > \Theta$, and the material is at rest for $x > X(t)$.

The Fourier equation is assumed to govern heat transfer over the half-space, $0 < x < \infty$:

$$\rho c\frac{\partial \theta}{\partial t} - \lambda\frac{\partial^2 \theta}{\partial x^2} = \tau\frac{\partial v}{\partial x} = \mu\left(\frac{\partial v}{\partial x}\right)^2,\tag{16}$$

where c and λ are specific heat capacity and thermo-conductivity, acordingly. These coefficients are considered to be constant as well as μ. The right hand part of this equation corresponds to the heat production by the work of shear viscosity stresses in the region $0 < x < X(t)$. The following additional conditions are imposed:

the stress or the velocity at $x = 0$:

$$\tau = \tau_0 \quad [v(0,t) = v_+ < v_0] \quad \text{or} \quad v(0,t) = v_0 \quad [\tau(0,t) < \tau_0],\tag{17}$$

the velocity at $x = X$:

$$v(X,t) = 0,\tag{18}$$

the heat flow through the boundary, $x = 0$, if $v_+ < v_0$; otherwise, there is no heat flow at $x = 0$:

$$-\lambda\frac{\partial \theta}{\partial x} = \tau_0(v_0 - v_+)H(v_0 - v_+) \quad (x = 0),\tag{19}$$

temperature at the moving interface

$$\theta = \Theta \quad [x = X(t)],\tag{20}$$

continuity of temperature at $x = X$:

$$[\theta] = 0 \quad (x = X),\tag{21}$$

temperature at infinity

$$\theta = 0 \quad (x = \infty). \tag{22}$$

Thus, for the determination of solutions to these two equations, each of the second order (one of them is defined in one region, and the other is for two regions) and the coordinate of the wavefront, $X(t)$, there are seven conditions: Eqs. (17) – (22) and the energy-release-rate relation (13):

$$\lambda \left(\frac{\partial \theta(X + 0, t)}{\partial x} - \frac{\partial \theta(X - 0, t)}{\partial x} \right) = \rho L \frac{dX}{dt}. \tag{23}$$

It could be shown that in the third period the fields of stresses and temperature tend to the corresponding fields for the related self-similar solution. Taking into account the first and second periods turn out to be very short, it can be saied that such a self-similar solution gives us an adequate representation of the melting wave.

3.2. SELF-SIMILAR SOLUTION FOR THE MELTING WAVE

In the case when the constitutive equation for the melt-down material (12) is assumed to be valid independently of the level of stresses, the self-similar solution exists which satisfies all the equations and additional conditions

$$v = v(\eta), \quad \theta = \theta(\eta), \quad \eta = \frac{\rho x^2}{4\mu t}, \quad t > 0, \tag{24}$$

In these terms, stresses, the governing equations and the additional conditions take the form

$$\tau = \sqrt{\frac{\mu \rho \eta}{t}} v' \quad \left(v' = \frac{dv}{d\eta} \right), \tag{25}$$

$$v'' + \left(1 + \frac{1}{2\eta} \right) v' = 0 \tag{26}$$

$$\theta'' + \left(\kappa + \frac{1}{2\eta} \right) \theta' = -\frac{\mu}{\lambda} (v')^2, \quad \kappa = \frac{c\mu}{\lambda}. \tag{27}$$

Further

$$v = v_0 \quad (\eta = 0), \quad v = 0 \quad (\eta = Y), \tag{28}$$

where the point $\eta = Y$ corresponds to the wavefront (in the self-similar solution considered, $X(t) = 2\sqrt{\mu Y t / \rho}$),

$$\zeta = \sqrt{\eta} \theta' \to 0 \quad (\eta \to 0), \tag{29}$$

$$N_- - N_+ = \lambda(\zeta_+ - \zeta_-)\sqrt{\frac{\rho}{\mu t}} = \rho L\sqrt{\frac{\mu Y}{\rho t}} \quad (\eta = Y), \tag{30}$$

$$\theta = \Theta \quad (\eta = Y), \quad \theta = 0 \quad (\eta = \infty). \tag{31}$$

Equation (26), with conditions (28), leads to the solution

$$v = v_0 \left[1 - \frac{\mathrm{erf}(\sqrt{\eta})}{\mathrm{erf}(\sqrt{Y})}\right] \tag{32}$$

Equation (27) can be represented in the form

$$\zeta' + \kappa\zeta = -\frac{\mu}{\lambda}(v')^2\sqrt{\eta} = -\frac{v_0^2\mu\, exp(-2\eta)}{\pi\lambda\sqrt{\eta}\,\mathrm{erf}^2(\sqrt{Y})}H(Y - \eta). \tag{33}$$

This equation, with conditions (29) and (31), gives us the solution

$$\zeta = -\frac{v_0^2\mu}{\lambda\sqrt{2\pi(2 - \kappa)}}\frac{\mathrm{erf}[\sqrt{(2 - \kappa)\eta}]}{\mathrm{erf}^2(\sqrt{Y})}e^{-\kappa\eta} \quad (\eta < Y)$$

$$\zeta = -\frac{\Theta\sqrt{Y}\,exp(-\kappa\eta)}{\sqrt{\pi}\,\mathrm{erfc}(\sqrt{\kappa Y})} \quad (\eta > Y) \tag{34}$$

and the temperature field under condition (31)

$$\theta = \Theta + \frac{v_0^2\mu}{\lambda\sqrt{2\pi(2 - \kappa)}\,\mathrm{erf}^2(\sqrt{Y})}I \quad (\eta < Y)$$

$$I = \int_\eta^Y \mathrm{erf}\left(\sqrt{(2 - \kappa)\eta}\right)e^{-\kappa\eta}\frac{d\eta}{\sqrt{\eta}}$$

$$\theta = \Theta\frac{\mathrm{erfc}(\sqrt{\kappa\eta})}{\mathrm{erfc}(\sqrt{\kappa Y})} \quad (\eta > Y). \tag{35}$$

The rest of condition (30) gives us an equation with respect to the wave front coordinate Y

$$\frac{v_0^2}{\sqrt{2\pi(2 - \kappa)Y}}\frac{\mathrm{erf}(\sqrt{(2 - \kappa)Y})}{\mathrm{erf}^2(\sqrt{Y})} - \frac{\Theta\lambda}{\sqrt{\pi}\mu\,\mathrm{erfc}(\sqrt{\kappa Y})} = Le^{\kappa Y}. \tag{36}$$

Consider two asymptotic cases. The low-velocity case, when $v_0 \to 0$, corresponds to the vanishing of the melt-down-material zone: $Y \to 0$. In this case, it follows from (36) that

$$Y \sim \frac{\sqrt{\pi}\mu v_0^2}{2\sqrt{2}(\Theta\lambda + \sqrt{\pi}\mu L)} \approx 8 \cdot 10^{-8}v_0^2 \quad (v_0 \to 0), \tag{37}$$

and at $\eta < Y$:

$$v \sim v_0 \left(1 - \sqrt{\frac{\eta}{Y}} \right),\tag{38}$$

$$\tau \sim -\frac{v_0}{2} \sqrt{\frac{\rho\mu}{Yt}} \approx -\frac{6 \cdot 10^{-5}}{\sqrt{t}},\tag{39}$$

$$\theta \sim \Theta.\tag{40}$$

The high-velocity case ($v_0 \to \infty$) corresponds to $Y \to \infty$. In this case, it can be found that

$$Y = Y_1 - Y_2 + o(1),\tag{41}$$

where

$$Y_1 = \ln A - \frac{1}{\kappa} \ln\ln A, \quad A = \left[\frac{\mu v_0^2}{\Theta\lambda\sqrt{2\pi\kappa(2-\kappa)}} \right]^{1/\kappa},\tag{42}$$

$$Y_2 = \frac{1}{\kappa} \ln\left[1 + \sqrt{2\pi(2-\kappa)Y_1}\frac{L}{v_0^2}exp(\kappa Y_1) \right],\tag{43}$$

and at $\eta < Y$:

$$v \sim v_0[1 - \mathrm{erf}(\eta)],\tag{44}$$

$$\tau \sim -v_0\sqrt{\frac{\rho\mu}{\pi t}}e^{-\eta} \approx -\frac{1.8 v_0}{\sqrt{t}}e^{-\eta},\tag{45}$$

$$\theta \sim \Theta + \frac{v_0^2\mu}{\lambda\sqrt{2\pi(2-\kappa)}} \int_\eta^\infty \mathrm{erf}(\sqrt{(2-\kappa)\eta})e^{-\kappa\eta}\frac{d\eta}{\sqrt{\eta}}.\tag{46}$$

It can be seen that the velocity, v_0, of the order of 10^3 m/s corresponds to the low-velocity case ($Y \approx 0.08$). In this case, the shear stresses comprise only 6% of the yield limit τ_0 when $t \geq 10^{-6}$ s. As follows from (45), τ increases with the velocity, however, the ratio $\tau/(\rho v_0^2)$ decreases.

In applying these results to the projectile – target interaction, we assume that the moment $t = 0$ corresponds to the beginning of the plastic flow at a considered *material* coordinate. Thus, the distance from this initial point can be measured from the front point of the projectile. This distance can be expressed as $h = v_0 t$. Using the above-mentioned numerical values, it can be found that the shear stresses fall drastically at the distance of the order of $10^3 \cdot 10^{-6} = 10^{-3}$ m. Thus, for a projectile of the length of several cm, the shear resistance in the localized shear bend can really be neglected.

4. Acknowledgment

This research was supported by grant No. 94-00349 from the United States – Israel Binational Science Foundation (BSF), Jerusalem, Israel, and by grant No. 9673-1-96 from the Ministry of Science, Israel.

References

Alekseevskii, V. P. (1966) Penetration of a rod into a target at high velocity. *Combust., Explos., Shock Waves* **2**, 63-66.

Allen, W. A. and Rogers, J. W. (1961) Penetration of a rod into a semi-infinite target. *J. Franklin Inst.* **272**, 275-275.

Anand, L., Kim, K. H. and Shawki, T. G. (1986) Onset of shear localization in viscoplastic solids. *J. Mech. Phys. Solids* **35**, 407-429.

Bai, Y. and Dodd, B. (1992) *Adiabatic Shear Localization.* Pergamon. Oxford.

Barta, R. C. (1987) Effect of material parameters on the initiation and growth of adiabatic shear bands. *Int. J. Solid Struct.* **23**, 1435-1446.

Birkhoff, G., MacDougall, D. P., Pugh, E. M. and Tailor, G. I. (1948) Explosives with lined cavities. *J.Appl.Phys.* **19**, 563-582.

Gioia, G. and Ortiz, M. (1996) The two-dimensional structure of dynamic boundary layers and shear bands in thermoviscoplastic solids. *J. Mech. Phys. Solids* **44**, 251-292.

Hartley, K. A., Duffy, J. and Hawley, R.H. (1987) Measurement of the temperature profile during shear band formation in steels deforming at high strain rates. *J. Mech. Phys. Solids* **35**, 283-301.

Marchand, A. and Duffy, J. (1988) An experimental study of the formation process of adiabatic shear bands in a structural steel. *J. Mech. Phys. Solids* **36**, 251-283.

Meunier, Y., Roux, R. and Moureaud, J. (1992) Survey of adiabatic shear phenomena in armor steels with perforation. *Shock-Wave and High-Strain-Rate Phenomena in Materials* (ed. M. A. Meyers, L. E. Murr and K. P. Staudhammer), Marcel Dekker Inc., New-York, Basel, Hong Kong, 637-644.

Molinari, A. and Clifton, R. J. (1987) Analytical characterization of shear localization in thermoviscoplastic materials. J. Appl. Mech., Trans. ASME **54**, 806-812.

Nicolas T. and Rajendran, A. M. (1992) Material characterization at high strain rates. In: *High Velocity Impact Dynamics* (ed. Jonas A. Zukas), John Wiley & Sons, New York, 127-296.

Recht, R. F. (1964) Catastrophic thermoplastic shear. *J. Appl. Mech., Trans. ASME* **31E**, 189-193.

Slepyan, L. I. (1978) Calculation of the size of the crater formed by a high-speed impact. *Sov. Mining Sci.* **14**, 467-471.

Tate, A. (1967) A theory for the deceleration of long rods after impact. *J. Mech. Phys. Solids* **15**, 387-399.

Wright, T. W. and Walter, J.W. (1987) On stress collapse in adiabatic shear bands. *J. Mech. Phys. Solids* **35**, 701-720.

Wright, T. W. and Walter, J. W. (1996), The asymptotic structure of an adiabatic shear band in antiplane motion. *J. Mech. Phys. Solids* **44**, 77-97.

Zehnder, A. T. and Kallivayalil, J. A. (1991) Temperature rise due to dynamic crack growth in beta-C titanium. *Speckle Techniques, Birefringence Methods, and Applications to Solid Mechanics, SPIE* **1554A**, 48.

Zehnder, A. T. and Rosakis, A. J. (1992a) On the temperature distribution at the vicinity of dynamically propagating cracks in 4340 steel. *J. Mech. Phys. Solids* **39**, 385-415.

Zehnder, A. T. and Rosakis, A. J. (1992b) Temperature rise at the tip of dynamically propagating cracks: measurements using high-speed infrared detectors. *Experimental Techniques in Fracture* (ed. J. S. Epstein), VCH Publishers, 125-169.

Zhou, M., Rosakis, A. J., and Ravichandran, G. (1996a) Dynamically propagating shear bands in impact-loaded prenotched plates – I. Experimental investigations of temperature signatures and propagation speed. *J. Mech. Phys. Solids* **44**, 981-1006.

Zhou, M., Rosakis, A. J. and Ravichandran, G. (1996b) Dynamically propagating shear bands in impact-loaded prenotched plates – II. Numerical simulation.*J. Mech. Phys. Solids* **44**, 1007-1032.

Zukas, J. A. (ed.) (1990) *High Velocity Impact Dynamics.* John Wiley & Sons. New York.

FRICTION AND SINGULARITIES IN STEADY PENETRATION

D. DURBAN
Faculty of Aerospace Engineering
Technion
Haifa 32000, Israel

1. Introduction

It is conceivable that penetration of a rigid sharp indentor (e.g. a cone or a wedge) into a viscoplastic material will induce stress singularities near the tip of the penetrator. That singularity which models the tearing stresses along the penetration axis can be treated, for certain constitutive models, by available methods used in fracture mechanics to study crack propagation. However, unlike crack tip fields, the presence of wall friction and penetrator geometry may have an appreciable influence on the singular field during the penetration process. This paper reviews recent work on steady penetration, with neglect of inertia effects, of cones and plane wedges into power law viscoplastic media, with emphasis on the effects of friction and material orthotropy.

Consider first the axially symmetric field generated by a conical indentor (Fig. 1) with rough walls (Fleck and Durban, 1991). At the wall we assume that the normal velocity vanishes and that the resisting shear stress $\sigma_{r\theta}$ is a constant fraction of the effective Mises stress σ_e. Material behaviour is modelled by the power law

$$\mathbf{D} = \frac{3}{2}\left(\frac{\sigma_e}{\sigma_o}\right)^n \frac{\mathbf{S}}{\sigma_e} \qquad \text{with} \qquad \sigma_e = \sqrt{\frac{3}{2}\mathbf{S}\cdot\mathbf{S}} \qquad (1)$$

where $\mathbf{D}$ is the Eulerian strain rate, $\mathbf{S}$ is the stress deviator and (σ_o, n) are material parmeters. For the Newtonian fluid $n=1$ and for the Mises perfectly plastic solid $n = \infty$.

At the wall we have the boundary conditions $v_\theta = 0$ and $\sigma_{r\theta} = -(m/\sqrt{3})\sigma_e$ where $0 \le m \le 1$ is the friction factor. Thus, for a smooth wall $m=0$ while for a perfectly rough wall $m=1$. The mathematical formulation is completed with the equilibrium equation $\nabla\cdot\sigma = 0$ with σ denoting the Cauchy stress tensor.

Material incompressibility (1) implies that the velocity components can be derived from a stream function $\psi(r,\theta)$ by

$$v_r = \frac{\psi,_\theta}{r^2 \sin\theta} \qquad v_\theta = -\frac{\psi,_r}{r \sin\theta} \qquad (2)$$

141

D. Durban and J.R.A. Pearson (eds.),
IUTAM Symposium on Non-Linear Singularities in Deformation and Flow, 141-154.
© 1999 *Kluwer Academic Publishers. Printed in the Netherlands.*

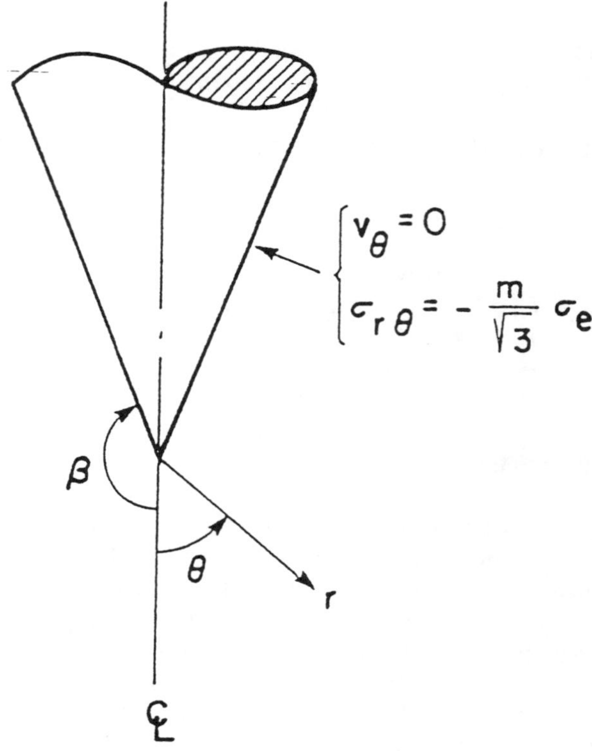

Figure 1. Notation for cone penetration. A spherical polar system is attached to the apex. Wall friction is modelled by friction factor m.

The local singular field is represented by a separation of variables solution in the form

$$\psi = r^s (\sin\theta)\phi(\theta) \tag{3}$$

where s is an eigenvalue and $\phi(\theta)$ is the associated eigenfunction. With (2) and (3) we have that

$$v_r = \left(\frac{d\phi}{d\theta} + \phi\cot\theta\right) r^{s-2} \qquad v_\theta = -s\phi r^{s-2} \tag{4}$$

and a further substitution in (1) shows that the strain rates depend on the radial coordinate like r^{s-3}. Thus, we expect the eigenvalues to be bounded by $2<s<3$ to ensure vanishing velocities and stress singularity as $r \to 0$. Inserting the stresses from (1) in the equations of equilibrium (Fleck and Durban, 1991) we arrive at a nonlinear eigenvalue problem whose numerical solution gives the strongest (smallest) permissible singularity s along with the associated eigenfunction $\phi(\theta)$. Surely, the stress singularity level (3-s) depends on (n,m,β) and becomes stronger with increasing β and decreasing m. Sharper and smoother cones induce a stronger singularity and penetrate more easily. Figure 2 illustrates the dependence of s on the friction factor m for a 90° cone (β=135°) and with different values of the hardening exponent n.

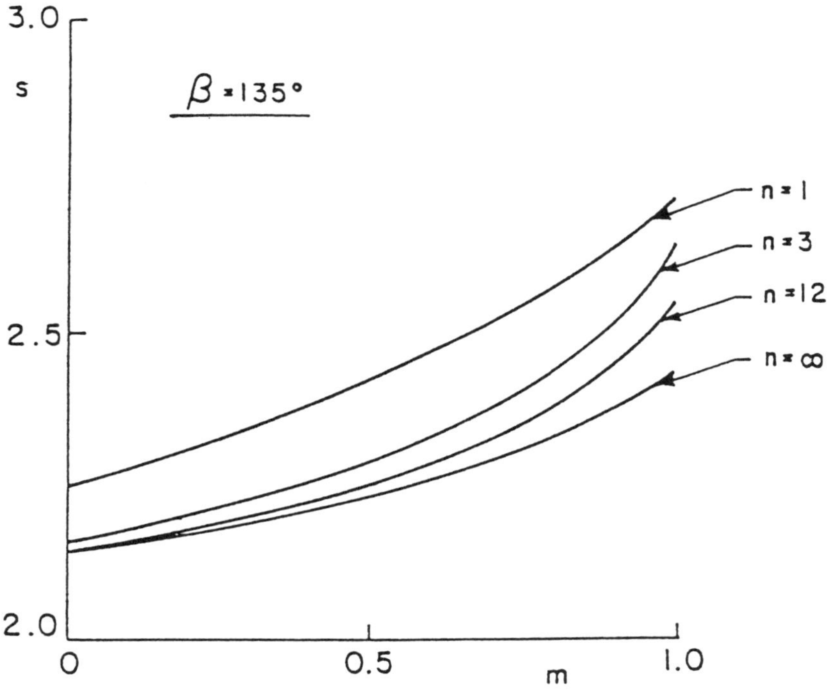

Figure 2. Variation of eigenvalue s with friction factor m, for different values of n, in cone penetration with β=135°.

Representative contours of constant effective stress σ_e are shown in Fig. 3 for β=135°. For all values of m and n, the contour of constant σ_e hugs the wall of the pentrator and extends ahead of the cone into the material. Deepest penetration of the σ_e contour occurs for m=1. There is also an obvious build-up of a boundary layer as both m and n increase. In fact, at the extreme case of m = 1 and n = ∞ (perfectly rough cone penetrating a perfectly plastic solid) the eigenfunction $\phi(\theta)$ admits, near the wall, the asymptotic expansion

$$\phi(\theta) \sim A(\beta-\theta)^{3/2} + B(\beta-\theta)+\cdots \tag{5}$$

where constants A, B depend on β.

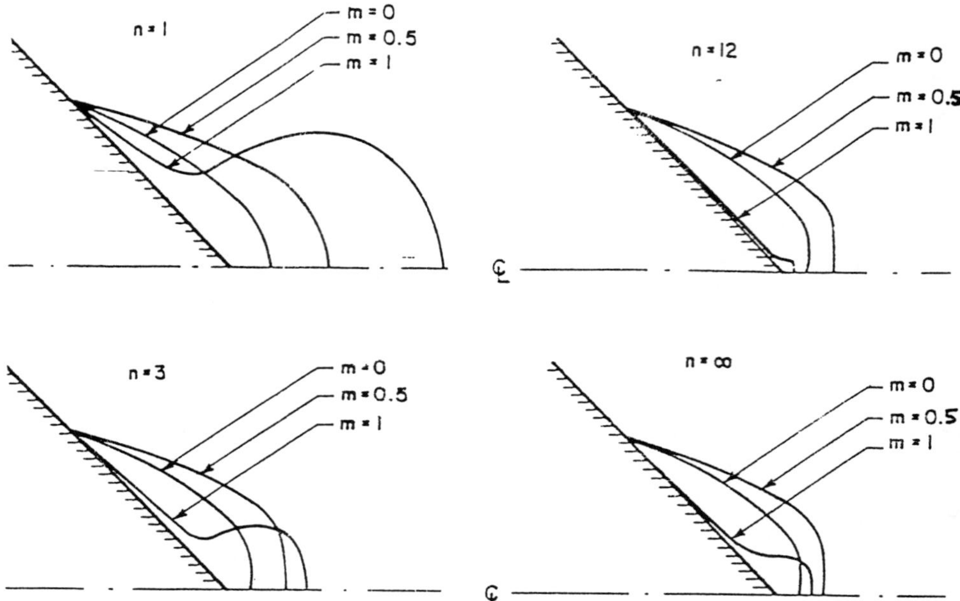

Figure 3. Contours of constant σ_e for different values of m and n with $\beta=135°$.

2. Wedge Penetration in Plane Strain

The analogous plane strain singular field near the tip of a penetrating wedge (Fig. 4) reveals essentially the same features (Durban and Rand, 1991) as the conical field. With the notation of Fig. 4 we introduce the Airy stress function $F(r,\theta)$ so that the stresses are expressed as

$$\sigma_{rr} = \frac{1}{r}F,_r + \frac{1}{r^2}F,_{\theta\theta} \qquad \sigma_{\theta\theta} = F,_{rr} \qquad \sigma_{r\theta} = -\left(\frac{1}{r}F,_\theta\right),_r \qquad (6)$$

and assume a separation of variables solution $F = r^s\phi(\theta)$. Thus, the stresses behave like r^{s-2} while the velocities vary with $r^{n(s-2)+1}$. Simple considerations show that here the eigenvalue s should be bounded by $2n/(n+1) < s < 2$.

The governing equations are now reduced to the HRR strain rate compatibility equation (Hutchinson, 1968; Rice and Rosengren, 1968)

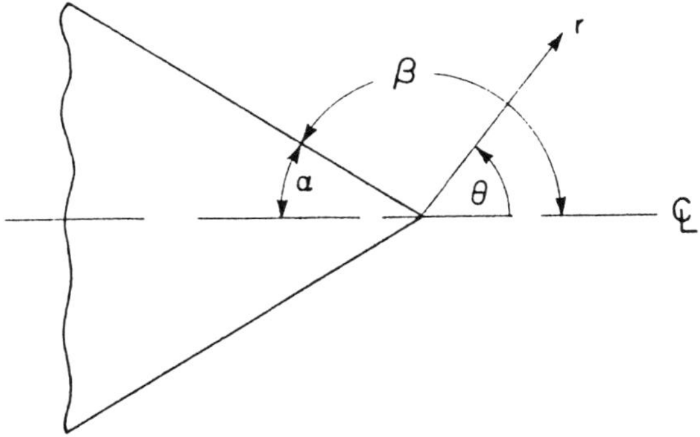

Figure 4. Notation for wedge penetration.

$$\left\{\frac{d^2}{d\theta^2} - n(s-2)[n(s-2)+2]\right\}\left\{\tilde{\sigma}_e^{\,n-1}\left[\frac{d^2\phi}{d\theta^2} - s(s-2)\phi\right]\right\}$$

$$+ 4(s-1)[n(s-2)+1]\frac{d}{d\theta}\left(\tilde{\sigma}_e^{\,n-1}\frac{d\phi}{d\theta}\right) = 0 \tag{7}$$

where $\tilde{\sigma}_e(\theta)$ is the circumferential profile of the effective stress given by

$$\tilde{\sigma}_e = \left\{\left[\frac{d^2\phi}{d\theta^2} - s(s-2)\phi\right]^2 + 4(s-1)\left(\frac{d\phi}{d\theta}\right)^2\right\}^{1/2} \tag{8}$$

The fourth order system (7)-(8) is supplemented by the boundary data

$$\sqrt{3}\sigma_{r\theta} = \left\{\begin{matrix} -m_1 \\ m_2 \end{matrix}\right\}\sigma_e \quad \text{at} \quad \theta = \pm\beta \quad \text{and} \quad v_\theta = 0 \quad \text{at} \quad \theta = \pm\beta \tag{9}$$

which can be expressed as homogeneous relations for ϕ and its derivatives (Durban and Rand, 1991). Notice that the friction factor in (9) is not necessarily identical on both walls of the wedge. The numerical solution of the eigensystem (7)-(9) for the eigenvalue s and the associated eigenfunction is straightforward and typical results for s are displayed in Fig. 5.

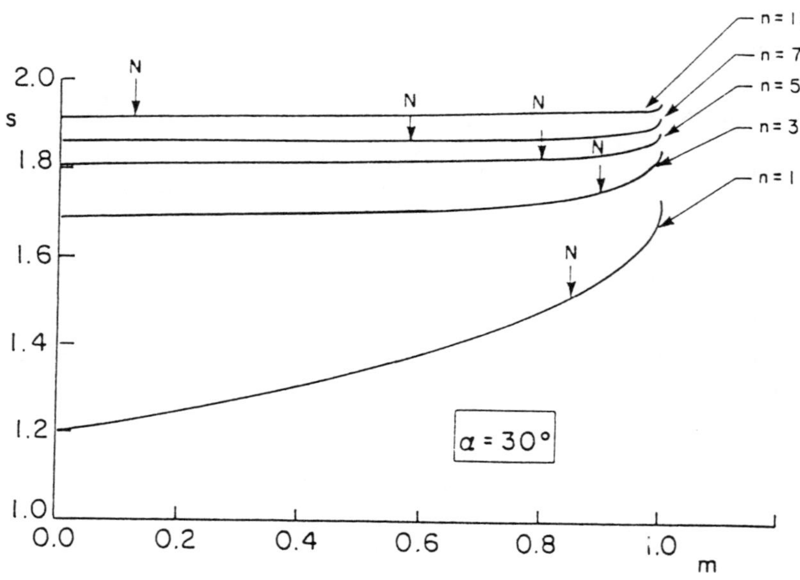

Figure 5. Variation of eigenvalue s with friction factor m, for different values of n, in wedge penetration with $\alpha = 30°$. Corresponding free notch eigenvalues are indicated by an arrow (N).

By comparison with the cone eigenvalues, in Fig. 2, we find that in both fields the level of singularity (3-s and 2-s, respectively) decreases with m, but unlike the conical fields increasing the strain rate exponent n reduces the singularity in plane strain. Reducing the wedge angle 2α will of course decrease s and thus increase the near tip singularity.

An instructive asymptotic analysis for the Newtonian fluid (n=1) results in the power expansion for s

$$s \sim \frac{\pi}{\beta}\left[1+\left(\frac{m}{2\beta}\right)+\frac{\pi-2\beta+\tan 2\beta}{\tan 2\beta}\left(\frac{m}{2\beta}\right)^2+\cdots\right] \tag{10}$$

which is a good approximation as long as $m^2 \ll (2\beta)^2 |\tan 2\beta|$. It can be deduced from (10) that when m is small there is an upper bound on the wedge angle, given by $2\alpha < \pi - m$, ensuring singular near tip behaviour. In steady penetration of a rigid knife with zero thickness ($\beta=\pi$) we have the exact relation

$$s = 1 + \frac{1}{\pi} \arctan \frac{m}{\sqrt{1-m^2}} \tag{11}$$

For a perfectly rough knife (m=1, s=3/2) the stress singularity is identical with that of a crack in a Hookean solid but the contours of effective stress are entirely different (Fig. 6), particularly the near wall boundary layer. Another distinction is that while the hydrostatic stress in steady penetration of a rigid knife is negative there is a tension hydrostatic environment at the tip of a crack.

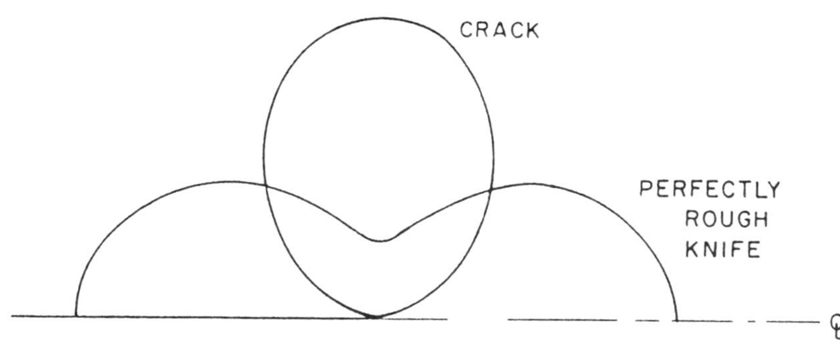

Figure 6. Contours of constant effective stress for a perfectly rough knife penetrating a Newtonian fluid, and for a crack in a Hookean solid.

Numerical mapping of constant effective stress contours are shown in Fig. 7 for a wedge with semi-angle $\alpha = 30°$. The extent of the contours have been normalized with respect to their span along the wall and the boundary layer build-up as m increases is clearly visible. For high values of n there is an apparent convergence of the contours shape to a narrow shear plane which is common to perfectly plastic plane strain fields.

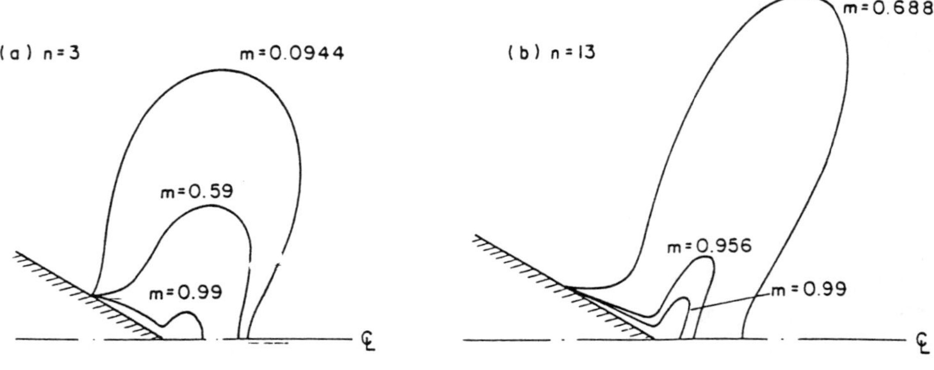

Figure 7. Contours of constant effective stress for a wedge with $\alpha = 30°$.

3. "Shear Shock" in Cone Penetration

The concentrated shear plane model has been studied further in (Durban and Fleck, 1992) for cone penetration into a perfectly plastic medium. With the notation of Fig. 8 we assume a rigid body motion with velocity V ahead of the rigid/plastic interface ($\theta = \lambda$) and fully plastic flow immediately after the instantaneous "shear shock". The constitutive relation in the plastic zone is given by

$$\mathbf{S} = \sqrt{\frac{2}{3}} \mathbf{Y} \frac{\mathbf{D}}{\sqrt{\mathbf{D} \cdot \cdot \mathbf{D}}} \tag{12}$$

where Y is the yield stress in simple tension. The proposed velocity field is independent of r which in turn implies, by (12) and the equilibrium equations, that $\sigma_{r\theta}$ depends only on coordinate θ while the normal stress component σ_{rr}, $\sigma_{\theta\theta} = \sigma_{\phi\phi}$, admit a logarithmic singularity of the form $(Y/\sqrt{3})D\ell nr$ where D is a constant. At the wall $\theta = \beta$ we have the conditions $\sigma_{r\theta} = -mY/\sqrt{3}$ and $v_\theta = 0$, while at the interface $\theta = \lambda$ we require instantaneous yield in shear with $\sigma_{r\theta} = Y/\sqrt{3}$.

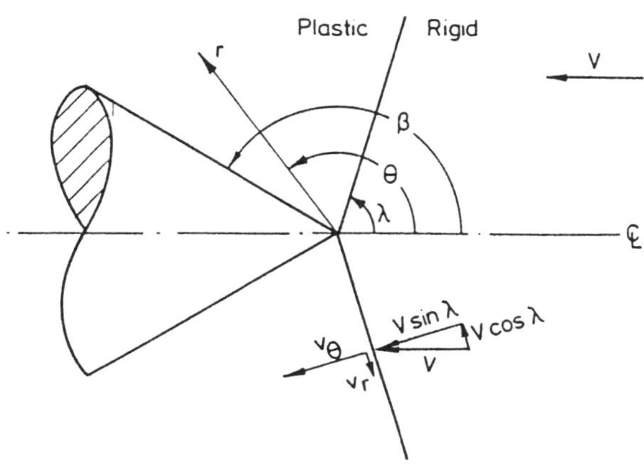

Figure 8. "Shear shock" discontinuity, at $\theta = \lambda$, in cone penetration.

The "shear shock" discontinuity at the rigid/plastic interface is supposed to simulate the narrow elastic-plastic transition zone which is expected to develop near $\theta = \lambda$. This model is constructed in the spirit of early studies on plane strain machining (Johnson and Mellor, 1973) based on the maximum shear plane hypothesis. In reality, one may expect a relatively large curvature of the streamlines in the vicinity of the idealized shear shock. Continuity of the normal velocity component at the interface requires that

$$v_\theta(\lambda) = V \sin \lambda \tag{13}$$

Now we introduce the ad-hoc requirement that the instantaneous shear strain at the interface will be as small as possible. The radial velocity jump at the interface (Fig. 8) is given by

$$\Delta v_r = v_r(\lambda) - (-V \cos \lambda) \tag{14}$$

which facilitates the definition of a shear strain $\tan\gamma$ at the interface by

$$\tan \gamma = \frac{\Delta v_r}{V \sin \lambda} \tag{15}$$

where γ is the shear angle. We minimize (15) with respect to λ, for given values of β and m, in order to determine the orientation of the interface. The variation of the shear strain $\tan\gamma$ with interface orientation λ, as evaluated from (15), is illustrated in Fig. 9 for a cone angle of 30° and with different values of m. As expected, there exists a clear minimum for the shear strain at a certain value of λ. That value of λ is taken here as the preferred orientation of the rigid/plastic interface within the framework of our model.

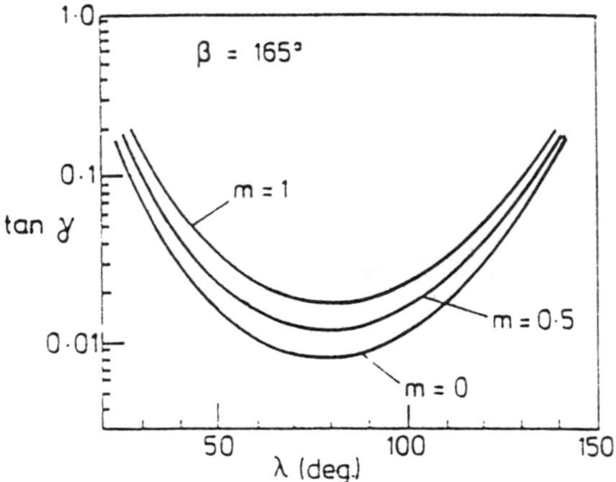

Figure 9. Variation of shear strain with interface angle for different friction factors m.

The interface orientation varies almost linearly with β but shows little sensitivity to wall friction (Fig. 10). An approximate relation for the interface angle is $\lambda \approx \beta/2$ provided that the cone is not very sharp. This result resembles known studies for plane strain orthogonal machining (Johnson and Mellor, 1973). A recent generalization for cone penetration into pressure sensitive solids was given by Balashov and Zvolinskii (1996).

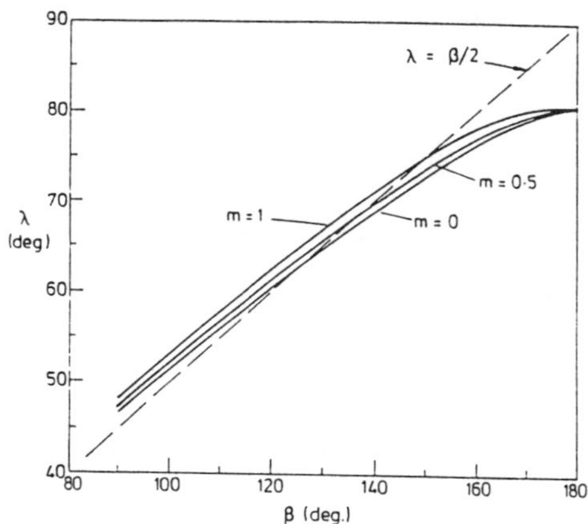

Figure 10. Variation of interface orientation λ with wall angle β.

4. Wedge Penetration into Orthotropic Viscoplastic Media

A recent study (Leser, 1992) on steady penetration, in plane strain conditions, of a sharp wedge into orthotropic viscoplastic media has revealed the interesting coupling between wall friction and plastic orthotropy. While the full report on that research is in preparation we shall conclude the present paper with a brief description of the essential results in (Leser, 1992).

With the notation of Fig. 11 we employ the plane strain effective stress definition

$$\frac{1}{3}\sigma_e^2 = A\left(\sigma_{xx} - \sigma_{yy}\right)^2 + 2N\sigma_{xy}^2 \tag{16}$$

where (x,y) are the material orthotropy axes and (A,N) are orthotropy parameters. For the Mises solid $A = \frac{1}{4}$ and $N = \frac{1}{2}$. Measuring the polar coordinate anti-clockwise from the x-axis, we denote by θ_o the wedge orientation (Fig. 11).

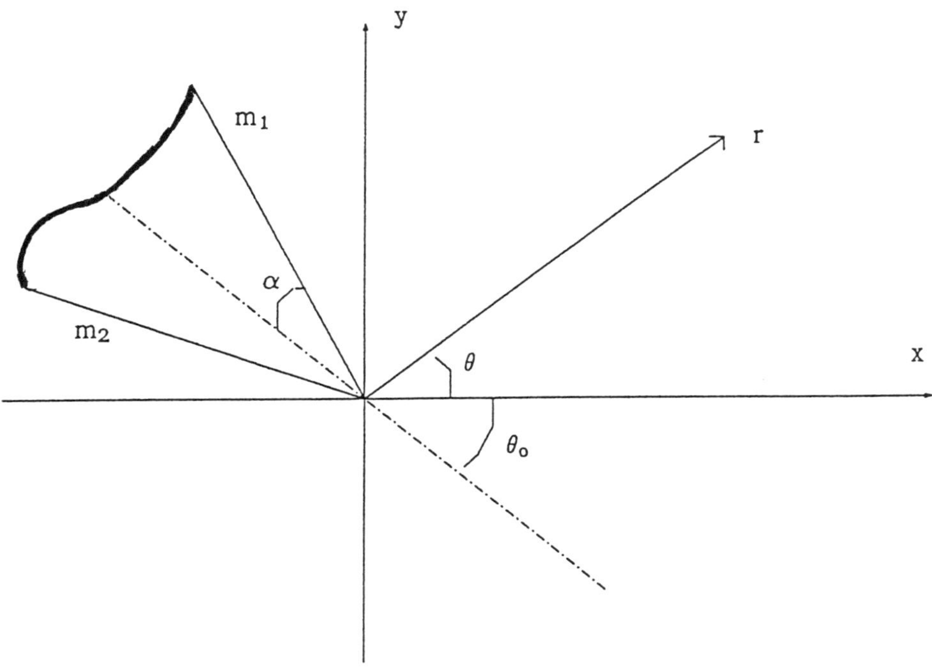

Figure 11. Wedge penetration into an orthotropic medium.

152   D. DURBAN

The strain rate components associated with (16) can be written in the polar system as

$$d_{rr} = -d_{\theta\theta} = k\sigma_e^{n-1}\left[(\sigma_{rr} - \sigma_{\theta\theta})(X\cos^2 2\theta + 1) - \sigma_{r\theta}X\sin 4\theta\right] \qquad (17a)$$

$$d_{r\theta} = k\sigma_e^{n-1}\left[-\frac{1}{2}(\sigma_{rr} - \sigma_{\theta\theta})X\sin 4\theta + 2\sigma_{r\theta}(X\sin^2 2\theta + 1)\right] \qquad (17b)$$

where $k = (3/2)N\sigma_o^{-n}$, with (σ_o, n) denoting material constants of (1) and

$$X = \frac{2A}{N} - 1 \qquad (18)$$

Similarly, the effective stress (16) takes here the form

$$\sigma_e = \sqrt{\frac{3}{2}}N\left\{[(\sigma_{rr} - \sigma_{\theta\theta})\cos 2\theta - 2\sigma_{r\theta}\sin 2\theta]^2 X + (\sigma_{rr} - \sigma_{\theta\theta})^2 + 4\sigma_{r\theta}^2\right\}^{1/2} \qquad (19)$$

The standard Mises viscoplastic material is recovered from (17)-(19) with $X = 0$.

The stress function formulation (6), along with a separation of variables representation of the singular field, generates here an eigenvalue equation, which is the orthotropic analogue of (7)-(8), along with the boundary data

$$v_\theta = 0 \quad \sqrt{6N}\sqrt{1 + X\sin^2 2\theta}\,\tau_{r\theta} = \left\{\begin{matrix} -m_1 \\ m_2 \end{matrix}\right\}\sigma_e \quad \text{at} \quad \theta = \left\{\begin{matrix} \pi - \alpha - \theta_o \\ -\pi + \alpha - \theta_o \end{matrix}\right\} \qquad (20)$$

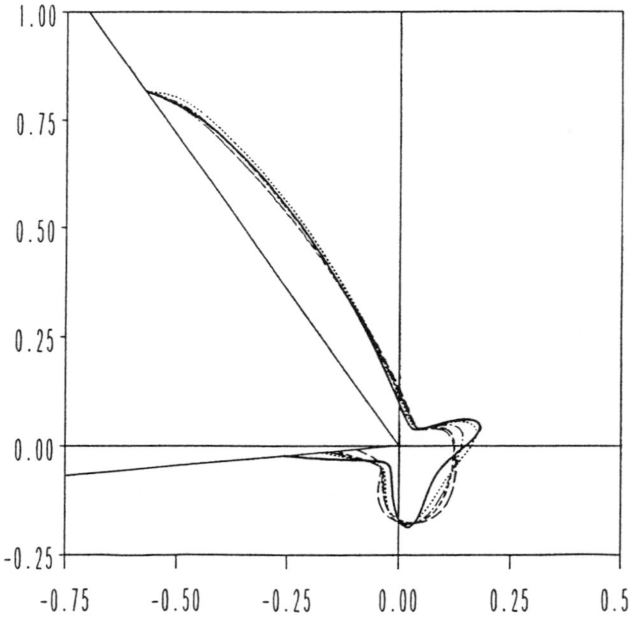

Figure 12. Constant effective stress contours for different values of orthotropy parameter X. $\theta_o = 25°$.

thus inducing a friction-orthotropy coupling. A typical constant effective stress contour is shown in Fig. 12 for high and nearly equal levels of wall function, and with different values of orthotropy parameter X. By comparison (Fig. 13) in "normal penetration" with $\theta_0 = 0$, and higher friction on the lower wall there is a stronger influence of material orthotropy.

Plastic orthotropy arises in directionally strengthened composite materials and is a major source of structural penetration resistance. There is a definite possibility that fiber reinforced materials can be designed to lower the level of singularity during a penetration process. The interaction and coupling among wedge geometry, wall friction and material properties should be further investigated along the lines suggested in the present preliminary work.

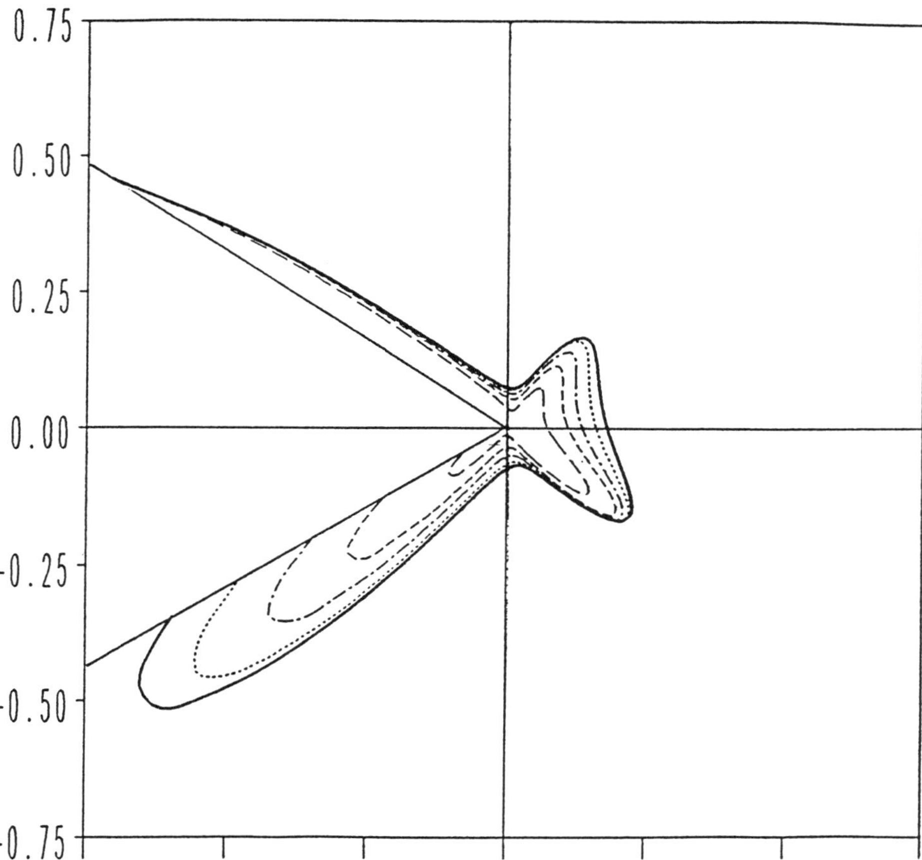

Figure 13. Constant effective stress contours for different values of orthotropy parameter X. $\theta_0 = 0°$.

Finally, it is worth recalling that singular fields induced during indentation of rigid penetrators into power law rigid/plastic solids can be examined by similar methods. The constitutive equations (1) and (12) remain essentially the same except for the exchange $\mathbf{D} \to \mathbf{E}$ where $\mathbf{E}$ is the infinitesimal strain tensor. Likewise, velocities are replaced by displacements and the solution is constructed within the framework of small strain plasticity theory. Studies of this nature for ideal boundary data are available for conical indentors in (Fleck and Durban, 1989) and for wedges in (Alexandrov and Grishin, 1987; Duva, 1988).

5. References

Alexandrov, V.M. and Grishin, S.A. (1987) State of stress and strain of a small neighbourhood of the apex of a wedge for a physical non-linearity and different boundary conditions, *PMM* (English Translation) **51**, 509-515.

Balashov, D.B. and Zvolinskii, N.V. (1996) Flow of a rigid-plastic medium around a cone, *Mechanics of Solids* (English Translation) **31**, 3, 39-45.

Durban, D. and Rand, O. (1991) Singular fields in plane strain penetration, *J. Appl. Mech.* **4**, 910-915.

Durban, D. and Fleck, N.A. (1992) Singular plastic fields in steady penetration of a rigid cone, *J. Appl. Mech.* **59**, 706-710. Authors' Closure to Discussion (1993) *J. Appl. Mech.* **60**, 1062.

Duva, J.M. (1988) The singularity at the apex of a rigid wedge embedded in a nonlinear material, *J. Appl. Mech.* **55**, 361-364.

Fleck, N.A. and Durban, D. (1989) Asymptotic fields at tip of conical notches and inclusions in a power hardening solid, *J. Mech. Phys. Solids* **37**, 233-263.

Fleck, N.A. and Durban, D. (1991) Steady penetration of a rigid cone with a rough wall into a power law viscous solid, *J. Appl. Mech.* **4**, 872-880.

Hutchinson, J. (1968) Singular behaviour at the end of a tensile crack in hardening materials, *J. Mech. Phys. Solids* **16**, 13-31.

Johnson, W. and Mellor, P.B. (1973) *Engineering Plasticity*, Van Nostrand, London.

Leser, R. (1992) Singular Fields in Steady Penetration of a Rigid Wedge into Orthotropic Viscoplastic Solids, M.Sc. Research Thesis, Technion, Haifa, Israel.

Rice, J.R. and Rosengren, G.F. (1968) Plane strain deformation near a crack tip in a power-law hardening material, *J. Mech. Phys. Solids* **16**, 1-12.

CREEP INDUCED COHESIVE CRACK
PROPAGATION IN MIXED MODE

F. BARPI, S. VALENTE
Politecnico di Torino, C.so Duca degli Abruzzi 24, 10129 Turin, Italy

F. CHILLE'
E.N.E.L.-C.R.I.S., Via Ornato 90/14, 20162 Milan, Italy

L. IMPERATO
I.S.M.E.S., Via Pastrengo 9, 24068 Seriate (Bergamo), Italy

1. Introduction

Westergaard (1939) gave the first solution to the problem of a Mode I crack in an elastic medium. He admits infinite stresses (stress singularity) at the crack tip, introducing the concept of *stress intensity factors* as a measure of stress field severity. This theory was called Linear Elastic Fracture Mechanics (*L.E.F.M.*).

Infinite stresses are not believed to exist in reality, and therefore new non linear fracture theories were formulated, based on the hypothesis that stresses have to be less than or equal to the yield strength of the material.

In this direction, Barenblatt (1959) and, independently, Dugdale (1960) proposed the use of a fictitious crack longer than the actual one. On this crack extension, *cohesive forces* are applied in order to eliminate the stress-singularity due to the application of mechanical loads.

The theories described up to this points were developed to study crack propagation in metals with limited ductility. The behaviour of such materials, called ductile, is characterised by marked *strain hardening* prior to the attainment of ultimate tensile strength. In this case, energy dissipation occurs mainly in a butterfly-shaped area. Entirely different is the behaviour of materials such as concrete, rock, ceramics, sea ice, etc. These materials are full of flaws, such as pores, air filled voids, lenses of bleed water under coarse aggregate, and shrinkage cracks, even before they are mechanically loaded. These flaws, and especially small cracks (microcracks) grow steadily under external loading, and coalesce with existing or newly-formed microcracks, until large fractures are formed and cause the collapse of the structure.

Due to such micro-mechanical phenomena, these materials exhibit a very modest degree of strain-hardening prior to the attainment of ultimate tensile strength. Therefore they are called quasi-brittle materials.

The *tension softening* phase is very large for these materials and cannot be neglected. In concrete it is due to localised microcracking and to bridging by coarse

155

D. Durban and J.R.A. Pearson (eds.),
IUTAM Symposium on Non-Linear Singularities in Deformation and Flow, 155-168.
© 1999 *Kluwer Academic Publishers. Printed in the Netherlands.*

aggregate. It represents a violation of Drucker's postulate about stable materials and produces the localisation of strains in a very narrow band, which is a continuation of the real crack. For this reason, the concept of a fictitious crack extending beyond the tip of the real crack, as initially proposed by Barenblatt and Dugdale for metals, was applied to three-point bending fracture of concrete by Hillerborg *et al.* (1976).

These authors called this crack extension, where cohesive forces are applied, Fictitious Process Zone (*F.P.Z.*, Figure 1a).

The closing forces applied are a descending function of the Crack Opening Displacement (*C.O.D.*, denoted with w in Figure 1c). Therefore, a critical value, w_c, exists beyond which no stress transfer occurs. This value represents an intrinsic length of the material which induces *size effects* in the structural response. In other words, a non linear condition on an evolving boundary achieves the full status of a constitutive law (Figure 1c). The area under the σ vs. w diagram is the energy dissipated during the process of formation of a unitary, stress-free crack surface (Figure 1c). This energy is called fracture energy ($\mathcal{G}_F$) and is assumed to be a property of the material. In this case, the tensile behaviour of the material is prevalently softening and energy dissipation occurs mainly in the fictitious process zone.

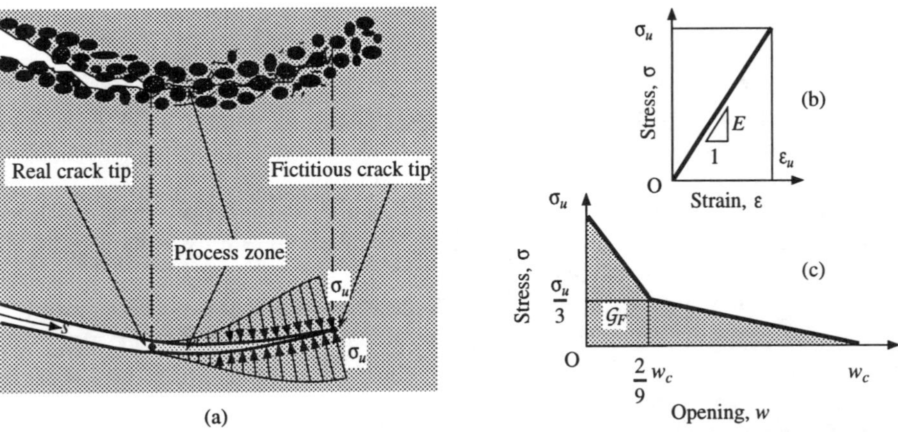

Figure 1. (a): process zone, (b): constitutive law for the undamaged material, (c): for the process zone.

In brittle structures (i.e., structures with small fracture energy or large sized), the load vs. loading point displacement diagram presents a snap-back branch. In order to follow this branch, Carpinteri (1985) proposed using the length of the fictitious crack as a control variable. This length is surely a monotonely increasing function of time during the irreversible fracture propagation process. This technique, called *fictitious crack length control scheme,* was generalised to mixed-mode problems by Carpinteri and Valente (1988), and was used for various boundary conditions (Bocca *et al.*, 1991; Valente, 1991, 1992, 1993; Valente *et al.*, 1994).

Keeping all the geometric ratios constant while varying the size produces a transition in both the numerical and the experimental response, which shifts from ductile behaviour (as is typical of large fracture energy or small sized structures) to brittle behaviour (small fracture energy or large structures). In the latter case, a strong snap-back phenomenon occurs and the cohesive crack model predicts the same maximum load as the Linear Elastic Fracture Mechanics method (Carpinteri, 1985; Carpinteri and Valente, 1988; Bocca *et al.*, 1991).

Therefore the material can be characterised through four parameters (Young's modulus E, Poisson's ratio v, ultimate tensile strength σ_u, fracture energy $\mathcal{G}_F$), and the numerical model explains why the behaviour observed in laboratory tests is usually ductile (small specimens) and the behaviour of real structures is usually brittle (large size). The cohesive crack model can also explain equilibrium path bifurcation phenomena, which are size-independent, such as losses of symmetry (for direct tension, see Rots *et al.*, 1987; for four-point shear tests on double notched specimens, see Valente, 1992), and bifurcation phenomena which are size-dependent, such as a second crack propagation during four-point shear tests on single notched specimens (Barpi and Valente, 1997).

Up to this point we have neglected the effects of time. As time elapses, the closing forces in the fictitious process zone reduce due to *creep*. Therefore, crack propagation occurs also under a constant load. This phenomenon is addressed, in this paper, in the context of the cohesive crack model. Since the study of time induced crack propagation is important in dam safety assessment, three dam models (scale $S_L = 1:40$) are analysed, from an experimental and a numerical point of view.

2. Static tests

As anticipated in the introduction, the phenomenon of creep induced crack propagation is investigated here through an extension of the static cohesive crack model. Therefore, the materials used in the laboratory tests have to be characterised from a static, time independent point of view.

Three types of concrete mix proportion are considered (Table 1): a micro-concrete (test series 2788), an ordinary concrete (test series 2857) and a mortar (test series 2859).

TABLE 1. Mix proportions for concrete

Test series	Cement Portland 325 (kg/m³)	Plasticizer (kg/m³)	Water (kg/m³)	Sand $\Phi < 3.15$ mm (kg/m³)	Gravel $3.15 < \Phi < 12$ mm (kg/m³)	Gravel $12 < \Phi < 25$ mm (kg/m³)
2788	300	3	190	915	915	0
2857	270	2.7	170	656	502	772
2859	400	4	240	1506	0	0

The specimens for static and creep tests were prepared together with the three dam models. The specimens and the models were stored at 20° C and 95% relative humidity. Mean compressive strength was measured on six 15×15×15 cm cubes, Young's modulus on two 15×15×60 cm prisms.

Stable bending tests were performed according to R.I.L.E.M. recommendations, on 10×10×84 cm beams with a 5 cm deep notch (testing span 80 cm), in order to determine the fracture energy.

Stable tensile tests were performed on 20 cm high, 10 cm diameter cylinders with a 1 cm deep notch in order to determine the ultimate tensile strength. Results, at 8 months from batch, are summarised in Table 2.

TABLE 2. Mean compressive strength, Young's modulus, fracture energy and tensile strength

Test series	Maximum aggregate size (mm)	Mean compressive strength (MPa)	Mean Young's modulus (MPa)	Mean fracture energy (N/m)	Mean tensile strength (MPa)
2788	12	28	28000	141.15	2.375
2857	25	35	35200	121.49	2.885
2859	3.15	32	22100	96.55	1.950

The testing machines were *Galdabini PM60* of electro-hydraulic and closed loop type, with a loading capacity of 60 tons. Deformations were measured using HBM W2 type LVDTs with a measuring range of ± 2 mm and accuracy of 0.2%. In tensile tests, C.M.O.D. (*Crack Mouth Opening Displacement*) was measured as the mean value of the readings obtained on three transducers, glued at 120° to one another.

In the numerical simulations, the mean values shown in Table 2 were used. The Poisson's ratio was 0.1.

3. Tensile creep tests

The creep law for the fictitious process zone (F.P.Z.) was determined by means of direct tensile tests, performed on 20 cm high, 10 cm diameter cylinders, with a 1 cm deep notch. The age of the concrete was 8 months.

The following phases (Figure 2) were carried out in the creep tests:

phase (1): load up to the maximum load P_{max} (C.O.D. rate of 0.2 μm/s),

phase (2): follow the softening branch until the load decreases to the prescribed value, P_D (C.O.D. rate of 0.2 μm/s),

phase (3): unload to P_E (about 10% of the maximum load P_{max}, load control),

phase (4): reload up to P_G corresponding to a certain percentage (80%, 85%, 90%) of P_D, load control,

phase (5): the load is kept constant (± 5 N) until creep rupture occurs.

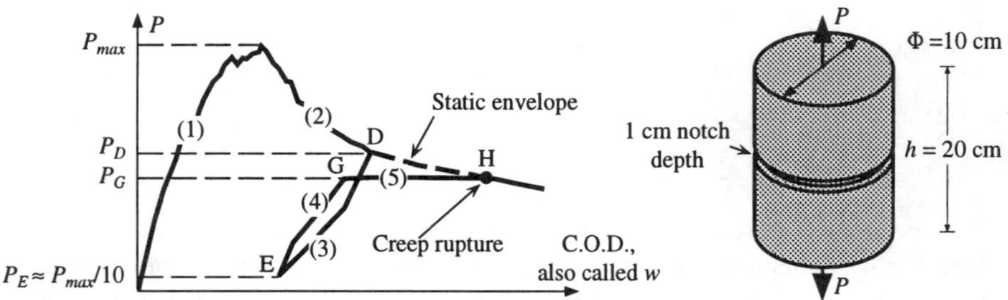

Figure 2. Tensile creep test procedure, test serie 2857 ($P_G/P_D = 0.80$).

Figure 2 shows that the curve obtained during a static test (phases (1), (2) and related extrapolation) can be assumed as a valid envelope criterion for creep fracture, as suggested by Carpinteri *et al.* (1995). The same conclusions were drawn from the observation of other types of concrete (see Table 1) and other sustained load levels $P_G/P_D = 0.80$, 0.85, 0.90 (where P_G and P_D are as defined in Figure 2). Barpi and

Valente (1996) give the complete serie of nine diagrams. This criterion implies that the fracture energy is completely independent of the time duration of the sustained load, as reported by Aassved Hansen (1991) for three-month-old concrete specimens.

During phases (1) and (2), a C.O.D. rate of 0.2 μm/s was imposed (C.O.D. control). Phases (3), (4) and (5) were executed under load control. Phases (1), (2), (3) and (4) are executed in a time which is small compared to the time of phase (5). Therefore they are considered time independent. Typical load and C.O.D. results are summarised in Tables 3, 4 and 5.

TABLE 3. Tensile creep results for test series 2788

Specimen	P_{max}	C.O.D. at P_{max}	P_D	P_G/P_D	P_G	$(C.O.D.)_G$	$(C.O.D.)_H$	Failure lifetime
	(kg$_f$)	(μm)	(kg$_f$)	(%)	(kg$_f$)	(μm)	(μm)	(s)
2788C	1136	7.1	800	90	720	11.5	15.7	1184
2788D	1221	8.3	800	90	720	16.3	24.5	4902
2788E	1314	9.5	800	85	683	17.2	27.0	11916
2788F	1125	6.5	800	85	680	10.0	15.4	4392
2788G	1260	8.0	800	85	680	13.5	22.8	16590
2788H	1050	7.5	800	80	644	9.1	18.2	62985
2788I	1062	8.4	800	80	640	10.3	17.4	33642
2788L	1235	8.8	800	80	640	14.8	24.8	64867

TABLE 4. Tensile creep results for test series 2857

Specimen	P_{max}	C.O.D. at P_{max}	P_D	P_G/P_D	P_G	$(C.O.D.)_G$	$(C.O.D.)_H$	Failure lifetime
	(kg$_f$)	(μm)	(kg$_f$)	(%)	(kg$_f$)	(μm)	(μm)	(s)
2857E	900	8.0	800	90	723	12.1	21.6	4693
2857F	1383	10.8	800	90	723	18.6	24.7	2626
2857H	1029	9.1	800	85	683	12.3	20.9	30184
2857I	1160	9.0	800	80	644	16.0	27.4	30058
2857L	1407	9.3	800	80	641	15.5	26.4	35926

TABLE 5. Tensile creep results for test series 2859

Specimen	P_{max}	C.O.D. at P_{max}	P_D	P_G/P_D	P_G	$(C.O.D.)_G$	$(C.O.D.)_H$	Failure lifetime
	(kg$_f$)	(μm)	(kg$_f$)	(%)	(kg$_f$)	(μm)	(μm)	(s)
2859C	1116	8.9	640	90	579	12.0	17.9	3051
2859G	858	8.4	640	90	575	9.7	12.1	877
2859D	615	8.3	640	85	543	8.3	16.0	9781
2859F	918	9.6	640	85	547	10.9	16.8	10589
2859E	1034	8.4	640	80	515	11.1	20.0	28206
2859I	649	7.5	640	80	515	7.4	12.7	56760

Let us now focus our attention on phase (5). Figures 3, 4 and 5 show normalised creep C.O.D. vs. normalised time. The first ratio is defined as $(w - w_G)/(w_H - w_G) = w^c/w^c_{ult}$ with reference to Figure 2, where C.O.D. is indicated simply as w. The second ratio is defined as time/time-until-failure. For each level of sustained load, experimental diagrams show a small scattering. Therefore the mean curves, reported in Figures 3, 4, 5 have been assumed as representative for the phenomenon.

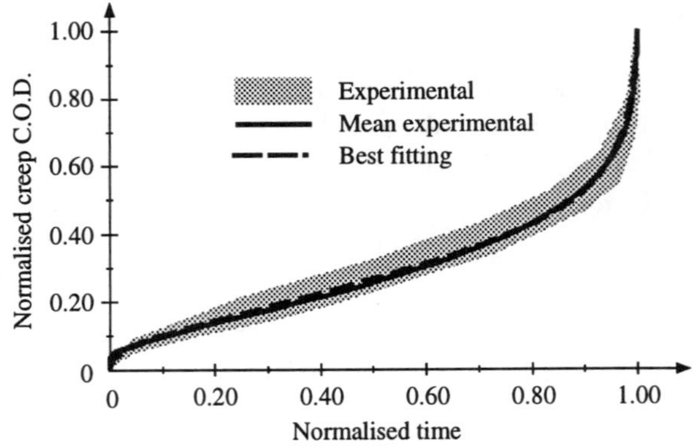

Figure 3. Experimental and best-fitting creep curves in test serie 2788.

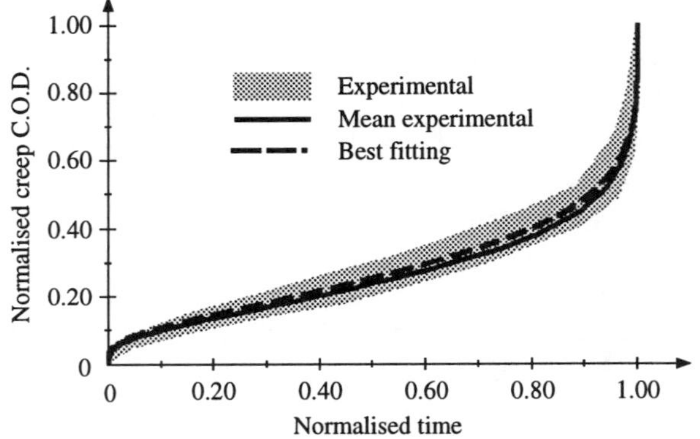

Figure 4. Experimental and best-fitting creep curves in test serie 2857.

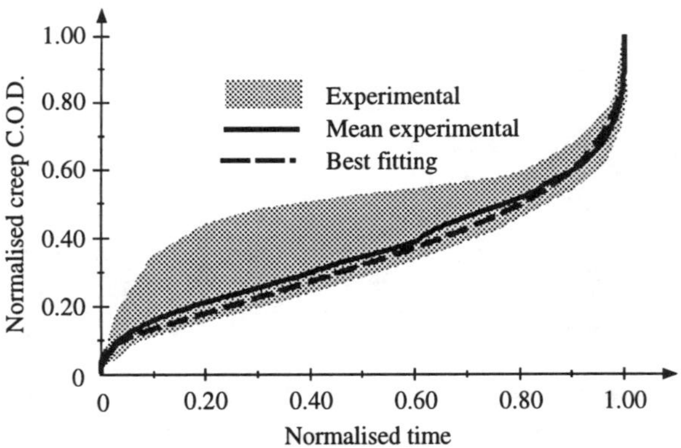

Figure 5. Experimental and best-fitting creep curves in test serie 2859.

They display a three-stage process, according to the change of creep rate. Creep rate decreases gradually in the *primary* stage, is almost constant in the *secondary* stage, and increases rapidly until failure in the *tertiary* stage. In terms of failure lifetime, the primary, secondary and tertiary stages take up about 5%, 75% and 20% respectively. In other words, the secondary stage dominates the failure lifetime in creep rupture.

Reinhardt and Cornellissen (1985) and Carpinteri *et al.* (1995) obtained similar experimental results. An interaction occurs between creep and micro-crack onset and coalescence. This is the reason for the non constant value of the C.O.D. rate at constant load. Numerical simulations were based on the following creep laws (for serie 2788, 2857, 2859 respectively):

$$\frac{\dot{w}^c}{w^c_{ult}} = 6.90 \times 10^{-3} \left(\frac{\sigma_G}{\sigma_D}\right)^{26} \left(0.40 + 25\left(\frac{w^c}{w^c_{ult}}\right)^5 + 9\exp\left(-50\frac{w^c}{w^c_{ult}}\right)\right), \quad (1a)$$

$$\frac{\dot{w}^c}{w^c_{ult}} = 16.72 \times 10^{-4} \left(\frac{\sigma_G}{\sigma_D}\right)^{19} \left(0.35 + 32\left(\frac{w^c}{w^c_{ult}}\right)^5 + 15\exp\left(-50\frac{w^c}{w^c_{ult}}\right)\right), \quad (1b)$$

$$\frac{\dot{w}^c}{w^c_{ult}} = 9.71 \times 10^{-3} \left(\frac{\sigma_G}{\sigma_D}\right)^{27} \left(0.45 + 12\left(\frac{w^c}{w^c_{ult}}\right)^5 + 35\exp\left(-50\frac{w^c}{w^c_{ult}}\right)\right), \quad (1c)$$

where:

(a) the first factor, depending on σ_G/σ_D is related to failure lifetime through the best fitting functions shown in Figures 6, 7, 8. A similar factor was proposed by Bazant and Gettu (1992) to reach agreement with experimental evidence that a small change in load level (for example from 0.80 to 0.90) causes a change in the failure lifetime of more than one order of magnitude. This sensitivity was also found by Karihaloo (1995) analysing the growth rate of a crack in a visco-elastic tension-softening material,

(b) the second factor, depending on w^c/w^c_{ult}, is related to three creep stages through the best fitting functions shown in Figures 3, 4, 5.

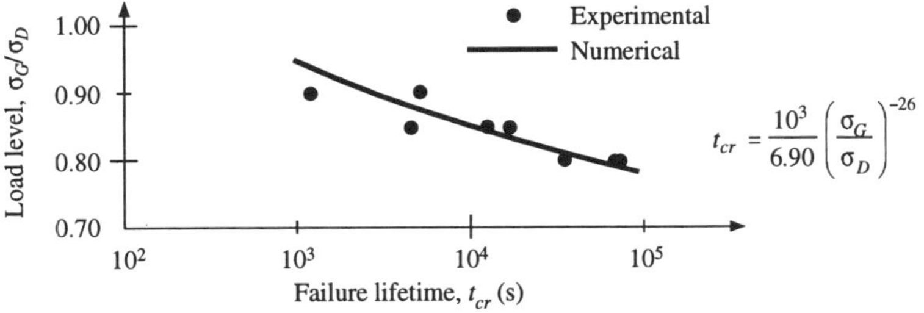

Figure 6. Sustained load level vs. failure lifetime curve in test serie 2788.

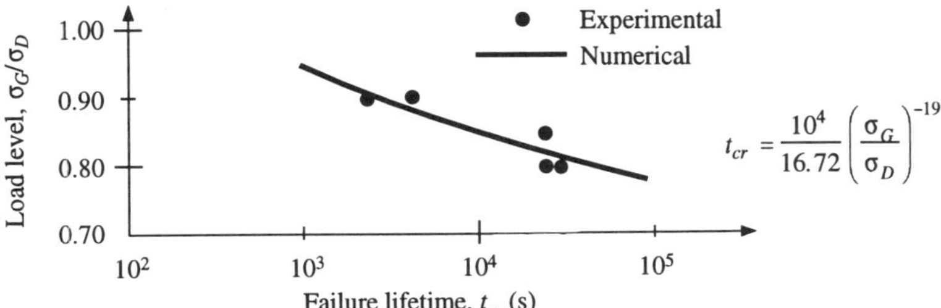

Figure 7. Sustained load level vs. failure lifetime curve in test serie 2857.

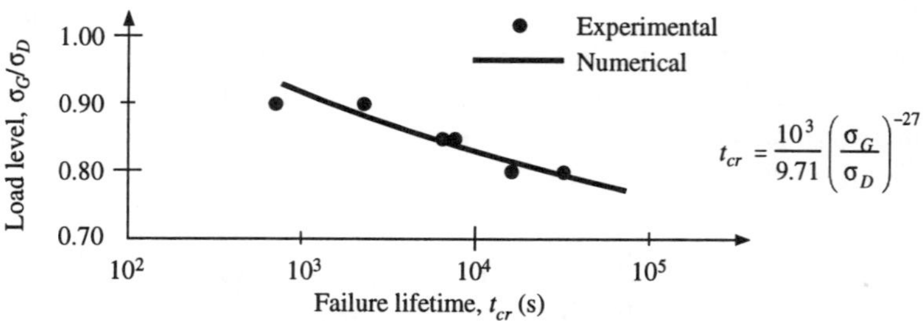

Figure 8. Sustained load level vs. failure lifetime curve in test serie 2859.

4. Constitutive law for the fictitious process zone

Figure 11 shows a typical finite element mesh used in the analysis. During the crack propagation process, some points are divided in two: one for each crack surface. A cohesive element is generated for each pair of such points, called *nodes,* as shown in Figure 11.

As proposed in (Carpinteri and Valente, 1988; Bocca *et al.*, 1991; Valente, 1992, 1993; Valente *et al.*, 1994):

(a) *tangential stresses* in the process zone are neglected,

(b) interaction between cohesive elements occurs only through the elastic bidimensional continuum, not directly (*uncoupled elements*).

The loading procedure described in Figure 2 with reference to a specimen (five phases) was also applied to the structure (gravity dam model).

The following observations may be of use in the definition of the incremental constitutive law:

- during phases (1), (2), (4), (5) $\dot{w} > 0$, $\forall\, \mathcal{P} \in S_c$,   (2a)
- during phase (3) $\qquad\qquad\quad \dot{w} < 0$, $\forall\, \mathcal{P} \in S_c$,   (2b)

where dot means derivation with respect to time and $\mathcal{P}$ is a generic point belonging to the cohesive surface S_c. Conditions (2a-b) are not imposed *a priori* in the numerical simulation, but they always turn out to be verified *a posteriori*. In other words, with reference to Figure 9, the following observation applies:

a) phases (3), (4) and (5) start at the same time for the overall structure and for every cohesive element. The respective starting points are D, L', G. In some cases, point L' is replaced by point L".

b) on the contrary, F (the intermediate point of phase (4)) and H (the ending point of phase (5)) can be reached at different times in two different cohesive elements.

c) during phases (1) and (2) all stress paths follow the static envelope shown in Figure 9 (AB and BC). The area internal to the closed polygon OABCO is the fracture energy of the material. The position of the knee point ($w_B = 2\, w_c/9$, $\sigma_B = \sigma_u/3$) was chosen according to Petersson (1981).

The shape of the unloading (DL') and reloading (L'F'G) paths shown in Figure 9 was proposed by Carpinteri *et al.* (1995), with reference to the three-point bending tests. In that case, it always was $w > f\,w_D$. On the contrary, in dam model analysis, it was $w < f\,w_D$ and hence branch LL" was added.

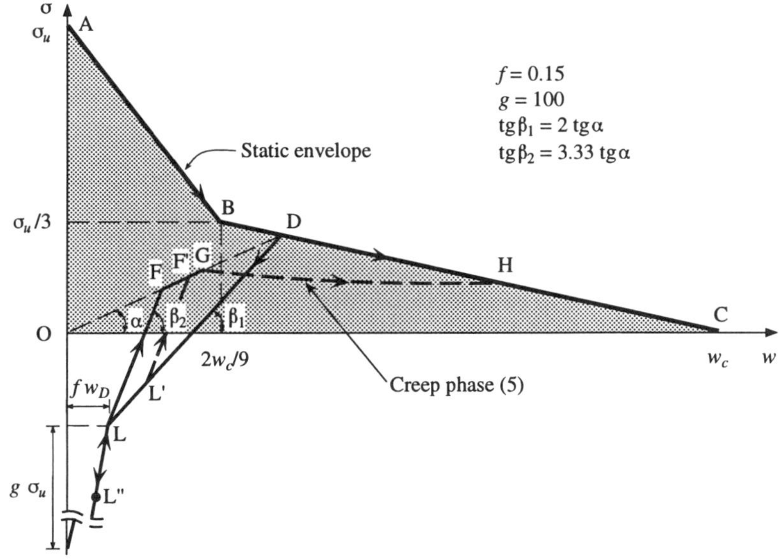

Figure 9. Piecewise linear constitutive law assumed for the F.P.Z.

The behaviour of the process zone analysed up to this point (phases (1), (2), (3) and (4)) is assumed as time independent. On the contrary, during phase (5), the external load is kept constant and the relaxation of the process zone moves the stress from point G to H (Figure 9).

As shown in Figure 10a, a generic time increment can be divided into two steps:

(a) during time dt, w is kept constant, and a stress relaxation $d\sigma^R = -F\,\dot{w}^c\,dt$ occurs in the process zone,

(b) the reduction in the closing stresses acting on the elastic continuum generates an opening increment dw and a stress increment $d\sigma^I$, in each cohesive element.

Figure 10b shows how w^c and w^c_{ult} are defined in the numerical simulation. Through Equation (1), this value enables us to determine $\dot{w}^c$, mentioned in point (a) above. Figure 10b also shows that the stress path, related to phase (5), is not pre-defined as a consequence of the starting point, G. It is computed as a sequence of steps, which are so small that $|d\sigma^R| < \sigma/100$ (in Figure 10b they are enlarged for the sake of clarity). Upon reaching the static envelope, the stress point is forced to follow it ($\dot{w} > 0$).

In dam model analysis, self-weight tends to close the crack. Therefore a constitutive law is also required for the real crack and the notch (Barpi and Valente, 1996).

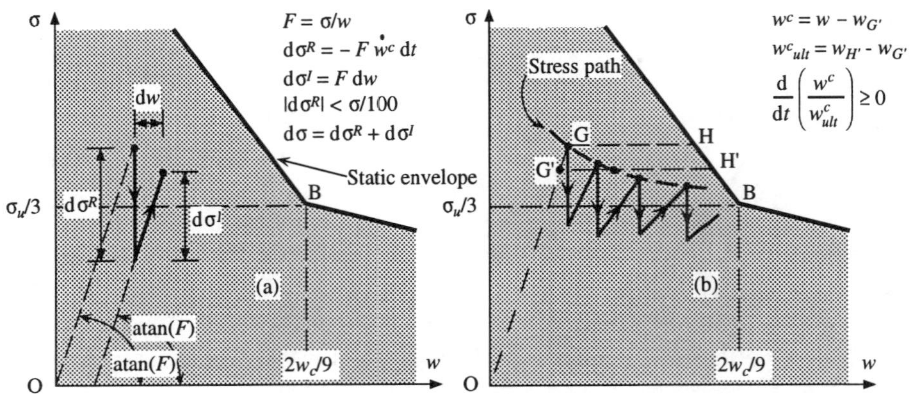

Figure 10.   (a): Piecewise linear constitutive law assumed for the F.P.Z.,
(b): Stress path in the F.P.Z. due to creep under constant load.

5. The cohesive crack model

Under high level sustained load, *creep* outside the process zone can be neglected compared to creep in the process zone (Bazant and Xiang, 1997). Therefore, the constitutive law for the continuum is linear elastic, time independent. According to the finite element method, keeping in mind that the constitutive law for the fictitious crack is assumed to be piece-wise linear, by taking the unknowns to be the n nodal displacements, u, it is possible to write the equilibrium condition, for each solution increment, through the virtual work equation as follows (Bocca *et al.*, 1991; Barpi and Valente, 1996):

$$(K_T + C_T)\, \mathrm{d}u = \mathrm{d}\lambda\, P + \mathrm{d}t\, Q, \tag{3}$$

where:

K_T: positive definite tangential stiffness matrix, containing contributions from linear elastic elements and possible contributions from cohesive elements with values (σ, w) inside the static envelope (phases (3), (4) and (5)),

C_T: negative definite tangential stiffness matrix, containing contributions from cohesive elements with values (σ, w) on the static envelope (phases (1), (2) and some elements during phase (5)),

P: vector of incremental load normalised in order to obtain $\sum P_i^2 = 1$,

$\mathrm{d}\lambda$: maximum load multiplier which is compatible with all active limits. For example, no stress increment can go through point B or C or F or L of Figure 9, because such a condition implies a modification of matrix K_T or C_T,

Q: vector of unbalanced load (or pseudo-load) due to relaxation in the process zone, related to a unitary time increment,

$\mathrm{d}t$: max time increment which is compatible with all limits active (see also $\mathrm{d}\lambda$).

The self-weight increment is applied at the first step, then it is kept constant. For this reason it does not appear in Equation (3).

During phases (1), (2), (3) and (4), the behaviour of the material is assumed to be time independent ($Q = 0$), the external load changes ($\mathrm{d}\lambda \neq 0$), and $\mathrm{d}t = 0$. On the contrary, during phase (5), the behaviour of the process zone is assumed to be time dependent ($Q \neq 0$), the external load is kept constant ($d\lambda = 0$), and $dt \neq 0$. The crack propagates, by a pre-defined length, perpendicularly to the principal tensile stress at the Fictitious Crack Tip ($\sigma_{F.C.T.}$, Figure 1), when $\sigma_{F.C.T.}$ reaches the tensile strength of the material (σ_u). Since the crack trajectory is unknown a priori, a remeshing technique is necessary at each crack growth step. For further details on this subject, see Bocca *et al.* (1991). The above conditions represent one of the limits which define $d\lambda$ or dt.

At the beginning of phase (5), all stress paths are inside the static envelope ($F > 0$), matrix $K_T + C_T$ is positive definite so that $du^T Q > 0$. During time evolution, F decreases, some stress paths reach the static envelope ($d\sigma/dw < 0$), new cohesive elements appear. Therefore, matrix $K_T + C_T$ will no longer be positive definite so that we may find $du^T Q < 0$. This condition determines the end of this phase and therefore the failure lifetime (t_{cr}) of the structure.

When remeshing takes place, the difference between the old and the new mesh produces small out-of-balance loads in the F.P.Z. If $\det(K_T + C_T)$ is positive but small, the process of elimination of these loads can be divergent. Usually this condition precedes the previous one and determines t_{cr}.

6. Experimental and numerical analyses of the three gravity dam models

Figure 11 shows the finite element mesh used in the numerical simulation of the experimental tests conducted on gravity dam models.

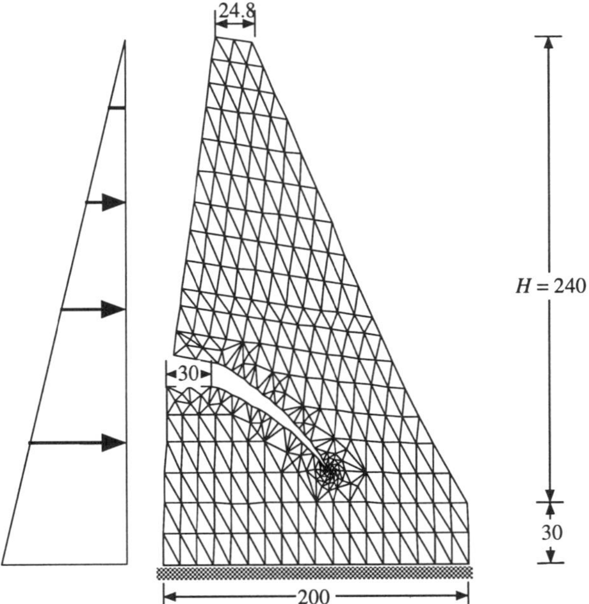

Figure 11. Finite element mesh and horizontal loads (dimensions are in cm).

The geometric scale ratio between prototype and model was assumed to be $S_L = 1:40$.

The scale adopted for the elastic modulus and for strength was $S_E = 1$ (same value for both the prototype and the model). Having established the scales of two independent quantities (lengths and stresses), the scales of the other quantities playing a role in static problems are determined accordingly (for instance, the scale for the volume forces must be: $S_q = S_E S_L^{-1} = S_L^{-1} = 40$). For this reason, it was necessary to simulate a self-weight increment by means of a system of discrete vertical forces, which were applied before the horizontal thrust (further details on the experimental set-up are given by Barpi and Valente (1996)).

As a first consequence of this loading procedure, a compressive stress concentration could occur at the notch tip. In order to prevent the related concrete damage, a mortar filled plastic bag was inserted into the notch. This is the reason of the small negative C.O.D. (the transducer base length was 8 cm) shown in Figures 12, 13 and 14.

As a second consequence of the loading procedure, a knee point appears in phases (1), (3) and (4) (see Figures 12, 13 and 14). This is the most evident difference in comparison with the behaviour recorded during tensile creep tests (Figure 2). As a third consequence, the C.M.O.D. due to creep is small compared to the corresponding quantity in tensile tests (Figure 2).

In the case of dam models, the incremented self-weight tends to close the notch, while the horizontal loads tend to open it. Therefore a load level exists for which C.M.O.D. vanishes and the diagram crosses the load-axis. Following the application of the incremented self-weight, the five subsequent local phases were seen to be the same as in tensile creep tests (Figure 2), except for the imposed C.M.O.D. rate, which was 2 μm/min.

Figures 12, 13 and 14 show that experimental and numerical load-C.M.O.D. diagrams are in good agreement and so are the values obtained for failure lifetime. Barpi and Valente (1996) showed that experimental and numerical crack trajectories also were in good agreement.

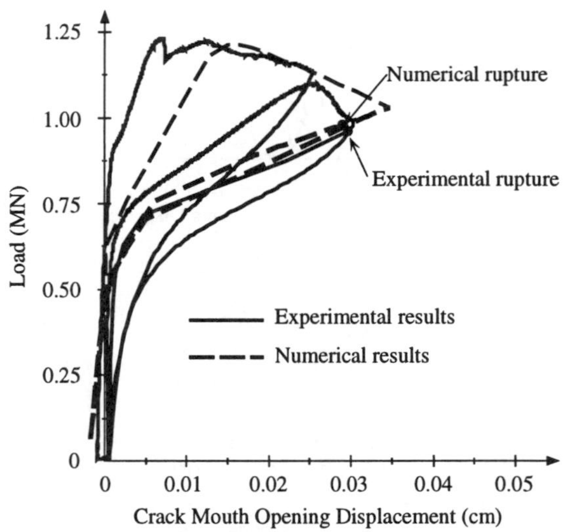

Figure 12. Load vs. Crack Mouth Opening Displacement for concrete 2788.

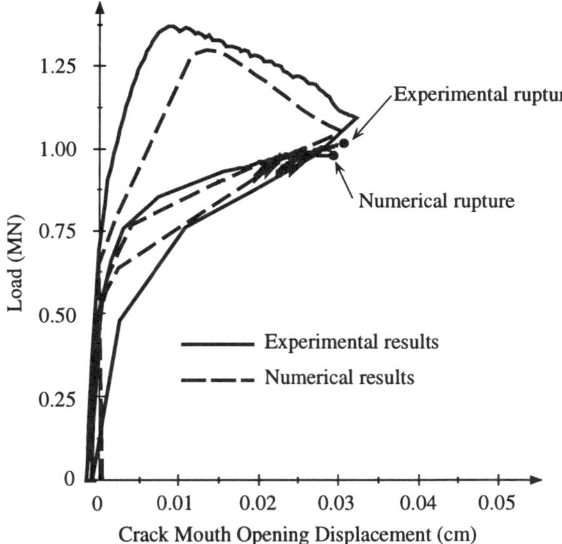

Figure 13. Load vs. Crack Mouth Opening Displacement for concrete 2857.

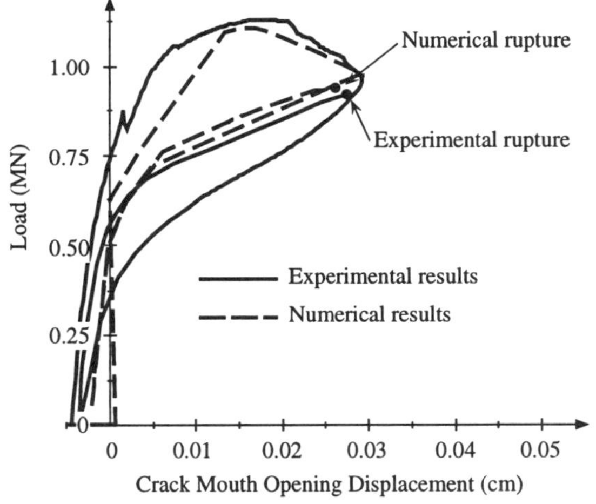

Figure 14. Load vs. Crack Mouth Opening Displacement for concrete 2859.

7. Conclusions

We can draw the following conclusions:

(a) Crack growth in quasi-brittle materials is the result of microcrack interaction and coalescence. The cohesive crack model replaces the details of such a process, which takes place on a very short length scale, with a discontinuity condition on the observation scale. The latter, acting on an evolving boundary, achieves the nature of a full constitutive law.

(b) Creep induced crack propagation in concrete structures can be analysed within the framework of the cohesive crack model, a well established approach to the fracture mechanics of quasi-brittle materials.

(c) The failure lifetime of pre-notched gravity dam models ($H = 240$ cm) can be predicted on the basis of the results of direct creep tensile tests ($h = H/12$).

(d) Experimental and numerical load vs. C.M.O.D. diagrams are in good agreement.

(e) The load vs. C.M.O.D. curves obtained from static tests can be assumed as a valid envelope criterion for creep fracture in mode I problems (direct tensile tests and three-point bending tests). In the mixed-mode problems analysed, creep fracture occurred before the static envelope was reached.

8. References

Aassved Hansen, E. (1991) Influence of sustained load on the fracture energy and the fracture zone of concrete, in van Mier J.G.M., Rots J.G., Bakker A. (eds.) *R.I.L.E.M./E.S.I.S. Conference on Fracture Processes in Brittle Disordered Materials: Concrete, Rock, Ceramics,* E&FN SPON, 829-838.

Barenblatt, G.I. (1959) The formation of equilibrium cracks during brittle fracture: general ideas and hypotheses. Axially-symmetric cracks, *Journal of Applied Mathematics and Mechanics,* **23,** 622-636.

Barpi, F. and Valente, S. (1996) Time induced crack propagation in concrete structures: cohesive crack model in mixed-mode conditions, Politecnico di Torino, Structural Engineering Department, *Research Report A845/95.*

Barpi, F. and Valente, S. (1998) Size-effects induced bifurcation phenomena during multiple cohesive crack propagation, *International Journal of Solids and Structures,* **35,** 1851-1861.

Bazant, Z.P. and Gettu, R. (1992) Rate effects and load relaxation in static fracture of concrete, *A.C.I. Materials Journal,* **89,** 456-468.

Bazant, Z.P. and Xiang, Y. (1997) Crack growth and lifetime of concrete under long time loading, *Journal of Engineering Mechanics,* **4,** 350-358.

Bocca, P., Carpinteri, A. and Valente, S. (1991) Mixed-mode fracture of concrete, *International Journal of Solids and Structures,* **27,** 1139-1153.

Carpinteri, A. (1985) Interpretation of the Griffith instability as a bifurcation of the global equilibrium, in Shah S.P. (ed.) *N.A.T.O.-Advanced Research Workshop on Application of Fracture Mechanics to Cementitious Composites,* Martinus Nijhoff Publishers, 287-316.

Carpinteri, A. and Valente, S. (1988) Size-scale transition from ductile to brittle failure: a dimensional analysis approach, in Mazars J. and Bazant Z.P. (eds.) *C.N.R.S.-N.S.F. Workshop on Strain Localisation and Size Effect due to Cracking and Damage,* Elsevier Applied Science, 477-490.

Carpinteri, A., Valente, S., Zhou, F.P., Ferrara, G. and Melchiorri, G. (1995) Crack propagation in concrete specimens subjected to sustained loads, in Wittmann F.H. (ed.) *Fracture Mechanics of Concrete Structures,* Aedificatio, 1315-1328.

Dugdale, D.S. (1960) Yielding of steel sheets containing slits, *Journal of Mechanics and Physics of Solids,* **8,** 100-104.

Hillerborg, A., Modeer, M. and Petersson, P.E. (1976) Analysis of crack formation and crack growth in concrete by means of fracture mechanics and finite elements, *Cement and Concrete Research,* **6,** 773-782.

Karihaloo, B. (1995) *Fracture Mechanics and Structural Concrete,* Longman Scientific and Technical.

Petersson, P.E. (1981) Crack growth and development of fracture zones in plain concrete and similar materials, *Doctoral Thesis Report TVBM-1006,* Division of Building and Materials, University of Lund, Sweden.

Reinhardt, H.W. and Cornellissen, H.W.A. (1985) Sustained tensile tests on concrete, *Baustoff,* **85,** 162-167.

Rots, J.G., Hordijk, D.A. and de Borst, R. (1987) Numerical simulation of concrete fracture in direct tension, in 4th *International Conference on Numerical Methods in Fracture Mechanics,* Pineridge Press, 457-471.

Valente, S. (1991) Influence of friction on cohesive crack propagation, in van Mier J.G.M., Rots J.G., Bakker A. (eds.) *R.I.L.E.M./E.S.I.S. Conference on Fracture Processes in Brittle Disordered Materials: Concrete, Rock, Ceramics,* E&FN SPON, 695-704.

Valente, S. (1992) Bifurcation phenomena in cohesive crack propagation, *Computers and Structures,* **44 1/2,** 55-62.

Valente, S. (1993) Heuristic softening strip model in the prediction of crack trajectories, *Theoretical and Applied Fracture Mechanics,* **19,** 119-125.

Valente, S., Barpi, F., Ferrara, G. and Giuseppetti, G. (1994) Numerical simulation of centrifuge tests on pre-notched gravity dam models, in Bourdarot E., Mazars J., Saouma V. (eds.) *Workshop on Dam Fracture and Damage,* Balkema, 111-119.

Westergaard, H.M. (1939), Bearing pressures and cracks, *Journal of Applied Mechanics,* **6,** 49-53.

ASYMPTOTIC ANALYSIS OF A SPONTANEOUS CRACK GROWTH. APPLICATION TO A BLUNT CRACK

D. LEGUILLON
Laboratoire de Modélisation en Mécanique - CNRS URA0229
Université P. et M. Curie, tour 66, case 162
4 place Jussieu, 75252 PARIS CEDEX 05, FRANCE

1. Introduction

The Griffith criterion is a universal approach to brittle fracture in elastic bodies. It is based on a critical value G_c of the energy release rate G defined as the derivative of the potential energy W with respect to the crack length ℓ (the present analysis is restricted to the bidimensional elasticity) :

$$G = - \frac{\partial W}{\partial \ell} \tag{1}$$

If G reaches the critical value $G_c = 2\gamma$, where γ is the fracture energy per unit length (surface in 3D), then the criterion predicts crack growth.

In general, the meaning and even the existence of the derivative (1) are unquestioned. However, there are at least two situations where this derivative does not exist : first, in case of an interface crack kinking out of the interface in a bimaterial (Rice 1988, Atkinson 1989, He *et al.* 1989, Miller *et al.* 1989, Leguillon 1993b), second in case of a crack lying in a stiff material with its tip located at the interface with a softer substrate (Erdogan *et al.* 1973, L. *et al.* 1992). In the first case it gives rise to redhibitory oscillations, whereas in the second one G increases to infinity as the tip approaches the interface.

Without considering these particular situations, the criterion fails in some other cases and especially at crack onset where G, calculated for $\ell = 0$, vanishes whatever the external applied loads. On one hand, experiments show that a crack can initiate at a corner point of a structure for instance, and on the other hand it is proved that G is zero (see (6) below) and thus cannot reach the critical value G_c. This paradox is mainly due to the me-

D. Durban and J.R.A. Pearson (eds.),
IUTAM Symposium on Non-Linear Singularities in Deformation and Flow, 169-180.

chanical meaning of the derivative (1). Let $\delta\ell$ denote an increment of crack length, the derivative (1) is defined by :

$$G = - \lim_{\delta\ell\to 0} \frac{W(\ell + \delta\ell) - W(\ell)}{\delta\ell} \tag{2}$$

The existence of this limit implies that for any $\delta\ell$, as small as needed, the situation can be described by a quasi-static state with a well known potential energy. This is a strong assumption which can be weaken considering spontaneous onset and growth of cracks at a ill-defined velocity. Such a process has been evoked by Kelly (1970), Aveston *et al.* (1973), and more recently by Hashin (1996) and Marigo (1996). Omiting the limit in (2) and assuming only the existence of two quasi-static states, the condition of propagation reads :

$$W(\ell) - W(\ell + \delta\ell) \geq 2\gamma\delta\ell \tag{3}$$

where $\delta\ell$ is now the minimum crack extension length such that the above inequality holds true.

2. Definition of a spontaneous crack extension length

Let us consider, within the linear bidimensional elasticity framework, a structure Ω^0 submited to prescribed displacements along a line Γ of its boundary $\partial\Omega^0$. A displacement field $\underline{U}^0$, a stress field σ^0 and a potential energy W^0 are associated with this equilibrium state, called (0) or unperturbed :

$$W^0 = \frac{1}{2} \int_{\Omega^0} \sigma^0 \nabla \underline{U}^0 dx \tag{4}$$

Next, let us consider the perturbed structure Ω^η such that there exists in addition a micro-flaw in the vicinity of a point O. We assume herein that it is a micro-crack with dimensionless length $\eta \ll 1$ (the ratio between the micro-crack length and a characteristic dimension of the structure). Under the same boundary conditions it corresponds new fields $\underline{U}^\eta$ and σ^η and a potential energy W^η :

$$W^\eta = \frac{1}{2} \int_{\Omega^\eta} \sigma^\eta \nabla \underline{U}^\eta dx \tag{5}$$

Using asymptotics with respect to the small parameter η, it is then shown that the most significant term of the potential energy change between the two states writes (Maz'ya *et al.* 1988, L. 1989) :

$$\delta W^{0\eta} = W^0 - W^\eta \simeq k^2 K \, \eta^{2\alpha} \tag{6}$$

This relation is true if the area Γ where prescribed displacements are acting is far from O in order to remain unchanged. $\alpha > 0$ in (6) is the characteristic exponent of the leading term of the elastic expansion around O expressed in polar co-ordinates r and θ as :

$$\underline{U}^0(x) = \underline{U}^0(O) + kr^\alpha \underline{u}(\theta) + ... \tag{7}$$

k is the intensity factor and $\underline{u}$ is the associated mode. If $\alpha < 1$ such a term is called a singularity. K in (6) is a geometrical coefficient taking into account the shape of the perturbation (its exact definition will be given in the forthcoming examples). The Griffith condition of growth can be written :

$$\delta W^{0\eta} \geq 2\gamma\eta \;\Rightarrow\; k^2 K \, \eta^{2\alpha-1} \geq 2\gamma \tag{8}$$

Remark 1 : If the point O coincides with a crack tip ($\alpha = 1/2$), the perturbation is a crack extension and (8) is equivalent to the Irwin criterion :

$$k \geq \sqrt{2\gamma/K_F} = k_c \tag{9}$$

where K_F denotes the particular coefficient K in (6) associated with a crack extension. If the crack is submitted to a pure opening mode I, $k = k_I$ and the critical value $k_c = k_{Ic}$ is the material toughness. ∎
It is clear from (8) that the reasonning fails whenever $\alpha > 1/2$. As $\eta \to 0$ (it is generally assumed that the micro-crack must grow continuously from 0 to its final length). applied loads and then k cannot be sufficiently high to pass the criterion.
Thus. as already mentioned, there is *a priori* a mismatch between this result and the experiments. although the reasonning based on an energy balance (8) seems to be unquestioned. Let us revisit the above conclusions with the additional information, provided for instance by the experiments, that there exists a critical load and thus a critical k denoted k' such that the growth criterion reads :

$$k \geq k' \tag{10}$$

Then. from (6) and (8), a critical fracture length η_s (the index s holds for "spontaneous") can be estimated :

$$\eta_s^{2\alpha-1} = \frac{2\gamma}{k'^2 K} \tag{11}$$

It is typically an unstable process such that there is no admissible position between the initial and final states.
Remark 2 : If the point O coincides with a crack tip ($\alpha = 1/2$), it comes from (8) that eqn. (11) is indeterminate. any length is admissible. As expected,

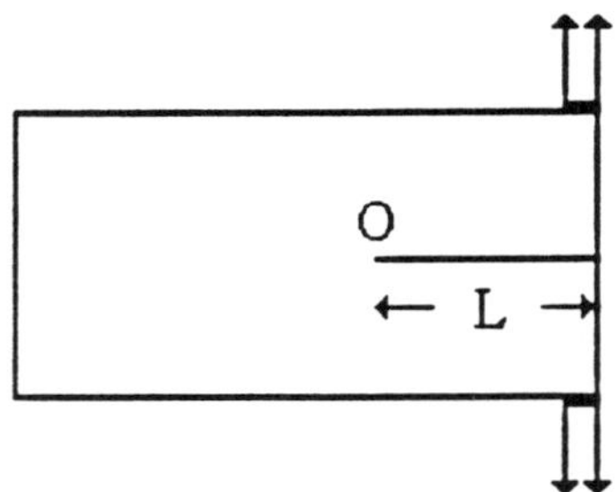

Figure 1.   The structure embedding a macro-crack

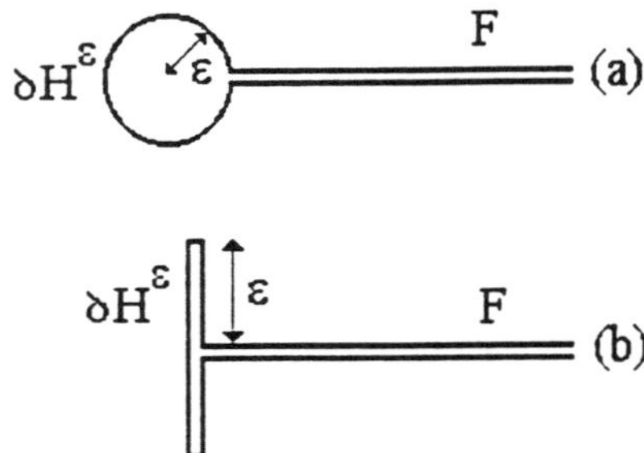

Figure 2.   Two kinds of blunting : circular hole (a), T-crack (b)

there is no characteristic fracture length associated with the classical quasi-static crack growth.   ∎

3. The blunt crack model

3.1. OUTER EXPANSION

Let us now consider two states (0) and (ε) of a structure, the first one corresponds to a homogeneous body embedding a crack with length L (fig. 1), the singular point O is located at the tip of this crack ($\alpha = 1/2$). The next state(ε) is a perturbation of the initial one, the crack tip is slightly blunt. Two examples are examined, a circular hole with radius $\varepsilon \ll 1$ centered at O (fig. 2a), two colateral branches with length ε, orthogonal at O to the initial crack (fig. 2b). Although they are two cases of blunting, the former will be denoted blunting and the later T-crack (Benveniste *et al.* 1989, Dollar *et al.* 1992). In the following, these two perturbations will be also denoted by the generic term of void. For the sake of simplicity, it is assumed that the prescribed displacements trigger a symmetric strain and a mode I crack opening.

Two displacements fields $\underline{U}^0$ and $\underline{U}^\varepsilon$ correspond to these states. $\underline{U}^0$, solution to the unperturbed problem (*i.e.* to a classical crack problem), fulfils the usual equilibrium and constitutive equations plus the boundary conditions :

$$\underline{U}^0 = \underline{\bar{U}} \quad \text{on } \Gamma \tag{12}$$

$$\sigma_{ij}(\underline{U}^0)n_j = 0 \quad \text{elswhere on } \partial\Omega^0 \text{ (including the crack lips } F) \tag{13}$$

$\underline{n} = \{n_j\}$ is the unit outer normal to Ω^0. Displacements $\underline{\bar{U}}$ are prescribed instead of forces in order to have a stable macroscopic crack growth.

$\underline{U}^0$ can be expanded in the vicinity of the crack tip in terms of the well known Williams series :

$$\underline{U}^0(x) = \underline{U}^0(0) + k_I\sqrt{r}\ \underline{u}_I^+(\theta) + ... \tag{14}$$

where $\sqrt{r}\ \underline{u}_I^+$ is the crack opening mode I and k_I the corresponding stress intensity factor.

$\underline{U}^\varepsilon$ is solution to a similar problem, but defined on the perturbed domain Ω^ε and including a condition along the new void boundary ∂H^ε, (13) is replaced by :

$$\sigma_{ij}(\underline{U}^\varepsilon)n_j = 0 \quad \text{elswhere on } \partial\Omega^\varepsilon \text{(including the crack lips } F \text{ and } \partial H^\varepsilon) \tag{15}$$

A description of $\underline{U}^\varepsilon$ can be obtained using two developments (Van Dyke 1964, Il'in 1992). First, an outer expansion in which each term is defined on the unperturbed domain Ω^0. Its specific form is a consequence of the matching conditions (L. 1993a) :

$$\underline{U}^\varepsilon(x) = \underline{U}^0(x) + \varepsilon\ k_I K\left(\frac{1}{\sqrt{r}}\underline{u}_I^-(\theta) + \underline{\hat{U}}^1(x)\right) + ... \tag{16}$$

$1/\sqrt{r}\ \underline{u}_I^-$ is the dual mode to the opening one $\sqrt{r}\ \underline{u}_I^+$.

The complementary term $\underline{\hat{U}}^1$ is solution to a well-posed problem with the specific boundary conditions :

$$\sigma_{ij}(\underline{\hat{U}}^1)n_j = -\sigma_{ij}(1/\sqrt{r}\ \underline{u}_I^-)n_j \quad \text{on } \partial\Omega^0 \text{ except } F \tag{17}$$

$$\sigma_{ij}(\underline{\hat{U}}^1)n_j = 0 \quad \text{on } F \tag{18}$$

In fact, (18) is included in (17) but it is important to single it out.

3.2. INNER EXPANSION

K in (16) is a coefficient derived from the second expansion, the inner one. It is obtained after a "stretching" of the space variables :

$$y = \frac{x}{\varepsilon} \; ; \; \rho = \frac{r}{\varepsilon} \tag{19}$$

As $\varepsilon \to 0$, after this change of variable, Ω^ε tends to an unbounded domain Ω^{in} with a semi-infinite crack and a circular hole with radius 1 or two unit colateral branches at its tip (fig. 2).

Matching conditions at infinity allow to define the inner expansion in the following form :

$$\underline{U}^\varepsilon(\varepsilon y) = \underline{U}^0(0) + \sqrt{\varepsilon} \, k_I \left(\sqrt{\rho} \, \underline{u}_I^+(\theta) + \underline{\hat{V}}^1(y) \right) + ... \tag{20}$$

$\underline{\hat{V}}^1$ is solution to a well-posed problem in Ω^{in}, with the usual equilibrium and constitutive equations (expressed with respect to the space variable y) plus boundary conditions and conditions at infinity :

$$\sigma_{ij}(\underline{\hat{V}}^1)n_j = -\sigma_{ij}(\sqrt{\rho} \, \underline{u}_I^+)n_j \;\; \text{on} \; \partial H \tag{21}$$

$$\sigma_{ij}(\underline{\hat{V}}^1)n_j = 0 \;\; \text{on} \; F \tag{22}$$

$$\underline{\hat{V}}^1 \to 0 \;\; as \; \rho \to \infty \tag{23}$$

∂H is the stretched void boundary. K in (6) and (16) is the stress intensity factor of the dual mode $1/\sqrt{\rho} \, \underline{u}_I^-$ in this problem. It expresses the decay of $\underline{\hat{V}}^1$ at infinity. Moreover, it is recalled that K_F is the particular value of K obtained when the perturbation under consideration is a straight crack extension.

Remark 3 : Eqn. (20) allows to recover a well known result by Inglis concerning the traction acting at the front face of a blunt crack. Using in (20) the derivation rules deduced from the change of variable (19) leads to conclude that this traction behaves like $1/\sqrt{\varepsilon}$ (*i.e.* like the inverse of the square root of the curvature radius of the blunting in the first case).   ∎

Remark 4 : As well as $\underline{U}^0$, the next term $\underline{\hat{U}}^1$ of the outer expansion (16) (and the following ones) can be expanded in a Williams series :

$$\underline{\hat{U}}^1(x) = \underline{\hat{U}}^1(0) + k_{I1}\sqrt{r} \, \underline{u}_I^+(\theta) + ... \tag{24}$$

Then, from (16), one can derive an expansion of the stress intensity factor of the opening mode I :

$$k_I^\varepsilon = k_I(1 + \varepsilon \, K k_{I1} + ...) \tag{25}$$

In a way, this is a fictitious factor since, taking into account the exact geometry at the crack tip, a true stress intensity factor no longer exists for the original problem $\underline{U}^\varepsilon$.   ∎

3.3. STRESS INTENSITY FACTORS COMPUTATION

The computation of the various intensity factors is based on contour integrals (L. *et al.* 1987, L. 1993a). K is independent of the applied loads and of the geometry of the structure Ω^0, it is a universal term (Leblond 1989) computed on a contour embedded in the inner domain Ω^{in} :

$$K = \frac{\psi(\hat{\underline{V}}^1, \sqrt{\rho}\, \underline{u}_I^+)}{\psi(1/\sqrt{\rho}\, \underline{u}_I^-, \sqrt{\rho}\, \underline{u}_I^+)} \tag{26}$$

For any fields $\underline{U}$ and $\underline{V}$ satisfying the equilibrium equation in a wedge and stress-free boundary conditions on the edges, ψ is a contour independent integral defined by :

$$\psi(\underline{U}, \underline{V}) = 1/2 \int_C [\sigma(\underline{U})\underline{N}\,\underline{V} - \sigma(\underline{V})\underline{N}\,\underline{U}]\, ds \tag{27}$$

C is any contour surrounding the singular point and starting and finishing at the stress-free edges. $\underline{N}$ is the unit normal to C pointing toward O.
k_{I1} is computed in the same way on a contour embedded in the outer domain Ω^0. However, it depends on the geometry of the structure and is not universal :

$$k_{I1} = \frac{\psi(\hat{\underline{U}}^1, 1/\sqrt{r}\, \underline{u}_I^-)}{\psi(\sqrt{r}\, \underline{u}_I^+, 1/\sqrt{r}\, \underline{u}_I^-)} \tag{28}$$

Remark 5 : These computations need to be performed with care, we emphasize mainly on the signs : K and K_F are positive while k_{I1} computed on the structure illustrated on fig. 1 is negative. Thus, as expected, in case of blunting, the fictitious stress intensity factor k_I^ε (25) smaller than the same coefficient estimated at a sharp crack tip.   ∎

4. Analysis of the crack extension

The third state to consider now, named (η), exhibits in addition to the previous one (ε), the starting of a new crack with relative length $\eta \ll 1$ in the front face of the blunting (a right notation for this state would be $(\varepsilon\eta)$ since both perturbations are present, it is avoided for simplicity). In case of a T-crack, it is assumed that the colateral cracks growth is firmly inhibited

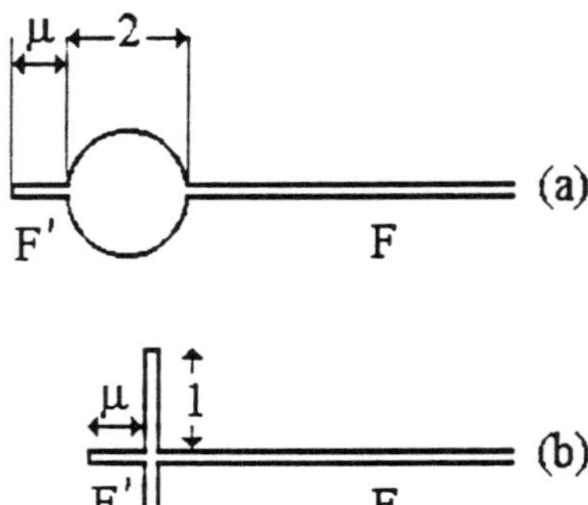

Figure 3.   The crack onset at blunting

by inclusions for instance. Otherwise, propagation will likely occurs at one or both tips of the T giving rise to a problem of crack branching (Leblond 1989, Amestoy *et al.* 1991, L. 1993a). There is no special assumption concerning the respective magnitudes of the initial defect (hole or colateral branches) and the micro-crack length, η can be as large or smaller than ε. To these three states correspond three potential energies W^0, W^ε and W^η (see (4) and (5)). The displacement field $\underline{U}^\eta$ can be described by two expansions :

$$\underline{U}^\eta(x) = \underline{U}^0(x) + \varepsilon\, k_I K(\eta) \left(\frac{1}{\sqrt{r}} \underline{u}_I^-(\theta) + \hat{\underline{U}}^1(x) \right) + \ldots \tag{29}$$

$$\underline{U}^\eta(\varepsilon y) = \underline{U}^0(0) + \sqrt{\varepsilon}\, k_I \left(\sqrt{\rho}\, \underline{u}_I^+(\theta) + \hat{\underline{V}}^{1\eta}(y) \right) + \ldots \tag{30}$$

In expansion (29) the coefficient $K(\eta)$ appears in replacement of the already known term K. Obviously, it depends on the new geometry and thus, on the restarting crack length η, moreover $K(0) = K$. (30) is obtained using as before the change of variable $y = x/\varepsilon$, hence, the resulting inner domain is identical to the previous one up to the perturbation shape (fig. 3). $K(\eta)$ has the same meaning than K defined by (16) (*i.e.* the stress intensity factor of $1/\sqrt{\rho}\, \underline{u}_I^-$ in the inner problem), but defined on a domain Ω^{in} depending now on the restarting crack length. In (30) $\hat{\underline{V}}^{1\eta}$ is solution to a well-posed problem in which condition (21) has been replaced by :

$$\sigma_{ij}(\hat{\underline{V}}^{1\eta})n_j = -\sigma_{ij}(\sqrt{\rho}\, \underline{u}_I^+)n_j \quad \text{on } \partial H \text{ and } F' \tag{31}$$

where F' denotes the new crack lips. We are now interested in evaluating the potential energy change $\delta W^{\varepsilon\eta}$ between states (ε) and (η) in order to predict the mechanism of restarting in case of blunting :

$$\delta W^{\varepsilon\eta} = W^\varepsilon - W^\eta \simeq \varepsilon\, k_I^2(K(\eta) - K) \tag{32}$$

The Griffith criterion (8) reads :

$$k_I^2(K(\eta) - K) \geq 2\gamma\,\eta/\varepsilon \tag{33}$$

Setting $\mu = \eta/\varepsilon$, it can be pointed out that the function $K(\eta)$ depends only on μ (it is the "stretched" restarting crack length). This function will still be denoted $K(\mu)$. Using the Irwin criterion and the toughness definition (9), it comes finally :

$$k_I^2(K(\mu) - K) \geq k_{Ic}^2 K_F \mu \tag{34}$$

The next section will focus on solutions to this inequality.

5. Estimation of a spontaneous crack extension length

We are now interested in the smaller value of μ such that (34) holds true. Obviously, at restart, the intensity factor k_I fulfils the inequality $k_I > k_{Ic}$ (it is suggested in remark 4). The existence of non trivial solutions to (34) is a consequence of the following remarks.

Remark 6 : There is an obvious solution $\mu = 0$ but it is not admissible. ∎

Remark 7 : For $\mu \ll 1$ ($\eta \ll \varepsilon$) $\hat{V}^{1\eta}$ in (30) can be expanded with respect to μ as :

$$\hat{\underline{V}}^{1\eta}(y) = \hat{\underline{V}}^{10}(y) + \mu^2 A\,\hat{\underline{V}}^{11}(y) + \dots \tag{35}$$

where A is a constant. The characterisitic exponent to take into account is $\alpha = 1$, it corresponds to a point of a straight boundary in a homogeneous body. It is clear in the example of the T-crack, in the other one the apparent curvature of the hole vanishes as $\mu \to 0$. Eqn. (35) implies :

$$\frac{K(\mu) - K}{\mu} = O(\mu) \to 0 \ \ \text{as } \mu \to 0 \tag{36}$$

the curve $K(\mu)$ has a horizontal tangent at $\mu = 0$. Hence. k_I being finite, there is no solution to (34) such that μ is infinitely small. *The restart occurs by a sudden crack growth, the extension length is at least of the same order than ε.* ∎

It is to be pointed out that this result does not hold in case of a T-crack if the restart takes place at one tip of the colateral branches, since the exponent to take into account in that case would be $\alpha = 1/2$.

Remark 8 : On the other hand. if μ becomes large ($\varepsilon \ll \eta \ll 1$) the cavity radius can be neglected at the first order and outer and inner expansions can be performed with respect to the small parameter η (instead of ε) using the change of variables :

$$y' = \frac{x}{\eta} \ \ ; \ \ \rho' = \frac{r}{\eta} \tag{37}$$

The main difference lies in the roles exchange. Coefficients depending on η (or more precisely on μ) are now constant while constant coefficients become functions of ε (or more precisely of $1/\mu$ and thus μ) :

$$\underline{U}^{\eta}(x) = \underline{U}^{0}(x) + \eta k_I \kappa(\mu) \left(\frac{1}{\sqrt{r}} \, \underline{u}_I^- + \hat{\underline{U}}^1(x) \right) + \ldots \tag{38}$$

$$\underline{U}^{\eta}(\eta y') = \underline{U}^{0}(0) + \sqrt{\eta} \, k_I \left(\sqrt{\rho'} \, \underline{u}_I^+ + \tilde{\underline{V}}^{1\varepsilon}(y') \right) + \ldots \tag{39}$$

In this last expression, $\tilde{\underline{V}}^{1\varepsilon}$ is solution to a problem settled on an inner domain, with a unit crack extension and slightly perturbated by a circular cavity (or two colateral cracks) with small radius (or length) $\varepsilon/\eta = 1/\mu$. Thus it can be expanded with respect to this parameter :

$$\tilde{\underline{V}}^{1\varepsilon}(y') = \tilde{\underline{V}}^{10}(y') + (1/\mu)^2 B \, \tilde{\underline{V}}^{11}(y') + \ldots \tag{40}$$

where B is a constant. Power two of the small parameter occurs in (40) for the same previous reason, the small hole or the colateral cracks form a perturbation of a straight line (the crack lips). With the help of (40), the stress intensity factors $\kappa(\mu)$ of $1/\sqrt{\rho'} \, \underline{u}_I^-$ in (39) is split into :

$$\kappa(\mu) = K_F + (1/\mu)^2 B K_{11} + \ldots \tag{41}$$

$\tilde{\underline{V}}^{10}$ in (39) corresponds to $\varepsilon = 0$ (unperturbed but with a unit crack extension) and thus undergoes the stress intensity factor K_F. K_{11} is associated with the second term of expansion (41). Taking into account the two outer expansions (16) and (38), it leads to the asymptotic expressions as $\mu \to \infty$:

$$K(\mu) = \mu \kappa(\mu) = \mu K_F + (1/\mu) B K_{11} + \ldots \tag{42}$$

$$\lim_{\mu \to \infty} \frac{K(\mu) - K}{\mu} = K_F \tag{43}$$

Thus, (34) *has always solutions for non zero ε provided $k_I > k_{Ic}$.* ∎
Coefficients K, K_F and function $K(\mu)$ have been computed using (26). Results are plotted on fig. 4 and 5.

To conclude let us consider now that the load must be increase by $\lambda \varepsilon^{\gamma}$ ($\lambda > 0$, $\gamma > 0$) to trigger a crack extension in case of blunting :

$$k_I' = k_{Ic}(1 + \lambda \varepsilon^{\gamma}) \tag{44}$$

From (34) and (44) it comes after some manipulations :

$$\frac{K(\mu) - K}{K_F \mu} \geq 1 - 2\lambda \varepsilon^{\gamma} \tag{45}$$

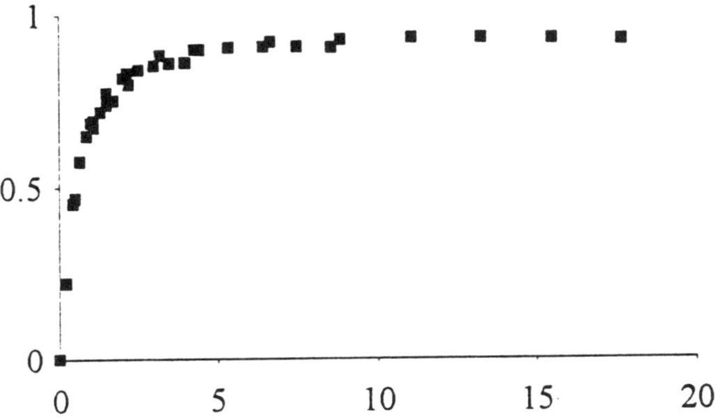

Figure 4. $(K(\mu) - K)/(K_F\mu)$ vs. μ for the blunt crack

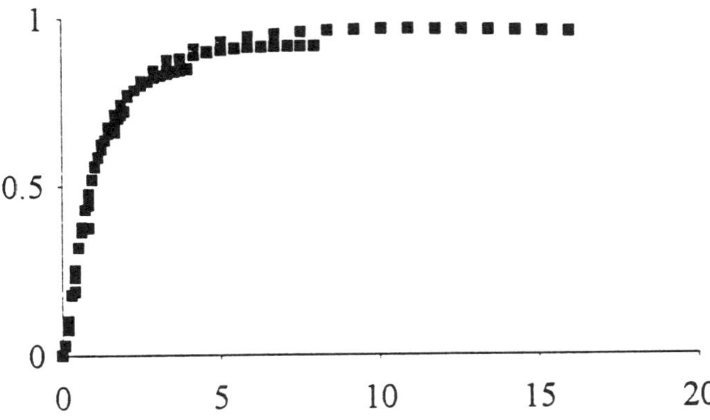

Figure 5. $(K(\mu) - K)/(K_F\mu)$ vs. μ for the T-crack

Obviously, as already mentioned in remark 7, according to (36) there is no solution if $\mu \ll 1$. Let us now consider $\mu \gg 1$, then (42) and (43) lead to :

$$\eta \geq \frac{1}{2\lambda} \frac{K}{K_F} \varepsilon^{1-\gamma} \tag{46}$$

Consistency with the assumptions implies $\gamma < 1$. Moreover, as suggested by Inglis result (remark 3), $\gamma = 1/2$ is a good candidate in (44) and then the magnitude of the extension length can be evaluate :

$$\eta_s = O(\sqrt{\varepsilon}) \tag{47}$$

References

Atkinson C. (1989) Cracks in bimaterial interface - an overview, *Proc. of the ICF7 - Advances in Fracture Research*, University of Houston, pp. 3053-3061.

Amestoy M., Leblond J.B. (1991) Crack paths in plane situation - II. Detailed for of the expansion of the stress intensity factors, *Int. J. Solids Structures*, **29**, 4, pp. 465-501.

Aveston J., Kelly A. (1973) Theory of multiple fracture of fibrous composites, *J. of Mat. Sci.*, **8**, pp. 352-362.

Benveniste Y., Dvorak G.J., Zarzour J., Wung E.C.J. (1989) On interacting cracks and complex crack configurations in linear elastic media, *Int. J. Solids Structures*, **25**, 11, pp. 1279-1293.

Dollar A., Steif P.S. (1992) Interface Blunting of Matrix Cracks in Fiber-Reinforced Ceramics, *J. of Appl. Mech.*, **59**, pp. 796-803.

Erdogan F., Biricicoglu V. (1973) Two bonded half planes with a crack going through the interface, *Int. J. Eng. Sci.*, 11, pp. 745-766.

Hashin Z. 1996 Finite thermoelastic fracture criterion with application to laminate cracking analysis, *J. Mech. Phys. Solids*, **44**, 7, pp. 1129-1145.

He M.Y., Hutchinson J.W. (1989) Kinking of a crack out of an interface, *J. of Appl. Mech.*, **56**, pp. 270-278.

Il'in A.M. (1992) *Matching of Asymptotics Expansions of Solutions of Boundary Value Problems*, Translations of Mathematical Monographs, 102, American Mathematical Society.

Kelly A. (1970) Interface effects and the work of fracture of a fibrous composite, *Proc. Roy. Soc. Lond.*, **A.319**, pp. 95-116.

Kipp M.E., Sih G.C. (1975) The strain energy density failure criterion applied to notched elastic solids, *Int. J. Solids Structures*, 11, pp. 153-173.

Leblond J.B. (1989) Crack paths in plae situation - I. General form of the expansion of the stress intensity factors, *Int. J. Solids Structures*, **25**, 1311-1325.

Leguillon D. (1989) Calcul du taux de restitution de l'énergie au voisinage d'une singularité, *C.R. Acad. Sci. Paris*, **309**, série II, pp. 155-160.

Leguillon D. (1993a) Asymptotic and numerical analysis of a crack branching in non-isotropic materials, *Eur. J. Mech., A/Solids*, **12**, 1, pp. 33-51.

Leguillon D. (1993b) Analysis of the brittle fracture in composites using singularities, in Mecamat 93, *International seminar on micromechanics of materials*, Eyrolles, Paris, 1993, pp. 60-71.

Leguillon D., Sanchez-Palencia E. (1987) *Computation of singular solutions in elliptic problems ans elasticity*, Masson, Paris, J. Wiley, New-York.

Leguillon D., Sanchez-Palencia E. (1992) Fracture in heterogeneous materials - Weak and strong singularities, *New Advances in Computational Structural Mechanics*, P. Ladevèze and O.C. Zienkiewicz ed., Studies in Applied Math, 32, Elsevier, Amsterdam, 1992, pp. 423-434.

Marigo J.J. (1996) Etude théorique et numérique d'une nouvelle formulation de la mécanique de la rupture fragile, rapport EDF/LPMTM n°174/ 1H740/ IMA232/ 174L17.

Maz'ya V.G., Nazarov S.A. (1988) The asymptotic behavior of energy integrals under small perturbations of the boundary near corner points and conical points, *Trans. Moscow Math. Soc.*, pp. 77-127.

Miller G.R., Stock W.L. (1989) Analysis of branched interface cracks between dissimilar anisotropic media, *J. Appl. Mech.*, **56**, pp. 844-849.

Rice J.R. (1988) Elastic fracture mechanics concepts for interfacial cracks, *J. Appl. Mech.*, **55**, pp. 98-103.

Van Dyke M. (1964) *Asymptotic methods in fluid mechanics*, Academic-Press, New-York.

EXPERIMENTAL INVESTIGATION OF DYNAMIC FAILURE MODE TRANSITIONS

D. RITTEL

Faculty of Mechanical Engineering
Technion
32000 Haifa, Israel

1. Introduction

The response of a crack subjected to quasi-static loading has been extensively investigated both theoretically and experimentally (see *e.g.* Kanninen and Popelar, 1985). The nature of the various (stress or strain) singularities has been identified for linear elastic (elastic singularities) as well as nonlinear materials. Emphasis has been put on stationary or steady state propagating cracks.

When dynamic loading is applied the theoretical problem becomes more complicated since inertia must be accounted for (Freund, 1990). Experiments aimed at identifying crack response under transient loading must involve not only a high spatial accuracy but also a high temporal resolution typically of the order of the microsecond (Klepaczko, 1985).

Even the "simple" case of the *stationary* crack in a *linear-elastic* material subjected to transient loading may prove to be quite complex to analyze. As an example, Bui and Maigre (1988) proposed a path-independent integral based on convolution between reference and the actual fields in the cracked solid. This approach was implemented in specially devised experiments by Rittel *et al.*(1992), and by Bui *et al.* (1992). These and later experiments provided a solid framework to investigate stationary crack behavior in linear elastic materials subjected to transient loading.

Ruling out for the moment cases where large scale material non-linearities develop (gross plastic yielding) it is interesting to extend the present framework to cases where the non-linearities remain confined, by analogy with the small scale yielding context of linear elastic fracture mechanics.

In this paper, we will report recent results on the transition of dynamic failure modes, such as reported by Kalthoff (1988) for maraging steel, Ravi-Chandar (1995) for polycarbonate and Zhou *et al.* (1996) for maraging steel and a titanium alloy. The basic idea is that when cracks (or notches) are loaded such as to generate dominant mode II, the failure mode may vary with the impact velocity. Typically, the initial kink angle formed by the crack as it starts to propagate varies from 70° to 40° for "low" impact velocities whereas it reaches almost 0° at" high" impact velocities. These authors

D. Durban and J.R.A. Pearson (eds.),
IUTAM Symposium on Non-Linear Singularities in Deformation and Flow, 181-192.

describe the high angle values as corresponding to local opening mode. At high velocities, shear dominated failure occurs with the development of adiabatic shear bands. Such results are quite interesting as they disclose a new kind of material response related to the development of a crack-tip material non-linearity.

2. Theoretical and experimental framework

The underlying framework is that of small scale yielding linear elastic fracture mechanics. The assumptions allow for limited non-linear material response, as previously mentioned. The crack is *stationary* and subjected to transient loading. The crack-tip stress fields, until the onset of propagation, are conveniently described by the stress intensity factors (SIF). The analysis is two-dimensional.

The experimental setup consists of an instrumented bar which is brought in contact with the cracked specimen (figure 1). A striker is fired against the bar using an air gun. The impact velocities are in the range of 8 m/s to 60 m/s. Upon impact a compressive stress wave progresses toward the sample and gets partly reflected partly transmitted to the specimen. The incident and reflected signals are collected on a high speed digital oscilloscope to be processed into the interfacial force applied to the specimen (Rittel and Maigre, 1993). The impacted specimen rests unsupported so that fracture is due to inertia only. Two specimen geometries are used which enforce in each case mixed-mode loading. The compact compression specimen (CCS) (Rittel *et al.*, 1992) is used for dominant mode I (minor mode II) loading experiments. The cracked plate (side impact) is used to enforce dominant mode II loading with a minor mode I component (Lee and Freund, 1990). In both cases the interfacial force pulse is used as boundary conditions for a finite element model of the specimen. This hybrid experimental-numerical approach (Kobayashi, 1987) is quite convenient to assist this kind of experiments as an alternative to direct crack-tip monitoring by means of optical methods (see *e.g.*, Mason *et al.*, 1992). The calculated crack opening displacement is converted into the stress intensity factors. The onset of crack propagation can be detected using single wire fracture gages located ahead of the crack-tip.

In our experiments, we *systematically compare* mixed-mode experiments in which mode I is *either* the dominant *or* the minor mode. Furthermore, all specimens contain fatigue precracks to assess the influence of friction between the crack flanks with respect to notches where closure can develop unrestrained. Consequently, the numerical model allows for Coulomb friction between the crack flanks as a first geometrical non-linearity of the problem.

Two polymeric materials are used and compared in this study: "brittle" polymethylmethacrylate (PMMA) and the more "ductile" commercial polycarbonate (PC).

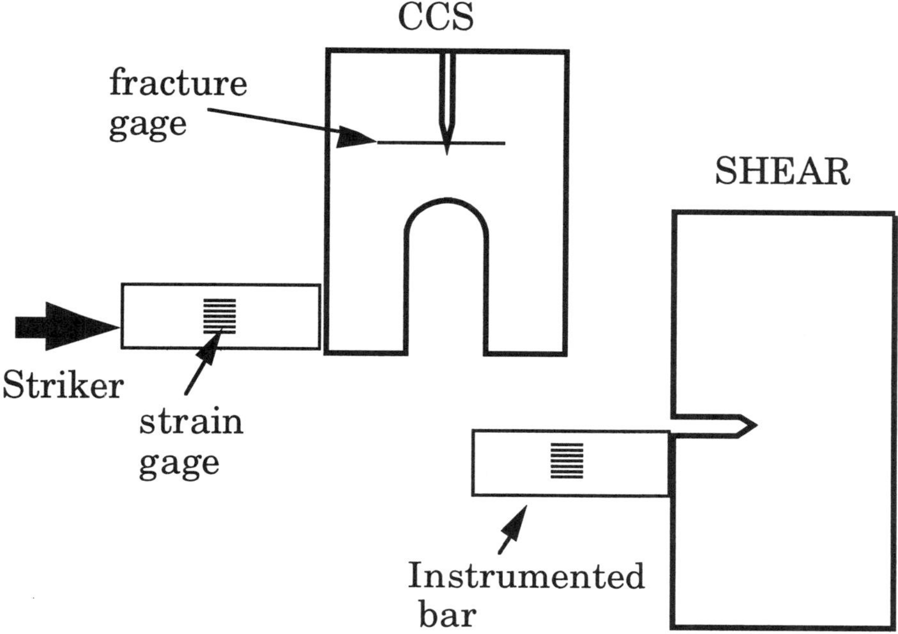

Figure 1: The experimental setup

3. Experimental results and discussion

3.1 FOREWORD

The results to be presented address two issues related to the dynamic failure mode transition.

The first is that of the relation between the failure mode transition and the mixity of the loading. The second issue relates to the nature of the thermomechanical crack-tip phenomena.

3.1.1 Failure mode transition and mode mixity – polycarbonate

The failure mode transition has been observed in dominant mode II experiments only. One may therefore wonder whether the operation of a shear failure mechanism is restricted high velocity impact loading or if it can be addressed in the more general context of mixed-mode loading. Side impact experiments carried out by Rittel *et al.* (1997) confirmed previous reports of the failure mode transition in polycarbonate (Ravi-Chandar, 1995). At impact velocities below 20 m/s, shear loaded cracks proceeded by forming a kink angle of typically 40°. When the impact velocity was noticeably increased, the cracks did not propagate. Rather, we observed a damage zone extending along the original crack direction over some 2 mm. This "flame-like" region is similar to an adiabatic shear band (figure 2).

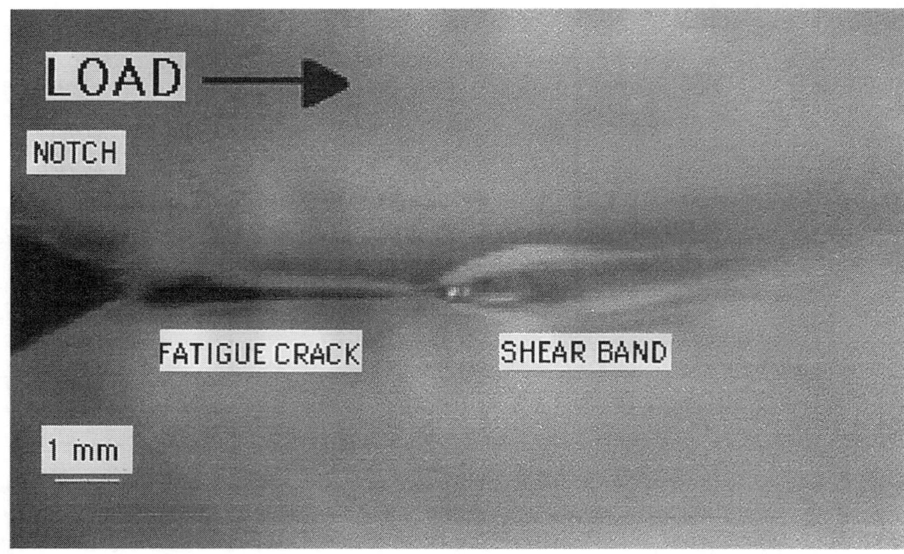

<u>Figure 2</u>: Adiabatic shear band ahead to the fatigue crack-tip.

The experiments also showed that fatigue cracks and notches reacted the same way with respect to the impact velocity. The numerical simulation of the experiments showed that as long as the crack remains stationary, the main effect of contact between the crack flanks is to impede interpenetration (negative mode I) which occurs for notches. Friction in itself restricts the relative motion of the crack faces, thus reducing the intensity of the mode II component.

A detailed characterization of the microfractographic features using light and scanning electron microscopy was carried out to identify the fracture micromechanisms which operate in each range of velocities (Rittel and Levin, 1997). At low impact velocity the fracture surface topography is relatively featureless at low magnifications. At higher magnifications, periodic craze markings indicate the operation of an *opening* fracture mechanism. Filament-like features are observed at the origin of the fracture. By contrast, the fractography of high impact velocity specimens reveals a wealth of details, among which the typical initial shear band in which elongated dimples are noticeable at higher magnifications. These features indicate the operation of a *shear* fracture mechanism by contrast with the crazes observed further away along the fracture path.

Considering now the CCS specimens (dominant mode I), a great similarity of fracture mechanisms could be observed with respect to the impact velocity. Low velocity impact causes the reported featureless and filament-like fracture while high velocity impact yields the characteristic shear band followed by crazing markings.

In fact the resemblance in failure mechanisms does not allow to distinguish a CCS from a side impacted plate, both impacted in the same range of velocities, on the sole evidence of fractographic analysis (Rittel and Levin, 1997).

These observations indicate that the operation and competition between a shear and normal fracture mechanism is not limited to dominant mode II experiments. Rather, our results show that the selection of a failure mechanism depends on the mixity and more specifically on the relative amount of the mode II component. While a CCS cannot be expected to fail by shear as the crack will ultimately open, the presence of a minor – yet non negligible – shear component will cause initial ductile shear failure.

The initial kink angle value of typically 40° can now be explained using LEFM concepts applied to a zone which encompasses the nonlinear shear fracture zone. Typically, we assume that (kink angled) fracture with its characteristic pattern of crazing markings obeys a maximum normal ($\sigma_{\Theta\Theta}$) stress criterion. Consequently, following determination of $K_I(t)$ and $K_{II}(t)$, $\sigma_{\Theta\Theta}(t)$ is calculated along with its the maximum value and the corresponding angle. Figure 3 illustrates the approach and one can note that a 40° kink angle corresponds to mixed mode loading. Higher angular values indicate opening failure during the mode II loading phase.

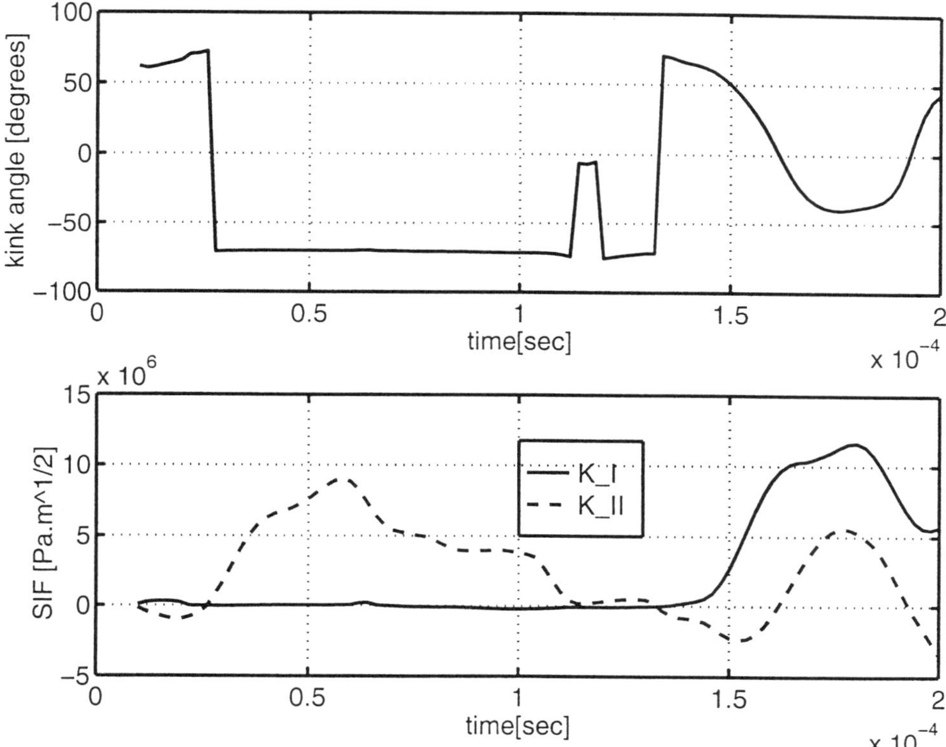

Figure 3: Calculated kink angle as a function of time for mixed-mode loading. Actual results of a side impact experiment. Note that initial part of the loading occurs in quasi-pure mode II followed by mixed-mode.

It must be mentioned at this stage, that all the room temperature side impact experiments carried out with PMMA samples caused fracture to propagate at high angular values (typically 60°) and more regardless of the impact velocity. We observed no failure mode transition in this material.

Therefore, the hybrid experimental-numerical approach can be successfully used to investigate the response of a crack to mixed mode dynamic loading where crack-tip material non-linearities can develop by analogy with the small scale yielding assumptions. In the next section we investigate further the nature of this material non-linearity.

3.1.2 On the role of temperature at the tip of a dynamically loaded crack

The observation of adiabatic shear band at high impact velocities indicates that significant temperatures are attained which cause material softening. For polymers, a characteristic softening temperature is the so-called "glassy temperature" (Tg). For the investigated polymers Tg(PC)$\approx$150°C while Tg(PMMA)$\approx$100°C.

Direct measurements of the temperature fields at the tip of a crack are very delicate to perform and require sophisticated equipment (see *e.g.* Zhender and Rosakis, 1991). However, the fractographic knowledge presented in the previous section can be used to this purpose as shown next.

Keeping in mind that high velocity mode II impact causes shear banding in PC, a series of such experiments was carried out on specimens which were globally cooled in the range of temperatures -120°C to +70°C. The initial kink angle was observed to decrease from $\alpha\approx$60° at T=-40°C (and below) to about $\alpha\approx$40° at T=-25°C. At higher temperatures the kink angle reached α=0° in accord with previous experiments. Fractographic observation of the sample fractured at T=-25°C clearly shows the elongated dimples characteristic of shear failure (figure 4). It is therefore assessed that the local temperature increase is at least of $\Delta T\geq$175°C. In other words the *thermally induced material non-linearity* develops at 0.59Tg.

When similar experiments are carried out on PMMA plates, it is observed that in order to modify the failure mechanism (in this case suppress crack initiation) the specimen has to be globally brought to 1.12Tg! In other words, the crack-tip material does not heat in proportions similar to PC and must be globally heated to temperatures well in excess of the glassy temperature (Rittel, 1997).

Assuming that PMMA represents the "brittle" material as opposed to PC, the "ductile" material, it is interesting to investigate the thermomechanical response of the crack-tip material, as shown next.

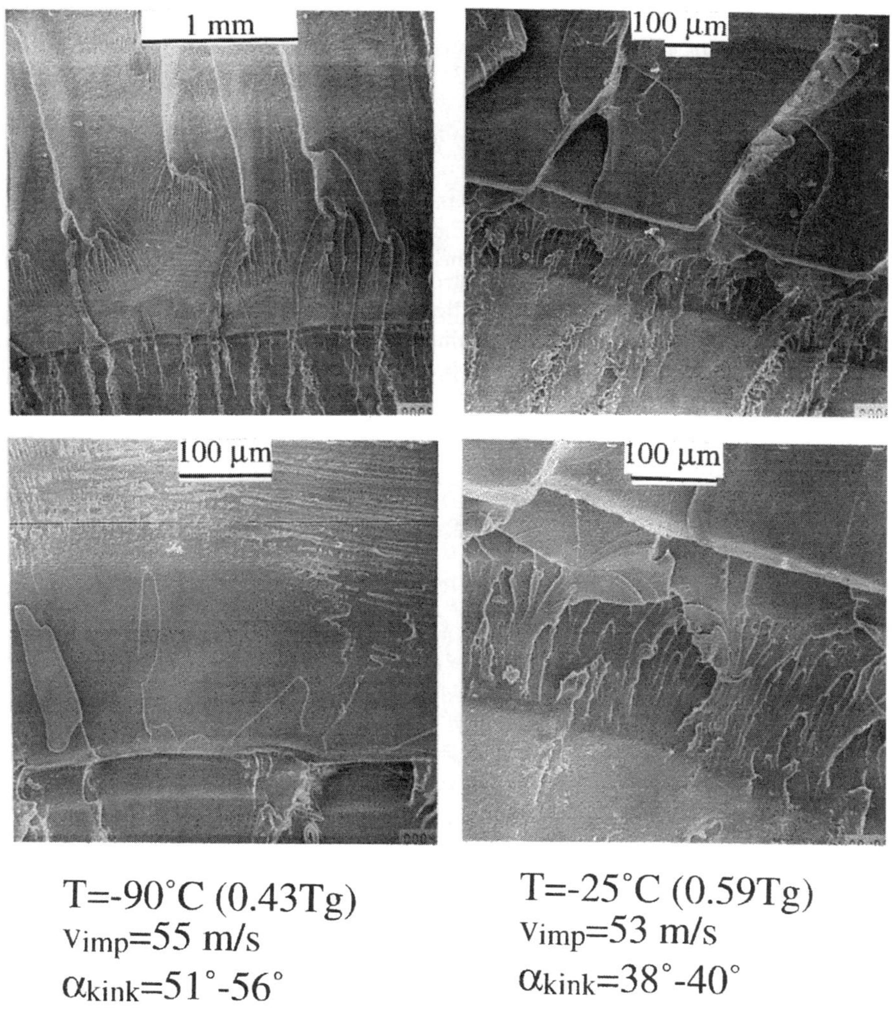

T=-90˚C (0.43Tg)
V_{imp}=55 m/s
α_{kink}=51˚-56˚

T=-25˚C (0.59Tg)
V_{imp}=53 m/s
α_{kink}=38˚-40˚

<u>Figure 4</u>:   Fracture surface topography of polycarbonate samples. High velocity mode II impact. Note the shear dimples in the specimen impacted at -25˚C, indicative of ductile failure.

3.1.3 Thermomechanical behavior of the crack-tip material

Early work by Taylor and Quinney (1934) has shown that the mechanical energy of the plastic deformation is converted essentially into heat (β= 0.6 to 0.9 depending on the material). As a consequence, most of the experimental and theoretical work has concentrated on aspects related to temperature rise and subsequent material softening. The transient heat equation is given by (Boley and Weiner, 1960):

$$kV^2T - \alpha(3\lambda + 2\mu)T_0\dot{\varepsilon}^e_{kk} + \beta\sigma_{ij}\dot{\varepsilon}^p_{kk} = \rho c\dot{T} \tag{1}$$

where k is the heat conductance and α is the thermal expansion coefficient. ρ, c, λ and μ stand for the material's density, heat capacity and Lamé constants respectively. T is the temperature and the strains ε are divided into elastic and plastic.

As mentioned earlier, the thermoelastic contribution is usually neglected and further simplification is gained by considering adiabatic conditions (k=0) (Zhender and Rosakis, 1991). The result of such approach will show the expected temperature rise. Considering now a thermoelastic solid subjected to the same adiabatic conditions. The volume change is given by:

$$\varepsilon^e_{kk} = \frac{-p}{\kappa} + 3\alpha(T - T_0) \tag{2}$$

where T_0 is some reference temperature.

The rate of volume change $\dot{\varepsilon}^e_{kk}$ can be expressed in terms of the rate of change of hydrostatic pressure $\dot{p}$ and the bulk modulus κ to yield:

$$3\alpha T_0\dot{p} = \dot{T}\left[\rho c + \frac{3\alpha^2 E}{1-2v}T_0\right] \tag{3}$$

in which the righthandside of the bracketed expression can be neglected with respect to the ρc term. Eqn. (3) can now be solved for a linear thermoelastic solid initially at room temperature and stress free, yielding:

$$T(t) = T_0(1 + \frac{3\alpha}{\rho c}p(t)) \tag{4}$$

Considering a stationary crack, the expressions for the angular stress distribution can be substituted and combined with Tresca's yield criterion to obtain an expression for the hydrostatic pressure , p, at $\Theta=0°$ (ahead of the crack). Assuming plane strain deformation (without external confinement), p is given by :

$$p = \frac{2}{3}\frac{(1+v)}{(1-2v)}\sigma_y \tag{5}$$

where σ_y is the uniaxial yield strength. For the mode I crack, eqn. (4) becomes:

$$T(t) = T_0(1 - \frac{2\alpha(1+v)}{\rho c}\frac{K_I(t)}{\sqrt{2\Pi r}}) \tag{6}$$

This expression can easily be extended to other loading modes including their combinations. Remark that if limited crack-tip plasticity is allowed (the physical situation), the position of the elastic-plastic boundary can easily be determined.

At this stage, it is important to note that the amplitude of the cooling effect bears a $r^{-1/2}$ singularity as dictated by the framework of linear elastic fracture mechanics. Eqn. (4) can be used to assess the amplitude of the crack-tip cooling while eqn. (6) describes the temperature distribution as a function of the distance from the crack-tip, as discussed next.

The factor $\dfrac{3\alpha}{\rho c}p(t)$ in eqn. 4 sets the magnitude of the cooling effect and it can be expressed as $\dfrac{2\alpha}{\rho c}\left(\dfrac{1+\nu}{1-2\nu}\right)\sigma y$. It can be noted that for a given material, α, ρ, c and ν are strain-rate insensitive. By contrast, the yield strength of most materials is known to be influenced by the strain rate. Consequently, eqn. 4 can be modified to take this point into account by specifying $\sigma_y(\dot{\varepsilon})$ (Meyers, 1994). For most materials the product ρc is of the order of 10^6 J/m$^{3\,\circ}$K (Ashby, 1992). We will consider two kinds of materials: polymers for which α is typically of the order of $10^{-4}/\,^{\circ}$K and metals (e.g. steels) for which α is ten times smaller. Keeping in mind that the ratio of the (static) yield strength of a polymer to that of steel is typically of the order of $\dfrac{\sigma y\_polymer}{\sigma y\_metal}\approx\dfrac{1}{10}$, the "amplitude" of the cooling effect will thus be dictated primarily by the ratio $\left(\dfrac{1+\nu}{1-2\nu}\right)$ along with the elastic requirements set by eqn. 5. Figure 5 shows the temperature profile (eqn. 6) as a function of the normalized distance from the elastic-plastic (dynamic) interface for a typical steel and a polymer whose properties are listed in table 1 below. To allow for comparison, the applied stress intensity factor has been (arbitrarily) scaled to the static yield strength ($\dfrac{K_I}{\sigma_{y\_static}}=0.4[m^{1/2}]$). The dynamic yield strength has been assumed to be equal to 1.5 the static yield strength.

Table 1:   Typical mechanical and thermal properties of a polymer and a metal used in estimating thermoelastic cooling shown in Fig. 5

material	static σ_y	dynamic σ_y	linear exp. α	ρc	Poisson's ν
	[MPa]	[MPa]	[1/$^{\circ}$K]	[J/m$^{3\,\circ}$K]	
PMMA	100	150	10^{-4}	10^6	0.37
STEEL	1000	1500	10^{-5}	10^6	0.30

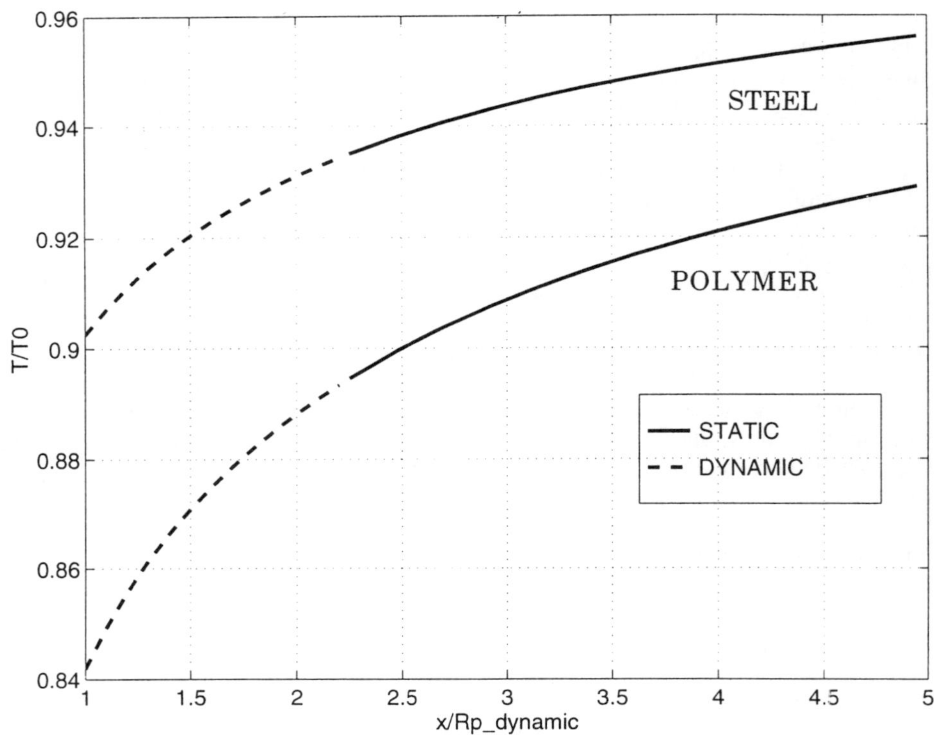

Figure 5:   Calculated thermoelastic cooling in a polymer and a steel as a function of
the normalized distance to the (dynamic) elastic-plastic interface. Solid line
calculated using static yield strength. Dashed line shows the effects of strain
rate sensitivity.

The results plotted in figure 5 first illustrate the previously addressed dependence of
the thermoelastic cooling on Poisson's ratio. Therefore, two widely differing materials
such as a metal and a polymer exhibit significant crack-tip cooling. Also, the reduced
dynamic plastic zone size allows for increased cooling closer to the crack-tip.
As an example, starting at room temperature (20°C), this figure shows that at the
elastic-plastic interface the temperature will drop to -27°C for the polymer and to about
-9°C for the metal. While these figures are *only indicative*, they nevertheless show that
thermoelastic effects must certainly be accounted for in the crack-tip vicinity when
transient loading is applied.

3.1.4 Linear versus non-linear material behavior

The above model and results show that thermomechanical crack-tip phenomena must certainly be taken into account in dynamic fracture mechanics investigations. Specifically, the "elastic" material is expected to adiabatically cool in the vicinity of the (mode I) crack-tip so that it will not soften unless plasticity develops in this area. By contrast, the "plastic" material will soften resulting in pronounced non-linear behavior.

Assuming now that PC can be treated as the "plastic" material by contrast with the "elastic" PMMA, one can tentatively propose that the global heating of PMMA is necessary to counteract local cooling below the transition temperature. This point needs further experimental evidence. It also seems that the observations of a transition from the linear to non-linear crack-tip behavior is probably related to the *ductile to brittle transition* and also perhaps to the observed trend (in polymers) for an *increase in the dynamic fracture toughness* with the stress intensity rate (Rittel and Maigre, 1996).

Finally, it must be noted that the thermal sequence consisting of crack-tip cooling followed by temperature rising has been actually observed by, *e.g.*, Fuller *et al.* (1975). These authors used sensitive thermal detectors which recorded the thermal signals related to steady state crack propagation (quasi-static loading rates). The adiabatic elastic precursor cooling phase is very clearly shown but is only very briefly addressed as the central point of their research was crack-tip heating.

4. Conclusions

- We have presented in this paper some results related to the transition of linear to non-linear behavior for the crack-tip material subjected to transient loading.
- The failure mode transition, i.e. the operation of a shear failure mechanism has been analyzed in terms of mixed-mode fracture in the framework of small scale yielding linear elastic fracture mechanics. Supported by scanning electron fractography, this result generalizes previous observations to arbitrary loading conditions.
- The transition fromlinear to nonlinear material behavior has been related to local crack-tip temperature elevation as a result of localized crack-tip adiabatic plastic deformations.
- This point has been further extended to include the thermoelastic response of the material. Whereas this response has been observed, it is generally neglected. Our results show that under transient conditions, thermoelastic phenomena must also be accounted for when examining the global crack-tip response.

Acknowledgment

This research is supported by the Israel Science Foundation (grant 030-039) and the Technion VPR Fund for the Promotion of Research.

5. References

Ashby, M.F. (1992) *Materials selection in mechanical design*, Pergamon Press, Oxford, England.

Boley, B.A. and Weiner, J.H. (1960), *Theory of thermal stresses*, J. Wiley and Sons, New York, NY.

Bui, H.D. and Maigre, H. (1988) Facteur d'intensité dynamique des contraintes tiré des grandeurs mécaniques globales, *C. R. Acad. Sc. Paris*. t. 306, Série II, 1213-1216.

Bui, H.D., Maigre, H. and Rittel, D. (1992) A new approach to the experimental determination of the dynamic stress intensity factor, *Int. J. Solids Structures*, **29**, No. 23, 2881-2895.

Freund, L.B. (1990) *Dynamic Fracture Mechanics*, Cambridge University Press, Cambridge.

Fuller, K.N.G., Fox, P.G. and Field, J.E. (1975) The temperature rise at the tip of a fast-moving crack in glassy polymers, *Proc. R. Soc.* , **A 341**, 537-557.

Kalthoff, J.F. (1988), Shadow optical analysis of dynamic fracture, *Optical Engng*, **27**, 835-840.

Kanninen, M.F. and Popelar, C.H. (1985) *Advanced Fracture Mechanics*, Oxford University Press, Oxford.

Klepaczko, J.R. (1985) Fracture initiation under impact, *Int. J. Impact Engng.*, Vol. **3** No. 3, 191-210.

Kobayashi, A.S. (1987), Hybrid experimental-numerical stress analysis,, in *Handbook on Experimental Mechanics*, Prentice-Hall, Inc., Englewood Cliffs NJ.

Lee, Y.J. and Freund, L.B. (1990) Fracture initiation due to asymmetric impact loading of an edge cracked plate. *J. Appl. Mech.* **57**, 104-111.

Mason, J.J., Lambros, J. and Rosakis, A.J. (1992) The use of a coherent gradient sensor in dynamic mixed-mode fracture mechanics experiments , *J. Mech. Phys. Solids*, **40**, No. 3, 641-661.

Meyers, M.A. (1994) *Dynamic behavior of materials*, J. Wiley and Sons, New York, NY.

Ravi-Chandar, K., (1995) On the failure mode transitions in polycarbonate under dynamic mixed mode loading, *International Journal of Solids and Structures*, Vol. **32** No. 6/7, 925-938.

Rittel, D. (1997) The influence of temperature on dynamic failure mode transitions, submitted for publication.

Rittel, D. and Levin, R. (1997) Mode-mixity and dynamic failure mode transitions in polycarbonate, submitted for publication.

Rittel, D. and Maigre, H. (1993) A new approach to dynamic fracture toughness testing, in *Novel Experimental Techniques in Fracture Mechanics*, ed. A. Shukla, ASME, (1993), 173-184.

Rittel, D. and Maigre, H. (1996) An investigation of dynamic crack initiation in PMMA, *Mechanics of Materials*, **Vol. 23** No. 3, 229-239.

Rittel, D., Maigre, H and Bui, H.D. (1992) A new method for dynamic fracture toughness testing, *Scripta Metallurgica et Materialia*, **26**, 1593-1598.

Rittel, D., Levin, R. and Maigre, H. (1997) On dynamic crack initiation in polycarbonate under mixed-mode loading, *Mechanics Research Communications*, Vol. **24**, No. 1, 57-64.

Taylor, G.I. and Quinney, H. (1934) The latent energy remaining in a metal after cold working, *Proc. R. Soc.* , **A 143**, 307-326.

Zehnder, A.T., and Rosakis, A.J. (1991) On the temperature distribution at the vicinity of dynamically propagating cracks in 4340 steel, *J. Mech. Phys. Solids,* **39** No. 3, 385-415.

Zhou, M., Rosakis, A.J. and Ravichandran, G. (1996) Dynamically propagating shear bands in impact-loaded prenotched plates. I- Experimental investigations of temperature signatures and propagation speed, *J. Mech. Phys. Solids,* **44**, No. 6, 981-1006.

ENERGY RELEASE IN FRACTURE OF RATE-DEPENDENT MATERIALS

L.I. SLEPYAN

Department of Solid Mechanics, Materials and Structures
Tel Aviv University, Ramat Aviv 69978 Tel Aviv, Israel

Abstract. A general, steady-state crack propagation problem is considered for both periodically structured and homogeneous media with rate-dependent moduli and/or density. The influence of the surface structure described by a specific dynamic condition at the crack faces is examined as well, and the energy release associated with the surface structure is determined. A general, structure- and crack-speed-dependent expression of the wave resistance is presented. The conditions for the existence of energy release in a 'refined' model of the medium are shown and a procedure for the proper homogenization of a structured material (to be adequate for fracture dynamics) is found.

1. Introduction

In crack propagation in a medium with a structure, wave resistance arises because of excitation of structure-associated waves which carry a part of the global energy release away from the crack tip. The local energy release as energy flux into the propagating crack tip is the remainder of the total energy released, and the local-global energy release ratio is a structure-dependent function of the crack velocity. In addition, the true surface of the crack faces can differ from the smoothed surface which is usually assumed in the theoretical formulation of crack dynamics, and this further decreases the ratio discussed.

A simple lattice model of an elastic or viscoelastic material gives an insight into what occurs during crack propagation. However, such a model has at least one disadvantage: it is strongly inhomogeneous and anisotropic on the 'microlevel'. This presents problems for a quantitative estimation of the role of the radiation in the formation of the surface structure. For example, conditions for micro-branching were found to depend on the shape of the lattice of possible crack trajectories. Therefore, the results of crack

193

D. Durban and J.R.A. Pearson (eds.),
IUTAM Symposium on Non-Linear Singularities in Deformation and Flow, 193-204.
© 1999 *Kluwer Academic Publishers. Printed in the Netherlands.*

dynamics analysis in a discrete lattice are lattice-structure-dependent, and seem to differ from those for a homogeneous structured material. At the same time, the crack dynamics phenomena considered are more likely to be associated with a macroscopic scale rather than the atomic one, and hence – with a continuous material. In this connection, it is of interest to consider homogeneous materials with an internal structure.

In the present paper, a general, steady-state crack propagation problem is considered for both periodically structured and homogeneous media with rate-dependent moduli and/or density. Some principles of the mathematical description of such media (to be adequate for fracture dynamics) are introduced. The corresponding technique valid for both types of structure and some results of the analysis are presented.

2. Some Results in Fracture of Discrete Lattices

Discrete lattice models were repeatedly used for determination of dynamic phenomena caused by a material structure. Note here the books by Kunin (1982, 1983) and Askar (1985) devoted to this subject.

A number of works exist, devoted to the analysis of dynamics of dislocations and cracks in discrete lattices as simple models of structured media: Atkinson and Cabrera (1965), Celli and Flytzanis (1970), Thomson *et al.* (1971), Weiner and Pear (1975), Ashurst and Hooper (1976), Maslov (1976, 1980), Sieradziki *et al.* (1988), Machova (1992).

Exact analytical solutions to such problems (2D-problems in fracture dynamics in lattices and 1D-problems in phase transformation dynamics in a chain of the simplest structure) were published in the 1980s. Investigations in this field were carried out to derive a relation between the structure and the speed of the crack or a phase separation interface, on the one hand, and the dissipation, on the other hand. In addition, these investigations made it possible to describe a number of phenomena which could not be discovered in the classical non-structured elastic medium.

Mode III of dynamic crack propagation in a square lattice was considered by Slepyan (1981a, 1981b, 1982a) for the sub-critical and super-critical crack speeds and for a general case, respectively. Modes I and II for a triangular lattice were studied by Kulakhmetova *et al.* (1984). The ratios, $r(v)$, of the "local" energy release rate, G_0 (the energy which is spent on the fracture itself – in accordance with the solution for the lattice), to the "global" energy release rate, G, on the macro-level – defined by the long wave approximation of this solution – were found for these three modes as functions of the ratios of the crack velocity, v, to the corresponding critical velocity (Slepyan, 1981a, 1993a; Kulakhmetova *et al.*, 1984). Similar relations were obtained by Marder and Gross (1995) for the lattice strips (the lattice strip was also considered by Slepyan, 1986). The same problems for anisotropic lattices (lattices that correspond to anisotropic elastic media) were solved by Kulakhmetova (1985a, 1985b).

Some general conclusions concerning the wave resistance to a propagating singular point are presented by Slepyan (1982b). Mikhailov and Slepyan (1986) used the same technique for the investigation of crack propagation in a model of a composite material; Slepyan and Kulakhmetova (1986) made use of this approach for a model of rock joints. Finally, the papers by Slepyan and Troyankina (1984, 1988) were devoted to fracture waves in the piecewise-linear and piecewise-linear-shifted chain structures (in the latter, the right stable branch of the stress-strain diagram is shifted with respect to the origin of the coordinates). These chain structures were the models for the phase transition dynamics in structured media, and all major phenomena that accompany the crack propagation in a lattice are presented in these more simple models which are susceptible to more detailed analysis. Many of the above mentioned results can be found in Slepyan, 1990, 1993a, 1996).

Under the condition of crack velocity oscillations, the energy radiation can accompany the crack propagation in a non-structured homogeneous material as well (Rice, 1978; Slepyan, 1978). In particular, such oscillations can be caused by a nonuniform, wavy-shaped distributed toughness (this phenomenon was considered by Das and Aki, 1977; Freund, 1987; Gilles and Rice, 1994).

The influence of the structure-associated waves depends on the crack speed; an extremal, minimal-wave-dissipation speed exists which required the minimal total (macro-level) energy release rate: almost all the released energy transforms into the surface energy. This extremum is clearly visible on the plots of dissipation versus crack speed (Slepyan, 1981a, 1993a; Kulakhmetova *et al.*, 1984). When the crack speed is less than the extremal one, the fracture does not coincide with the maximal stress intensity but, at least, takes place after the maximum. This is possible in the case of a delay in fracture. If this is not the case, the extremal speed must coincide with the lower limit of the crack speed. In accordance with this, the lower speed limits for cracks and failure waves in discrete lattices were found based on the analysis of the exact solutions (Slepyan, 1981b; Slepyan and Troyankina, 1984; Marder and Gross, 1995).

When the crack speed exceeds the extremal value, the fracture takes place before the maximum of the stress intensity. In this case, however, the wave radiation from the propagating crack tip increases with the crack velocity. Indeed, the fracture can consume only a part of the energy release, and the ratio of the surface energy to the total energy release rate must tend to zero when the crack speed approaches the critical one (Slepyan, 1981a,

1993a; Kulakhmetova *et al.*, 1984). Therefore, under a constant surface energy, the wave resistance increases unboundedly when the crack speed approaches the critical wave speed. This high-amplitude wave motion leads to damage to the material in the vicinity of the crack tip. This phenomenon, which defines the upper crack speed limit, was noted by Slepyan (1990) and investigated by Marder and Gross (1995).

Taking into account the fact that, in dynamics, the stress intensity for some inclined directions approaches that for the crack line continuation, it can be seen that the upper limit cannot be too far from the lower one. Note that in the case of "weak bonds", when the crack continuation line is much weaker than the surrounding material, crack speed can achieve the critical one (Lee and Knauss, 1989), and this does not contradict the above simple considerations.

The surface structure created during the crack propagation (see Ravi-Chandar and Knauss, 1984; Fineberg *et al.*, 1991, 1992; Marder, 1995) with roughness of the crack surfaces, micro-cracking and micro-branching are common for crack dynamics, and this increases energy dissipation dramatically. As a result, in dynamics, cracks do not obey any limited-energy-release-rate criterion, and under ordinary conditions crack propagation appears to correspond to a maximum dissipation rate per unit time (Slepyan, 1992, 1993b).

A micro-nonuniform crack propagation was studied theoretically, in particular, by Rice *et al.* (1994). Movchan *et al.* (1996) present asymptotic analysis of out-of-plane perturbations of 2D and 3D cracks. Gilles and Rice (1994) studied oscillations in crack dynamics in a medium with nonuniform distribution of the surface energy. In a sense, this corresponds to a structured material. A specific case of crack speed oscillations in a lattice was considered by Slepyan (1986). Irregular brittle fracture was considered by Frantziskonis (1994).

A number of works were devoted to stability of crack propagation in discrete lattices (Fineberg *et al.*, 1991, 1992; Marder, 1991; Marder and Xiangmin Liu, 1993; Marder and Gross, 1995; Langer, 1996; Ching *et al.*, 1996; Adda-Bedia *et al.*, 1997). Xu and Needleman (1994) considered crack dynamics in a homogeneous body, but with a lattice of possible crack trajectories.

3. Wave Resistance in Crack Propagation

In this section, a general expression for the wave resistance to crack propagation is derived as a structure-dependent function of the crack speed (Slepyan, 1982b, 1990). Suppose that a component of the displacement distribution, $u_-(\eta)$, for $\eta = x - vt < 0$ (i.e. on the crack propagating with velocity v) and the corresponding component of the traction distribution, σ_+, on the crack continuation, $\eta > 0$, are connected by the relation

$$S_u(0 + ikv, k)u_-^F(k) + S_\sigma(0 + ikv, k)\sigma_+^F(k) = Q^F(0 + ikv, k). \qquad (1)$$

Here Q is an external force, the upper index, F, means the Fourier transform (with the exponent $ik\eta$) of the corresponding function of η, and $0 + ikv = \lim \epsilon + ikv$ ($\epsilon \to +0$). This limit corresponds to the causality principle when the steady-state solution is considered as a limit ($t \to \infty$) of the corresponding transient problem solution with zero initial conditions.

Assume that the function $S(0 + ikv, k)$ has the asymptotes

$$S = A(0 + ik)^\alpha(0 - ik)^\beta + O(k^{\alpha+\beta+\gamma}) \quad (k \to 0),$$
$$S = Be^{\pm i\pi d}|k|^\nu + O(k^{\nu-\gamma}) \quad (k \to \pm\infty); \quad A, B, \gamma = \text{const} > 0, \qquad (2)$$
$$d = \text{Ind } S \equiv [\text{Arg } S(0 + i\infty, \infty) - \text{Arg } S(0 - i\infty, -\infty)]/(2\pi).$$

Let us introduce a function S_* according to the relation

$$S = BS_*(0 + ikv, k)(0 + ik)^\alpha(0 - ik)^\beta N_+ N_- \qquad (3)$$
$$N_\pm = \left[\left(\frac{A}{B}\right)^{1/(2n_\pm)} + (0 \mp ik)^2\right]^{n_\pm},$$
$$n_+ = -\frac{1}{2}\left(d + \beta - \frac{\nu}{2}\right), \quad n_- = \frac{1}{2}\left(d - \alpha + \frac{\nu}{2}\right).$$

Under this definition,

$$S_* = 1 + O(k^\gamma) \ (k \to 0), \ S_* = 1 + O(k^{-\gamma}) \ (k \to \pm\infty), \ \text{Ind } S_* = 0 \quad (4)$$

and the conventional technique can be used for the factorization of the function S_* and hence the function S.

Using the energy-release-rate relation in the form (Slepyan, 1984)

$$G = \lim_{s \to \infty} s^2 \sigma_+^F(is)u_-^F(-is) \qquad (5)$$

the true energy release rate, G_0, can now be found.

At the same time, using asymptotes of the functions in the right-hand part of (12), $k \to 0$, which correspond to the long-wave (low-frequency) approximation, the corresponding energy release rate on the 'macro-level', G, can be found.

In this way, we reach the following conclusions. An energy release rate exists only if the index is integer; it is positive for a positive even index and it is negative for a positive odd or negative index. Finite, nonzero energy release rates on both levels, G_0 and G, exist simultaneously if $d = D$, where

D is the macro-level index, $D = (\alpha - \beta)/2$. For the 'classical' fracture when the energy flux from infinity exists in the form of a non-oscillating wave, the ratio of the energy fluxes is

$$r(v) = \frac{G_0}{G} = \lambda^2 = \exp\left(\frac{2}{\pi}\int_0^\infty \text{Arg } S_*(0 + ikv, k)\frac{dk}{k}\right). \tag{6}$$

The wave resistance, $R(v)$, is the difference between the total energy release rate, G, and the energy release on the micro-level (as the level of the structure), G_0:

$$R = G - G_0 = [1 - r(v)]G. \tag{7}$$

Note that for the determination of the wave resistance, there is no need to solve a specific crack dynamic problem, but only the function $S(k)$, which depends on the structure of the medium and the crack speed, is important.

The derived results are essential for the problem of proper homogenization of a structured material to be adequate for fracture dynamics. Since $D = 0$ for a sub-critical crack speed, the index, d, must be equal to zero for a structured, 'refined' model of the medium as well. Note that for a discrete lattice, $d = 0$ (Slepyan, 1982b, 1990).

4. Homogeneous, Rate-Dependent Media

To avoid difficulties in the formulation of the boundary conditions for a structured continuous medium, consider media with a structure introduced by strain-rate-dependent moduli or/and effective density. Such a medium can be conservative or nonconservative. In either case, the relation between the total energy release rate, G (defined for the corresponding non-structured elastic medium), and the fracture energy itself, G_0 (defined for the considered continuous medium with a structure) is expressed by the same formula (6), and the only condition $d = D = 0$ must be satisfied.

To find a proper way for introducing a structure into the structure-less model of a medium (or changing the structure), we appeal to relations (2) and (3) and consider the problem of the index-invariant transformation of the function S. Assume that the medium considered is stable in the sense that the function S has no zero or singular points in the right half-plane of s, that is, for $\epsilon > 0$ [see the definition (1)]. In this case, Ind S is independent of ϵ ($\epsilon > 0$). Suppose that this function depends on parameter $\phi : S = S(\epsilon + ikv, k, \phi)$ and has an invariable index if $\phi \in \Omega$, and $\phi = \phi_0$ is an internal point of Ω.

To change the model of the medium, substitute parameter ϕ by a function $\mathcal{R}(s), \text{Re } s = \epsilon > 0, \text{Im } s = kv$. We require (a) that this transformation preserve the stability of the medium and (b) that $\mathcal{R}(s)$ tend uniformly to

ϕ_0 when $\epsilon \to \infty$. In this case, the indices are independent of ϵ (see Gakhov, 1966), Ind $\mathcal{R} = 0$ and Ind S remains invariable as required. Consider two examples.

4.1. VISCOELASTIC MATERIAL

First consider the standard solid model of a viscoelastic material (see Christensen, 1982). In such a model the bulk, K, and shear, μ, elastic moduli are replaced by the operators

$$K = K_0 \frac{1 + \tau_1 s}{1 + \tau_2 s}, \quad \mu = \mu_0 \frac{1 + \tau_3 s}{1 + \tau_4 s}, \tag{8}$$

where $K_0, \mu_0 = \text{const}$ (they represent the low-rate moduli), s is the time-derivative symbol, $0 < \tau_2 < \tau_1$ and $0 < \tau_4 < \tau_3$ ($K_0 \tau_1/\tau_2$ and $\mu_0 \tau_3/\tau_4$ are the high-rate moduli).

This representation satisfies the above-mentioned conditions. Therefore, both the energy criterion of fracture can be satisfied and the wave resistance can be obtained based on this model. For steady-state crack propagation, the elastic moduli, which appear in the function S (for the corresponding elastic body), can be substituted by the expressions (8) in which $s = 0 + ikv$. In this way, for a loading applied far from the crack tip, the total energy release corresponds to the elastic body, and the wave resistance corresponds to energy dissipated in the volume of the body.

Note that only ratios of the 'relaxation' parameters $\tau_2/\tau_1, ..., \tau_4/\tau_1$ are important as far as there is no other time-unit in the problem formulation. A number of works on viscoelastic fracture are referred in Freund (1990).

4.2. RATE-DEPENDENT DENSITY

Consider an elastic material which interacts with a medium of uniformly distributed 'spring-mass' elements. The dynamic equations are

$$\frac{\partial \sigma_{mn}}{\partial x_m} - \rho_0 \frac{\partial^2 u_n}{\partial t^2} = \gamma(u_n - w_n), \quad \rho_1 \frac{\partial^2 w}{\partial t^2} = \gamma(u_n - w_n), \tag{9}$$

where u_n, w_n and ρ_0, ρ_1 are displacements and densities of the elastic and spring-mass media, respectively, and γ is the spring rigidity. Note that a wave in such a one-dimensional system with a distribution of the spring-mass medium within a range of frequencies was considered by Slepyan (1967). Using the Laplace transform, one obtains

$$\frac{\partial \sigma_{mn}^L}{\partial x_m} - \rho s^2 u^L = 0, \quad \rho = \frac{(\rho_0 + \rho_1)\gamma + \rho_0 \rho_1 s^2}{\gamma + \rho_1 s^2}. \tag{10}$$

Here ρ is the density operator. For a low-rate process $(s \to 0)$, $\rho \sim \rho_0 + \rho_1$, and for a high-rate process $(s \to \infty)$, $\rho = \rho_0$. This representation also satisfies the conditions required.

4.3. MODEL OF A SURFACE STRUCTURE

In the framework of a homogeneous material, surface structure, as an additional waveguide, can be introduced by a specific dynamic condition at the crack faces. In the relation

$$u_-^{LF} = -S(s,k)(\sigma_+^{LF} + \sigma_-^{LF}), \tag{11}$$

assume that $\sigma_-^{LF} = g(s,k)u_-^{LF}$. One now has

$$[1 + g(s,k)S(s,k)]u_-^{LF} + S(s,k)\sigma_+^{LF} = 0. \tag{12}$$

For the steady-state crack propagation, this equation coincides with (1). In addition to the conditions mentioned above, we now require that the product $g(s,k)S(s,k) \to 0$, when $s,k \to 0$. This provides for relation (12) to be surface-structure-free for a low-rate process. For example, all the conditions are satisfied by the function $g = qs^2/(\tau^{-2} + s^2)$, where the surface structure intensity $q > 0$ and the relaxation time τ are constants. It can be found that energy release rate associated with the surface structure is

$$G_s = \int_{x/v}^{\infty} \sigma_- \frac{\partial u_-}{\partial t}\, dt = \frac{q}{2\tau^2 v^2}|u_-^F(k)|^2, \quad k = \frac{1}{\tau v}. \tag{13}$$

It would be reasonable to define the surface structure intensity, q, to correlate with energy flux density in the wave field at the crack faces.

4.4. SOME RESULTS OF CALCULATIONS

An investigation of a micro-nonuniform crack tip motion in such a medium with the spatial and surface structures is exceptionally interesting. However, we now restrict ourselves to the steady-state crack propagation. Some dependences of the wave resistance, $R/G = 1 - r(v)$, versus the crack speed for Mode I crack propagation are shown in Fig. 1. These results are obtained using the general relations (6) and (7), the expression of the function S for the non-structured elastic medium (Freund, 1990, Slepyan, 1990) and the introduced rate-dependent moduli (8), density (10) and surface structure (12). (Poisson's ratio, $\nu = 1/4$, and Rayleigh wave speed, $c = c_R = \sqrt{2 - 2/\sqrt{3}}\,c_2$, corresponds to the low-rate moduli and density.)

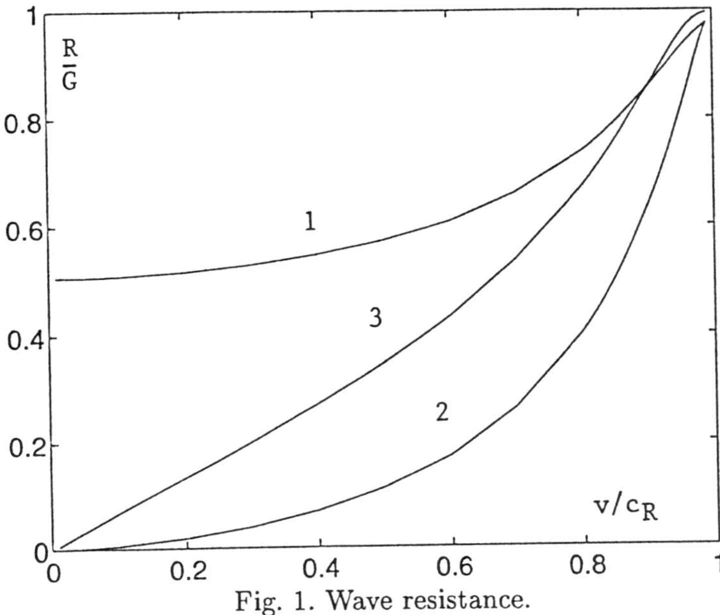

Fig. 1. Wave resistance.

1. Viscoelastic material (18); $\tau_1 = \tau_3 = 2, \tau_2 = \tau_4 = 1$
2. Strain-rate-dependent material (20); $\rho_0 = 1, \rho_1 = 1, \gamma = 1$
3. Material with surface structure (22); $q = 1, \tau = 1$.

4.5. QUASI-STATIC LIMIT OF THE WAVE RESISTANCE

Consider an isotropic material with a rate-dependent modulus and the rate-independent Poisson ratio:

$$\mu = \mu_0 \frac{1 + \tau_1 s}{1 + \tau_2 s}, \quad \lambda = \lambda_0 \frac{1 + \tau_1 s}{1 + \tau_2 s}, \quad c_1^2 = \frac{\lambda}{\rho}, \quad c_2^2 = \frac{\mu}{\rho}. \tag{14}$$

Now, when $v \to 0$

$$ArgS \sim Arg\frac{1 + i\tau_2 kv}{1 + i\tau_1 kv} = Arctan\frac{(\tau_2 - \tau_1)kv}{1 + \tau_1 \tau_2 k^2 v^2} \tag{15}$$

uniformly over k-axis, and the local/global energy release ratio has the following limit:

$$\lim_{v \to 0} r = exp\left(-\frac{2}{\pi}\int_0^\infty Arctan\frac{\alpha k}{1 + k^2}\frac{dk}{k}\right)$$

$$= exp\left(-\frac{4}{\pi}\int_0^1 Arctan\frac{\alpha k}{1 + k^2}\frac{dk}{k}\right), \quad \alpha = \sqrt{\frac{\tau_1}{\tau_2}} - \sqrt{\frac{\tau_2}{\tau_1}}. \tag{16}$$

Thus the normalized wave resistance, $R = 1 - r$, is nonzero even for zero crack speed if $\tau_1 > \tau_2$, and it becomes infinitely large in comparison with the local energy release when $\tau_1/\tau_2 \to \infty$.

4.6. WAVE RESISTANCE TO CRACK PROPAGATION IN A LAYER

Consider an elastic layer, $-\infty < x < \infty, -h < y < h, -\infty < z < \infty$, with a uniformly propagating crack at $x < vt, y = 0, -\infty < z < \infty$ loaded by stress uniformly distributed along its faces: $\sigma_{yq} = \sigma_- = -p$, where q can be x, y or z depending on the Mode of fracture. The layer surfaces are assumed to be immobile, at least in the direction of the load, or, generally, the boundary conditions must provide a *bounded* displacement under such a load.

Under this condition, the energy release rate is equal to half the work of the applied stresses per unit area. This work is twice the work at one surface. The local/global energy release ratio formula (6) is valid in the considered case as well.

Consider Mode III fracture of a viscoelastic layer with immobile surfaces at $y = \pm h$. In this case,

$$S(k) = \Lambda \frac{th(hk\sqrt{1 - \Lambda v^2 c_1^{-2}})}{\mu_0\sqrt{1 - \Lambda v^2 c_1^{-2}}}, \quad \Lambda = \frac{1 + i\tau_2 vk}{1 + i\tau_1 vk}, \quad c_1^2 = \frac{\mu_0}{\rho}. \tag{17}$$

The most interesting conclusion can be made concerning critical crack speeds. In the case of $h = \infty$, the lower wave velocity in the material, c_1, is critical: the corresponding local/global energy release ratio $r = 0$ if $v \geq c_1$. It means that the crack cannot propagate with such speeds. This conclusion follows from the fact that $ArgS(0) \neq 0$ and integral (16) does not exist. However, in the case of a finite h, as can be seen in (17), the point $k = 0$ is regular. In the latter case, the higher wave velocity, $c_2 = c_1\sqrt{\tau_1/\tau_2}$, is critical because the point $k = \infty$ becomes singular: the integral tends to infinity when $v \to c_2$. Thus, in contrast to the space ($h = \infty$), a finite-thickness viscoelastic layer permits the crack to propagate with speeds in the interval $c_1 \leq v < c_2$. In this connection, see Ryvkin and Banks-Sills (1992).

5. Acknowledgment

This research was supported by grant No. 94-00349 from the United States – Israel Binational Science Foundation (BSF), Jerusalem, Israel, and by grant No. 9673-1-96 from the Ministry of Science, Israel.

References

Adda-Bedia, M., Ben Amar, M. and Pomeau, Y. (1997) Fracture Propagation and Onset of Dynamic Instability (submitted).

Ashurst, W. T. and Hoover, W. G. (1976) Microscopic Fracture Studies in the Two-Dimensional Triangular Lattice. *Phys. Rev.* **B14**, 1465-1473.

Askar, A. (1985) *Lattice Dynamical Foundations of Continuum Theories*, World Sci.

Atkinson, W. and Cabrera, N. (1965) Motion of a Frenkel-Kontorova Dislocation in a One-Dimensional Crystal. *Phys. Rev.* **138**, A763-A776.

Celli, V. and Flytzanis, N. (1970) Motion of a Screw Dislocation in a Crystal. *J. Appl. Phys.* **41**, 4443-4447.

Ching, E. S. C., Langer, J. S. and Nakanishi, H. (1996) Instabilities in Mode I Fracture (submitted).

Christensen, R. M. (1982) *Theory of Viscoelasticity.* Academic Press, New York.

Das, S. and Aki, K. (1977) Fault Plane with Barriers: a Versatible Earthquake Model. *J. Geophys. Research* **82**, 565-570.

Fineberg, J., Gross, S. P., Marder, M. and Swinney, H. L. (1991) Instability in Dynamic Fracture. *Phys. Rev. Letters* **67**(4), 457-460.

Fineberg, J., Gross, S. P., Marder, M. and Swinney, H. L. (1992) Instability in the Propagation of Fast Cracks. *Phys. Rev.* **B 45**(10), 5146-5154.

Frantziskonis, G. (1994) On Scaling Phenomena in Fracture of Heterogeneous Solids. *Eur. J. Mech., A/Solids* **13**, 73-92.

Freund, L. B. (1987) The Apparent Fracture Energy for Dynamic Crack Growth with Fine Scale Periodic Fracture Resistance. *J. Appl. Mech.* **54**, 970-973.

Freund, L. B. (1990) *Dynamic Fracture Mechanics*, Cambridge University Press.

Gakhov, F. D. (1966) *Boundary Value Problems.* Pergamon Press, London.

Gilles, P. and Rice, J. R. (1994) Disordering of a Dynamic Planar Crack Front in a Model Elastic Medium of Randomly Variable Toughness. *J. Mech. Phys. Solids* **42**, 1047-1064.

Kulakhmetova, Sh. A. (1985a) Influence of Anisotropy of a Lattice on an Energy Outflow from a Propagating Crack. *Vestnik LGU*, No. 22, 51-57 (in Russian).

Kulakhmetova, Sh. A. (1985b) Dynamics of a Crack in an Anisotropic Lattice. *Sov. Phys. Dokl* **30**(3), 254-255.

Kulakhmetova, Sh. A., Saraikin, V. A., and Slepyan, L. I. (1984) Plane Problem of a Crack in a Lattice. *Mechanics of Solids* **19**(3), 101-108.

Kunin, I. A. (1982) *Elastic Media with Microstructure*, Springer-Verlag.

Kunin, I. A. (1983) *Elastic Media with Microstructure II*, Springer-Verlag.

Langer, J. S. (1996) Instabilities in Dynamic Fracture (submitted).

Machova, A. (1992) Molecular Dynamic Simulation of Micro-crack Initiation by Impact Loading. *Mater. Sci. Engng.* **A149**, 153-165.

Lee, O. S. and Knauss, W. G. (1989) Dynamic Crack Propagation along a Weakly Bonded Plane in a Polymer. *Experimental Mech.* **29**(3), 342-345.

Marder, M. (1991) New dynamical equation for cracks. *Phys. Rev. Lett.* **66**, 2484-2487.

Marder, M. and Xiangmin Liu (1993) Instability in Lattice Fracture. *Phys. Rev. Lett.* **71**(15), 2417-2420.

Marder, M. and Gross, S. (1995) Origin of Crack Tip Instabilities. *J. Mech. Phys. Solids* **43**, 1-48.

Maslov, L. A. (1976) Model of a Fracture as an Emitter of Elastic Vibrations. *J. Appl. Mech. Techn. Phys* **17**(2), 274-279.

Maslov, L. A. (1980) Motion of a Crack in a Discrete Medium. *Mechanics of Solids* **15**(4), 106-109.

Mikhailov, A. M., and Slepyan, L. I. (1986) Steady-State Motion of a Crack in a Unidirectional Composite. *Mechanics of Solids* **21**(2), 183-191.

Ravi-Chandar, K., and Knauss, W. G. (1984), An Experimental Investigation into Dynamic Fracture: III. On Steady-State Crack Propagation and Crack Branching, *Int.*

J. Frac, **26**, 141-154.

Rice, J. R. (1978) Radiant Energy Dissipation in Crack Propagation. *Proc. ARPA Materials Research Council Meeting*, July 1977, University of Michigan, 60-77.

Rice, J. R., Ben-Zion, Y. and Kim, K.-S. (1994) Three-Dimensional Perturbation Solution for a Dynamic Planar Crack Moving Unsteadily in a Model Elastic Solid, *J. Mech. Phys. Solids*, **42**, 5, 813-843.

Sieradziki, K., Dienes, G. J., Paskin, A. and Massounzadeh, B. (1988) Atomistics of Crack Branching. *Acta. Metall.* **36**, 651.

Slepyan, L. I. (1967) The Strain Wave in a Bar with Vibration-Isolated Masses. *Mechanics of Solids* **2**, 57-64.

Slepyan, L. I. (1978) Mechanics of Cracks. In: *Mechanics of an Elastic Body*, Polish Publ. House, Warsaw, 329-346 (in Russian).

Slepyan, L. I. (1981a) Dynamics of a Crack in a Lattice *Sov. Phys. Dokl.* **26**, 538-540.

Slepyan, L. I. (1981b) Crack Propagation in High-frequency Lattice Vibrations. *Sov. Phys. Dokl.* **26**(9), 900-902.

Slepyan, L. I. (1982a) Antiplane Problem of a Crack in a Lattice. *Mechanics of Solids* **17**(5), 101-114.

Slepyan, L. I. (1982b) The Relation between the Solutions of Mixed Dynamic Problems for a Continuous Elastic Medium and a Lattice. *Sov. Phys. Dokl.* **27**(9), 771-772.

Slepyan, L. I. (1984) Dynamics of Brittle Fracture in Media with a Structure. *Mechanics of Solids* **19**(6), 114-122.

Slepyan, L. I. (1986) Dynamics of Brittle Fracture in Media with a Structure - Inhomogeneous Problems. In: *Mathematical Methods of Mechanics of Solids*, (Eds. R. V. Goldstain and O.A. Oleinik) Nauka, Moscow, 143-149 (in Russian).

Slepyan, L. I. (1990) *Mechanics of Cracks*, Sudostroenie, Leningrad (2nd edition; 1st edition: 1981) – in Russian.

Slepyan, L. I. (1992) The Criterion of Maximum Dissipation Rate in Crack Dynamics. *Sov. Phys. Dokl.* **37**(5), 259-261.

Slepyan, L. I. (1993a) Role of Microstructure in Fracture and Phase Transition Dynamics. *Proc. of Int. Seminar 'Mecamat'93: Micromechanics of Materials'*, EYROLLES, Paris, 246-257.

Slepyan, L. I. (1993b) Principle of Maximum Energy Dissipation Rate in Crack Dynamics. *J. Mech. Phys. Solids* **41**(6), 1019-1034.

Slepyan, L. I. (1996) Crack Dynamics. In: *FRACTURE: A Topical Encyclopedia of Current Knowledge Dedicated to Alan Arnold Griffith* (ed., G. Cherepanov), Krieger Publishing Company, Melbourne, USA, 631-670.

Slepyan, L. I., and Kulakhmetova, Sh. A. (1986) Crack Propagation in a Bed Consisting of Rigid Blocks with Elastic Interlayers. *Phys. Solid Earth* **22**(12), 977-982.

Slepyan, L. I. and Troyankina, L. V. (1984) Fracture Wave in a Chain Structure. *J. Appl. Mech. Techn. Phys.* **25**(6), 921-927.

Slepyan, L. I. and Troyankina, L. V. (1988) Shock Waves in a Nonlinear Chain. In: *Strength and Visco-plasticity*, Nauka, 1988, 301-305 (in Russian).

Thompson, R., Hsieh, C. and Rana, V. (1971) Lattice Trapping of Fracture Cracks. *J. Appl. Phys.* **42**, 3154-3160.

Weiner, J. H. and Pear, M. (1975) Crack and Dislocation Propagation in an Idealized Crystal Model. *J.Appl.Phys.* **46**(6), 2398-2405.

Willis, J. R. and Movchan, A. B. (1997) Three-dimensional Dynamic Perturbation of a Propagating Crack. *J. Mech. Phys. Solids* **45**, 591-610.

Xu, X.-P. and Needleman, A. (1994), Numerical Simulation of Fast Crack Growth in Brittle Solids, *J. Mech. Phys. Solids* **42**, 1397-1434.

A COMBINED ELEMENT-FREE GALERKIN METHOD/ ARBITRARY LAGRANGIAN-EULERIAN FORMULATION FOR DYNAMIC CRACK PROPAGATION

J.-P. PONTHOT
LTAS-Continuum Mechanics and Thermomechanics
University of Liège, 21 Rue E. Solvay, B4000 Liège, Belgium

AND

T. BELYTSCHKO
Department of Civil and Mechanical Engineering,
Northwestern University, Evanston, Il 60208-3109, U.S.A.

1. Introduction

The Element Free Galerkin (EFG) method is an alternative to the finite element method (FEM) which is an industrial standard tool for solving a wide variety of mechanics problems. However, for some highly nonlinear problems, some difficulties are not yet totally overcome. These difficulties generally arise from the inherent structure of the FEM and the topological knowledge it requires, namely the rigid connectivity defined by elements. In fracture problems, for instance, finite element edges provide natural lines along which cracks can grow. However, if the crack path is not known a priori, FEM requires remeshing in order to follow an arbitrary crack path.

In contrast to the FEM, meshless methods do not have a rigid connectivity provided a priori. They require only nodes and a description of internal and external boundaries and interfaces, such as cracks. One meshless method (see [11, 14] for a review of existing meshless methods), the *element free Galerkin* (EFG) method, proposed by Belytschko and coworkers [2, 3, 4, 6] has been used extensively to solve fracture problems where it offers considerable simplifications. In EFG, the connectivity and shape functions are constructed by the method and they can be automatically modified during the computation to account for crack propagation.

However, as for any discretization method, acceptable solutions can only be obtained for a sufficiently refined discretization. In dynamic fracture

D. Durban and J.R.A. Pearson (eds.),
IUTAM Symposium on Non-Linear Singularities in Deformation and Flow, 205-216.
© 1999 *Kluwer Academic Publishers. Printed in the Netherlands.*

problems, where the crack path can be arbitrary, and is thus a priori unknown, this necessitates a refined discretization in large parts of the computational domain which can lead to prohibitive computation costs. This high cost can be avoided by using an Arbitrary Lagrangian-Eulerian (ALE) formulation. Indeed, ALE formulation allows to continuously relocate nodes on the computational domain. This versatility will be exploited here by adding to a Lagrangian discretization on the whole domain a few ALE nodes that move in a rigid body motion with respect to the propagating crack-tip. By combining EFG with ALE, it is thus possible, in a crack propagation problem, to locally and dynamically refine the spatial discretization in the neighborhood of a propagating crack-tip.

2. Element-Free Galerkin Approximation

The EFG method is a meshless or particle method. The objective of such a meshless method is to obtain an approximation to a function $u(\boldsymbol{x}, t)$ strictly in terms of parameters at a set of nodes and a description of the boundaries of the domain of interest. This boundary description includes any interior surfaces of discontinuity such as cracks or interfaces between different materials.

In the EFG method, the field variable $u(\boldsymbol{x}, t)$ is approximated by

$$u^h(\boldsymbol{x}, t) = \sum_{j=1}^{m} p_j(\boldsymbol{x}) a_j(\boldsymbol{x}, t) \equiv \boldsymbol{p}^T(\boldsymbol{x}) \boldsymbol{a}(\boldsymbol{x}, t) \tag{1}$$

where $\boldsymbol{p}(\boldsymbol{x})$ is a complete polynomial basis of arbitrary order and $\boldsymbol{a}(\boldsymbol{x}, t)$ are coefficients (to be determined) which are function of the space coordinate $\boldsymbol{x}$ and time t. For example, in two dimensions, one can choose the following basis:

$$\boldsymbol{p}^T(\boldsymbol{x}) = \begin{bmatrix} 1 & x & y \end{bmatrix} \qquad \text{(m=3, linear basis)} \tag{2}$$

$$\boldsymbol{p}^T(\boldsymbol{x}) = \begin{bmatrix} 1 & x & y & xy & x^2 & y^2 \end{bmatrix} \qquad \text{(m=6, quadratic basis)} \tag{3}$$

The unknown coefficients $\boldsymbol{a}(\boldsymbol{x}, t)$ in (1) are obtained at any point $\boldsymbol{x}$ by minimizing the following weighted, discrete error norm

$$J = \sum_{I=1}^{n} w(\boldsymbol{x} - \boldsymbol{x}_I)[u_I(t) - \boldsymbol{p}^T(\boldsymbol{x}_I) \boldsymbol{a}(\boldsymbol{x}, t)]^2 \tag{4}$$

where $w(\boldsymbol{x} - \boldsymbol{x}_I)$ is a weight function of compact support (often called the *domain of influence* of node I) and n is the number of nodes whose support includes point $\boldsymbol{x}$. The elements of the discrete set $\boldsymbol{x}_I, I = 1, ...n$ of nodes such that $w(\boldsymbol{x} - \boldsymbol{x}_I) > 0$ are called the *neighbor nodes* at point $\boldsymbol{x}$

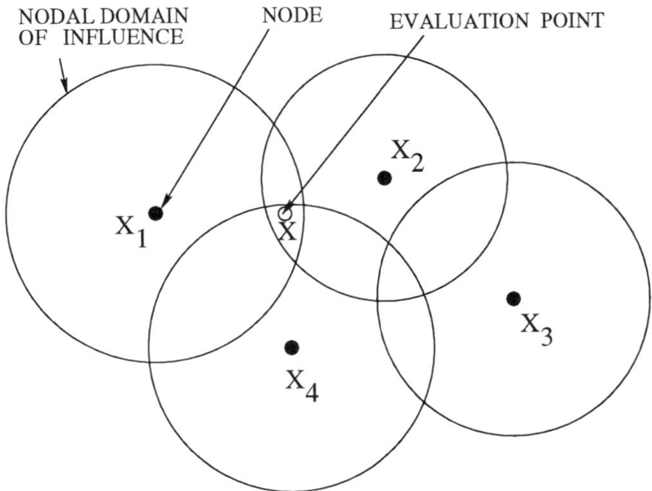

Figure 1. Illustration of nodal domains of influence in two dimensions. The neighbor list for evaluation at point x includes nodes 1, 2 and 4 since their domain of influence contains point x. Node 3 is excluded from the neighbor list for x.

(see Fig. 1). In expression (4), $u_I(t)$ is the parameter associated with node I of the approximated field; it is not the nodal value because $u^h(x, t)$ is an approximation, not an interpolation. Minimization of (4) with respect to $a(x, t)$ then yields to the following system of linear equations for the coefficient $a(x, t)$:

$$A(x)a(x, t) = B(x)u(t) \tag{5}$$

where

$$A(x) = \sum_{I=1}^{n} w(x - x_I)p(x_I)p^T(x_I) \tag{6}$$

$$B(x) = \left[\; w(x - x_1)p(x_1) \quad \cdots \quad w(x - x_n)p(x_n) \; \right] \tag{7}$$

$$u^T(t) = \left[\; u_1(t) \quad u_2(t) \quad \cdots \quad u_n(t) \; \right] \tag{8}$$

where $u(t)$ is the vector of nodal unknowns. If A is invertible (a necessary but not sufficient condition is that $n \geq m$; i.e. the number of neighbors is a least equal to the basis dimension), the coefficients $a(x, t)$ can be expressed as

$$a(x, t) = A^{-1}(x)B(x)u(t) \tag{9}$$

By substituting (9) into (1), the approximates are finally given by

$$u^h(\boldsymbol{x}, t) = \boldsymbol{p}^T(\boldsymbol{x})\boldsymbol{a}(\boldsymbol{x}, t) \equiv \Phi^T(\boldsymbol{x})\boldsymbol{u}(t) \tag{10}$$

or

$$u^h(\boldsymbol{x}, t) = \sum_{I=1}^{n} \Phi_I(\boldsymbol{x})u_I(t) \tag{11}$$

The EFG shape function $\Phi(\boldsymbol{x}, t)$ is defined as

$$\Phi^T(\boldsymbol{x}) = \boldsymbol{p}^T(\boldsymbol{x})[\boldsymbol{A}^{-1}(\boldsymbol{x})\boldsymbol{B}(\boldsymbol{x})] \tag{12}$$

Thus, in EFG, the list of neighbors and the resulting shape functions can be automatically evaluated once a given weight function w has been choosen.

In the absence of a crack, the weight function associated with the Ith node is choosen to be non-zero in a ball

$$B_I = \left\{ \boldsymbol{x} \in R^2 : |\boldsymbol{x} - \boldsymbol{x}_I| \leq r_I \right\} \tag{13}$$

where r_I is the radius of support of the $I^t h$ node, and $|\boldsymbol{x} - \boldsymbol{x}_I|$ is the distance between $\boldsymbol{x}$ and $\boldsymbol{x}_I$ in any suitable norm. The ball B_I defines the largest set on which the shape functions Φ_I may be non-zero, i.e. it defines the support of Φ_I. In other words, u_I influences the approximation of u only on B_I. If there is a crack, a more complex procedure to define the ball has to be used [11]. Details pertaining to the choice of the weight function $w(r)$ can be found in [11] as well as in [14].

3. ALE Basic Equations

The Lagrangian or *material coordinate* $\boldsymbol{X}$ is considered as a "marker", attached to a given material particle. It describes its motion through a function φ which gives the position of the particle as a function of time t:

$$\boldsymbol{x} = \varphi(\boldsymbol{X}, t) \tag{14}$$

where $\boldsymbol{x}$ is the *spatial coordinate* (the current position of the material particle identified by $\boldsymbol{X}$) and φ is the material motion function or the mapping function between $\boldsymbol{X}$ and $\boldsymbol{x}$ at a given time t.

Alternatively, it is also possible to describe the motion of a continuum by a set of coordinates χ, called the *referential coordinates*. Generally, the motion of the referential coordinate point can be arbitrary and therefore not coincide with the material coordinates. With this set of coordinates, the motion of the continuum can be described by a relation analogous to (14), i.e.

$$x = \hat{\varphi}(\chi, t) \tag{15}$$

where $\hat{\varphi}$ is the referential motion or the mapping between x and χ. The **referential derivative** of some continuous function f is then defined by:

$$\overset{\circ}{f} = \left. \frac{\partial f(\chi, t)}{\partial t} \right|_{\chi} \tag{16}$$

The material and referential derivatives are not independent. The fundamental relationship between material and referential time derivative can be established readily, for any quantity f as [9, 7, 13]:

$$\dot{f} = \overset{\circ}{f} + c \cdot \nabla f = \overset{\circ}{f} + c_i \frac{\partial f}{\partial x_i} \tag{17}$$

where c is the *relative or convective velocity* between the material v and referential $\hat{v}$ velocity:

$$c = v - \hat{v} \tag{18}$$

In the referential description, the equations of motion can be shown to read [14]:

$$\sigma_{ij,j} + \rho b_i = \rho \left(\overset{\circ}{v}_i + c_j v_{i,j} \right) \qquad \text{in } \Omega \tag{19}$$

The weak form corresponding to the conservation equation is obtained classically [14]. The discretized system of equations is obtained by substituting the following expression for the material displacement:

$$u(\chi, t) = \sum_{I=1}^{n} \Phi_I(\chi) u_I(t) \tag{20}$$

where Φ_I are the EFG shape functions given by (12) where the spatial coordinates x have been replaced by the referential coordinates χ. Test functions, convective velocity, material velocity, referential acceleration and initial velocity are discretized in a similar fashion. Substitution of these expressions in the weak forms results in:

$$M \overset{\circ}{v} + Cv + f^{int} = f^{ext} \tag{21}$$

with $f^{int} = Ku$ in the linear elastic case and

$$M_{IJ} = \int_\Omega \rho \Phi_I \Phi_J \, I \, d\Omega \tag{22}$$

$$C_{IJ} = \int_\Omega \rho c_j \Phi_I \Phi_{J,j} \, I \, d\Omega \tag{23}$$

$$K_{IJ} = \int_\Omega B_I^T D B_J \, d\Omega \tag{24}$$

$$f_I^{int} = \int_\Omega B_I^T \sigma \, d\Omega \tag{25}$$

$$f_I^{ext} = \int_\Omega \rho \Phi_I b \, d\Omega + \int_{\Sigma_t} \Phi_I t \, d\Sigma_t \tag{26}$$

where M is the mass matrix, C the convective matrix, K the stiffness matrix, f^{int} are the internal forces and f^{ext} are the applied forces.

4. Numerical Integration

One of the major difficulties associated with the time integration of (21) results from the presence of the non-symmetric convective term Cv (meaning the problem is a non self-adjoint one). This term stems from the relative motion between the material and referential system of reference. In a Lagrangian formulation, this relative motion vanishes, since the material and referential systems are identical. Therefore, no convection occurs and the C matrix is identically zero.

In fact, two ways are possible in implementing and solving the ALE equations. These two ways also correspond to the two approaches taken in implementing an Eulerian solver from the viewpoint of fluid mechanics.

The first approach solves the fully coupled equations as they are written. This approach is rather complex. Alternatively, one can use an operator splitting to integrate the equations in time. We have adopted this technique here. Typically, a Lagrangian step is performed first. During this phase, the nodes move with the material during the step and its configuration at the end of the step is different of a (to be defined) reference grid. The solution obtained from the Lagrangian phase is then mapped (convected) to the reference grid in order to complete the ALE step. One of the most salient features of the developed methodology is that in the EFG framework, some amount of upwinding can be easily and naturally introduced in the model. A detailed description of the procedure that is actually used can be found in [14].

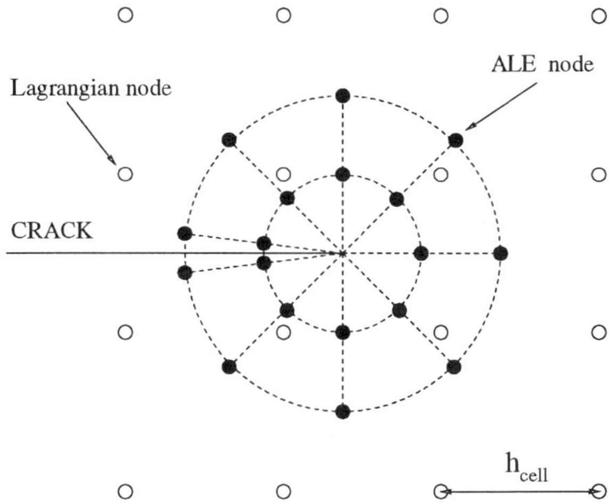

Figure 2.   A "star-like" configuration for ALE nodes around the crack tip

5. Numerical Examples

5.1. MODE I CRACK PROPAGATION

As already mentioned, ALE formulation is an attractive alternative to the
Lagrangian formulation in the sense that it is possible to automatically
maintain a dynamically high density of nodes in a limited region evolving
with time such as a crack tip neighborhood. This can be achieved by su-
perposing to a rather coarse Lagrangian discretization a few ALE nodes
which move at the same grid velocity $\hat{v}$ as the crack tip. One possible way
to do this is to generate a few ALE nodes in a star-like configuration (see
Fig. 2) around the initial position of the crack tip. As a first example of this
technique, we present the following problem of mode I crack propagation.
Consider (see Fig. 3) a semi-infinite crack in an infinite body subjected to
a pulse of magnitude $\sigma_0 = 1$ parallel to the crack plane. The body is under plane strain conditions with elastic constants $E = 211\,GPa$, $\nu = 0.30$
and $\rho = 7800\,kg/m^3$. The dilatational, shear and Raleigh wave speeds are
denoted as c_d, c_s, and c_r, respectively.

Using the central difference method for time integration and a linear
basis function in the EFG formulation, the transient response of the plate
in plane strain is calculated for 90 time steps with $\Delta t = 10^{-5}$. Details of the
implementation can be found in [6] which presents an EFG Lagrangian analysis. This problem has also been been previously analyzed numerically using
Lagrangian finite elements in [12]. Solutions were generated with a background Lagrangian grid composed of 100 x 41 (a much coarser grid than the
one used in [1, 6, 3, 2] where a 200 x 81 grid was used) cells with Lagrangian

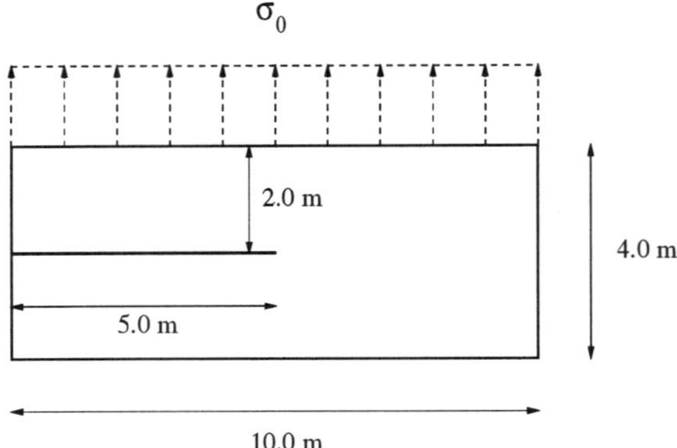

Figure 3. Schematic of the discrete model for the infinite plate with a semi-infinite crack subjected to a stress wave of magnitude σ_0 applied as a step function in time.

nodes placed at the corners plus 18 ALE nodes in a star-like configuration (2 nodes x 9 angular locations at $0^o; \pm 45^o; \pm 90^o; \pm 135^o; \pm 175^o$). On each radius of the star, the nodes location was at $0.3333\,h_{cell}$ and $0.6666\,h_{cell}$ where h_{cell} is the average dimension between two Lagrangian nodes. The shape functions were constructed with a linear basis and a Gaussian weight function with a domain of influence based on $r_I = 2.0\,h_{cell}$. Both Lagrangian and ALE solutions are compared to the analytical solution investigated by Freund [8]. Fig. 4 displays the initial grid and crack configuration. Two types of solution were generated and compared:

1. a stationary crack, $v = 0$
2. a crack that begins moving at a constant velocity of $v = 0.4c_s$ when the stress wave reaches the crack

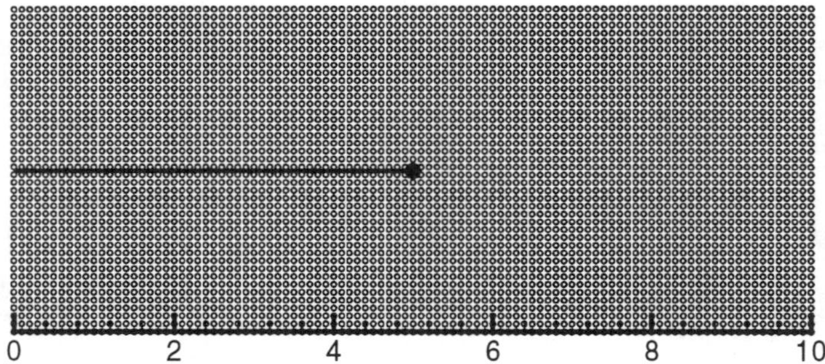

Figure 4. Typical discrete numerical model for the semi-infinite edge-cracked body. The grid contains 100 x 41 Lagrangian nodes plus 18 ALE nodes around the crack tip.

The energy release rates for these two cases are shown in Fig. 5. The stress wave arrives at $t_1 = H/c_d$ ($2H = 4m$ is the height of the plate–time t_1

corresponds to the unit reduced time on the figure), after which the energy release rate begins increasing linearly. The solution is then either along the slope corresponding to $v = 0$ or $v = 0.4c_s$. As can be noted on the figures, the ALE solution for moving cracks is much smoother than the Lagrangian one and ALE results are closer to the analytical solution.

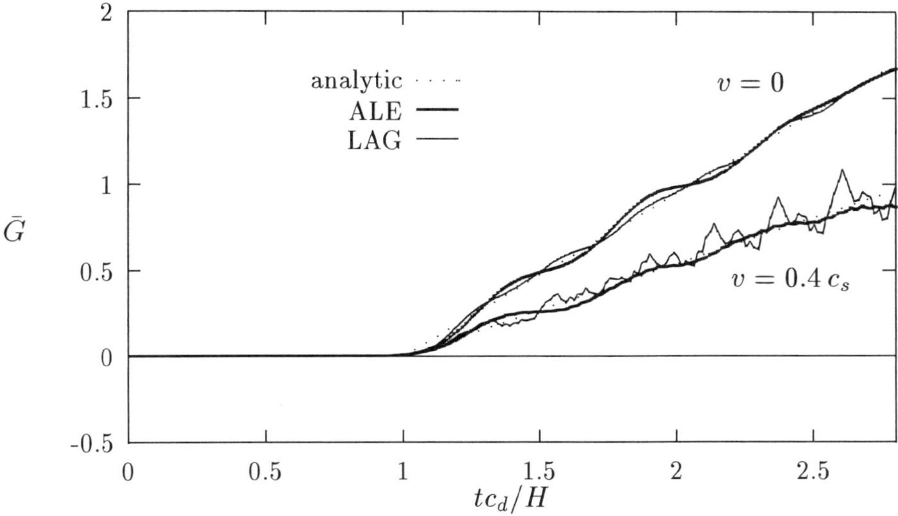

Figure 5.   Energy release rates for the infinite plate with a semi-infinite crack for the 50×21 cell mesh. The energy release rate is normalized by $H\sigma_0^2/E$. $r_I = 2.0h_{cell}$

5.2. PLATE WITH TWO PARALLEL EDGE CRACKS

Kalthoff and Winkler [10] have developed an experimental technique for investigating fracture behavior in high strength maraging steel subjected to in-plane shear loading. The experimental setup consists of a free plate with two parallel edge notches. The plate, initially at rest and stress–free, is impacted by a projectile as shown in Fig. 6. For a representative impact velocity of $v_0 = 33\,m/s$, experimental failure occurs in a brittle fracture mode and the crack propagates at an angle of about 70° with respect to the original notch.

Due to symmetry, only the top half of the plate is modeled in the simulation. The material properties are for high strength maraging steel (18Ni1900): $E = 190\,GPa$; $\nu = 0.30$ and $\rho = 8000\,kg/m^3$. To simulate the crack growth in an arbitrary direction, the maximum hoop–stress criterion is used for crack initiation [15]. When the maximum hoop–stress σ_{crit} reaches a critical value, the crack growth takes place in a direction

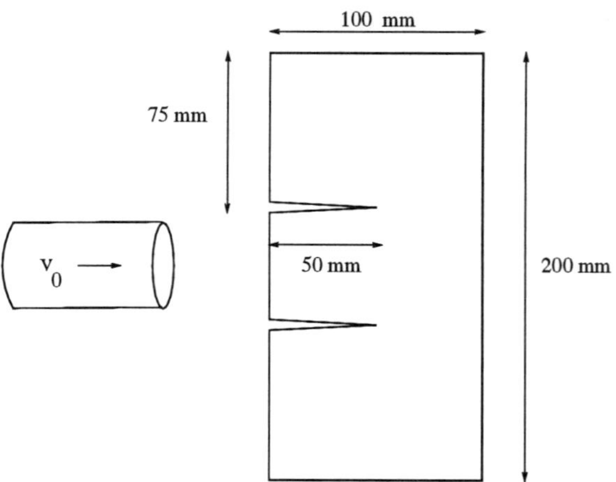

Figure 6.   Experimental setup for crack growth in edge-cracked plate.

perpendicular to σ_{crit}. Once growth is initiated, the crack is restricted to propagate at a constant velocity $v = 0.15\ c_s$, where c_s is the shear waves velocity.

EFG solutions were obtained with a discretization of 50 x 50 cells with Lagrangian nodes placed at the corners. For ALE simulations, a star-like configuration of 16 nodes, similar to the previous example, was added. The discretization, including ALE nodes and initial crack position is shown in Fig. 7. Central difference time integration was used with $\Delta t = 4.\,10^{-7}$ s until a final time of 80 μs, i.e. 200 time steps. This is in contrast with previously reported EFG/Lagrangian calculations [6, 5] which used a 100 x 100 cells grid and a time step of $\Delta t = 1.\,10^{-7}$ s. Fig. 8 displays an intermediate (after 100 steps) and the final configuration (200 steps) with the whole crack-path. The average computed slope of the propagating crack is 78° for the Lagrangian formulation (100 x 100 discretization) and 73° for the ALE formulation (50 x 50 discretization), a value which is much closer to the experimental value of 70°.

References

1. T. Belytschko, Y. Krongauz, M. Fleming, D. Organ, and W. K. Liu. Smoothing and accelerated computations in the element-free Galerkin method. *J. Computational and Applied Mathematics*, submitted for publication.
2. T. Belytschko, Y. Y. Lu, and L. Gu. Crack propagation by element-free Galerkin methods. *Engineering Fracture Mechanics*, 51(2):295–315, 1995.
3. T. Belytschko, Y. Y. Lu, L. Gu, and M. Tabbara. Element-free Galerkin methods for static and dynamic fracture. *International Journal of Solids and Structures*, 32(17-18):2547–2570, 1995.
4. T. Belytschko, Y.Y. Lu, and L. Gu. Element-free Galerkin methods. *International*

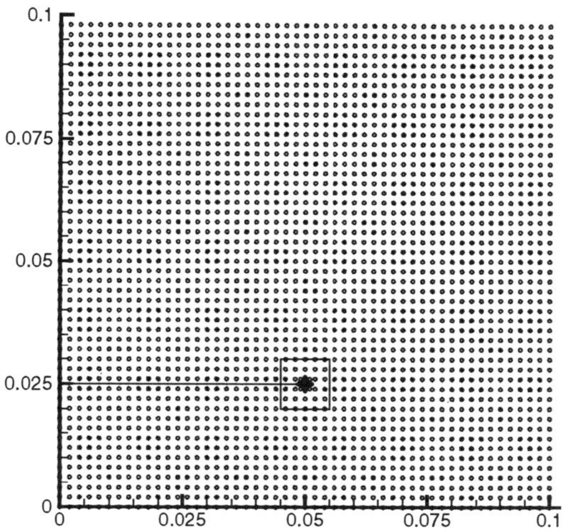

Figure 7. Typical grid for modeling crack growth in an edge-cracked plate. The discretization contains 51 × 51 Lagrangian nodes + 16 ALE nodes. The initial crack and J-domain are also shown.

Journal of Numerical Methods in Engineering, 37:229–256, 1994.

5. T. Belytschko, D. Organ, and Y. Krongauz. A coupled finite element–element-free Galerkin method. *Computational Mechanics*, 17:186–195, 1995.

6. T. Belytschko and M. Tabbara. Dynamic fracture using element-free Galerkin methods. *International Journal of Numerical Methods in Engineering*, 39:923–938, 1996.

7. J. Donéa. Arbitrary Lagrangian-Eulerian finite element methods. In T. Belytschko and T.J.R. Hughes, editors, *Computational Methods for Transient Analysis*, chapter 10, pages 474–516. North Holland, 1983.

8. L. B. Freund. *Dynamic Fracture Mechanics*. Cambridge University Press, 1990.

9. T.J.R Hughes, W.K. Liu, and T.K. Zimmerman. Lagrangian Eulerian finite element formulation for incompressible viscous flows. *Computer Methods in Applied Mechanics and Engineering*, 29:329–349, 1981.

10. J.F. Kalthoff and S. Winkler. Failure mode transition at high rates of shear loading. In Chiem C.Y., Kunze H.D., and Meyer L.W., editors, *Impact Loading and Dynamic Behaviour of Materials*, volume 1, pages 185–195, 1987.

11. P. Krysl and T. Belytschko. Element-free galerkin method: Convergence of the continuous and discontinuous shape functions. *Computer Methods in Applied Mechanics and Engineering*, accepted for publication.

12. T. Nakamura, C.F. Shih, and L.B. Freund. Computational methods based on an energy integral in dynamic fracture. *International Journal of Fracture*, 27:229–243, 1985.

13. J.P. Ponthot. *A Finite Element Unified Treatment of Continuum Mechanics for Solids Submitted to Large Strains*. PhD thesis, Université de Liège, Liège, Belgium, 1995. In French.

14. J.P. Ponthot and T. Belytschko. Arbitrary Lagrangian-Eulerian formulation for Element-Free Galerkin method. *Computer Methods in Applied Mechanics and Engineering*, submitted for publication.

15. D. Swenson and A.R. Ingraffea. Modeling mixed-mode dynamic crack propagation using finite elements: Theory and applications. *Comput. Mech.*, 3:381–397, 1988.

     J.-P. PONTHOT AND T. BELYTSCHKO

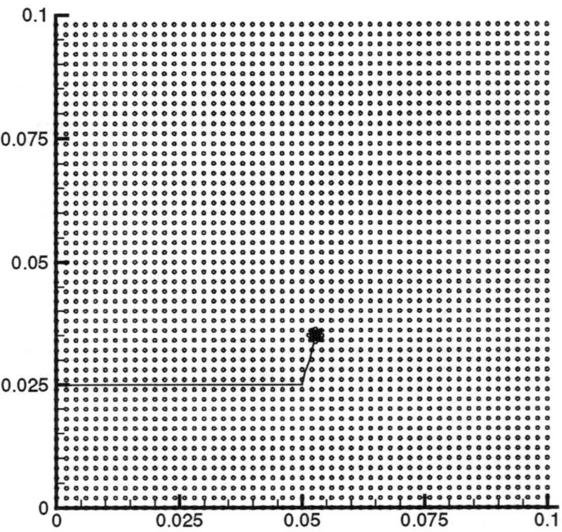

(a) configuration after 100 steps

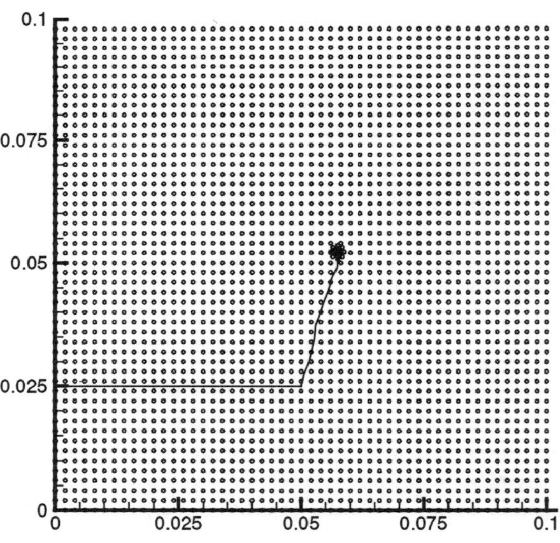

(b) configuration after 200 steps

Figure 8.   Numerical crack path in edge-cracked plate

BOUNDARY ELEMENT AND DISCRETE VORTICES METHOD

FOR IDEAL FLUID FLOW CALCULATIONS

D. V. YEVDOKYMOV
Dniepropetrovsk University,
Gagarin ave., 72, Dniepropetrovsk, 320625, Ukraine

1. Introduction

The fast development of computational hydrodynamics in recent years is mainly due to finite difference methods. However, that method cannot be used for small local flow effects investigation, and there are also difficulties in the application of the method for complex domains. A similar situation exists with finite element methods. A detailed analysis of finite difference and finite element methods' development is not the subject of the present work, but it should be noted that the above mentioned circumstances decelerate development of application of finite difference and finite element methods in hydrodynamics and stimulate interest in integral methods, such as the boundary element method and the discrete vortices method.

Among the various mathematical models of hydrodynamics the ideal fluid flow is the simplest, and potential flow problems are especially simple. There are two main mathematical models of incompressible fluid flow: ideal fluid flow and viscous fluid flow. The viscous fluid flow model is more exact, but ideal fluid flow model is more simple. The main effect of viscosity in flow is vorticity generation, but the mechanism of vorticity generation is absent in ideal fluid flow model, and for vortical flow it's necessary to introduce vorticity artificially. In fact, artificial vorticity introduction in ideal fluid flow model is a combination of the two mentioned models. For external flow vorticity is localised in a narrow domain, named the boundary layer. Since such domains are small within first approximation, their sizes are usually neglected, and, as a result, discrete vortices can be obtained.

The discrete vortices method is convenient for vortical ideal fluid flow and is based on the concept of a moving singular vortical source. According to this approach solid boundaries in flow are replaced by layers of discrete vortices, and the vorticity in flow is replaced by discrete vortices systems. The discrete vortices method was developed in studies by Belotserkovsky and his students [1-4] and by others [5-9].

The boundary element method has grown intensively during the last two decades, because the method is a most powerful tool for solution of linear homogeneous elliptic boundary value problems. The traditional boundary element method is described in several books, for example by Banerjee [10] and Brebbia [11], but boundary element method effective application is restricted to linear homogeneous problems. It is worth noting a number of papers by Wu [12-14], in which the boundary element method was

217

D. Durban and J.R.A. Pearson (eds.),
IUTAM Symposium on Non-Linear Singularities in Deformation and Flow, 217-230.
© 1999 *Kluwer Academic Publishers. Printed in the Netherlands.*

applied to viscous flow problems. A noteworthy development in the boundary element method is the Patterson [15] regular boundary element method.

The boundary element method is also based on the concept of a singular source, but the direct boundary element method includes two singularities, rendering it more accurate than the discrete vortices method. The main disadvantage of the boundary element method are difficulties with domain source consideration. Only in the particular case of external ideal fluid flow is the discrete vortices model a good model of physical behaviour of vorticity in flow.

2. Governing equations

Consider the classical fluid flow formulation:

$$\frac{\partial^2 \Psi}{\partial x^2} + \frac{\partial^2 \Psi}{\partial y^2} = -\omega \tag{1}$$

where Ψ is the stream function and ω the vorticity, along with the vorticity transfer equation

$$\frac{\partial \omega}{\partial t} + V_x \frac{\partial \omega}{\partial x} + V_y \frac{\partial \omega}{\partial y} = 0 \text{ (ideal fluid)} \tag{2}$$

and, for a viscous fluid,

$$\frac{\partial \omega}{\partial t} + V_x \frac{\partial \omega}{\partial x} + V_y \frac{\partial \omega}{\partial y} = \frac{1}{Re}(\frac{\partial^2 \omega}{\partial x^2} + \frac{\partial^2 \omega}{\partial y^2}) \tag{3}$$

where Re is the Reynolds number and V_x, V_y are the velocity components defined by

$$V_x = \frac{\partial \Psi}{\partial y}, \qquad V_y = -\frac{\partial \Psi}{\partial x}$$

Let the Dirichlet condition for the stream function be prescribed on the solid boundary Γ, viz.

$$\Psi\big|_\Gamma = \text{Const} \tag{4}$$

while at infinity:

$$\Psi\big|_\infty = y \tag{5}$$

It is important to note that equation (1) is identical for both ideal and viscous fluids flows while the mechanism of vorticity transfer and boundary conditions distinguishes between the mathematical models.

3. Boundary element method algorithm

This part of the paper is a short description of the boundary element method algorithm and a short review of author's earlier results in this field [16-19].

The following integral representation corresponds to the equation (1):

$$\chi(x_0,y_0)\Psi(x_0,y_0) = \oint_\Gamma P(x,y,x_0,y_0)\frac{\partial\Psi}{\partial n}dS$$
$$-\oint_\Gamma \Psi\frac{\partial P}{\partial n}(x,y,x_0,y_0)dS + \iint_D \omega(x,y)P(x,y,x_0,y_0)dxdy \tag{6}$$

with

$$\chi(x_0,y_0) = \begin{cases} 0,(x_0,y_0)\notin D,(x_0,y_0)\notin\Gamma \\ 1/2,(x_0,y_0)\in\Gamma \\ 1,(x_0,y_0)\in D \end{cases}$$

$$P = \frac{1}{2\pi}\ln\left(\frac{1}{\sqrt{(x-x_0)^2+(y-y_0)^2}}\right) \tag{7}$$

where D is the flow domain and Γ is the boundary of domain D.

The case of $\chi=1/2$ is the classical singular boundary integral equation of potential theory. The case of $\chi=0$ corresponds to the regular integral equation.

The boundary element method was applied to the problem of nonvortical ($\omega=0$) flow around a smooth body. In this case:

$$\chi(x_0,y_0)\Psi(x_0,y_0) = \oint_\Gamma P(x,y,x_0,y_0)\frac{\partial\Psi}{\partial n}dS - \oint_\Gamma \Psi\frac{\partial P}{\partial n}(x,y,x_0,y_0)dS \tag{8}$$

A new boundary element method algorithm, based on the regular boundary element method, was proposed. In this work Ψ and $\dfrac{\partial\Psi}{\partial n}$ on each boundary element are approximated by constants, viz.

$$0 = \sum_{n=1}^N \left(\frac{\partial\Psi}{\partial n}\right)_k \int_{\Delta\Gamma k} P_i(x,y)dS_k - \sum_{k=1}^N \Psi_k \int_{\Delta\Gamma k} \left(\frac{\partial P}{\partial n}\right)_i (x,y)dS_k$$

Values of $\displaystyle\int_{\Delta\Gamma k} P_i(x,y)dS_k$ and of $\displaystyle\int_{\Delta\Gamma k}\left(\frac{\partial P}{\partial n}\right)_i(x,y)dS_k$ were numerically calculated on the real boundary, with better accuracy at points near the boundary, in contrast with the traditional boundary element method algorithm.

The problem of flow around a circle contour is used as a test. The proposed boundary element method algorithm (without boundary approximation) gives an error of 0.03% at most, using 30 boundary elements per circle, and not more than 0.01%, using 60 boundary elements.

4. Circulation flow calculation

4.1. GENERAL PROBLEM

Since external ideal fluid flow is determined to within some circulation flow, it is necessary to consider algorithms of the problem completion.

Taking into consideration (4), and due to the substitution

$$\overline{\Psi} = \Psi - y \qquad (9)$$

we can obtain

$$\frac{1}{2}(C - y) = \int_{\Gamma} \frac{\partial \overline{\Psi}}{\partial n} P dS - C \int_{\Gamma} \frac{\partial P}{\partial n} dS + \int_{\Gamma} y \frac{\partial P}{\partial n} dS \qquad (10)$$

The second integral in (10) is not, in general, a Gauss integral. The discrete analog of equation (10) contains N equations and N+1 unknowns (N values of $\dfrac{\partial \overline{\Psi}}{\partial n}$ and C). For the system completion it's necessary to introduce an additional equation to determine the unknown stream function boundary value, which depends one-to-one on the circulation about the contour. We shall consider now several examples of that equation.

4.2. SMOOTH CONTOUR CASE

4.2.1. *Prescribed velocity*

At some point (x_c, y_c) on the contour the velocity is defined, that is the value of $\dfrac{\partial \overline{\Psi}}{\partial n}$ is defined on the boundary element containing this point:

$$(\frac{\partial \overline{\Psi}}{\partial n})_j = V_k \qquad (11)$$

4.2.2. *Prescribed separation point*
A point of streamline separation is prescribed. Two approaches are possible for this additional condition:
a) At the point of streamline separation the tangential velocity is equal zero:

$$(\frac{\partial \overline{\Psi}}{\partial n})_j = -n_y \qquad (12)$$

b) Some point, called Zhukovsky or Kutta point, is chosen on the separated streamline. The stream function at this point is equal to C. Thus:

$$C - y_c = \sum_{k=1}^{N}\left(\frac{\partial\overline{\Psi}}{\partial n}\right)_k \int_{\Gamma k} P(x_c, y_c)dS - C\int_{\Gamma}\frac{\partial P}{\partial n}(x_c, y_c)dS + \int_{\Gamma} y\frac{\partial P}{\partial n}(x_c, y_c)dS \qquad (13)$$

4.2.3. *Prescribed circulation*

Circulation about the contour is prescribed. Thus by definition an equation exists in discrete form:

$$\sum_{i=1}^{N}\left(\left(\frac{\partial\overline{\Psi}}{\partial n}\right)_i + n_{y_j}\right)S_I = \gamma \qquad (14)$$

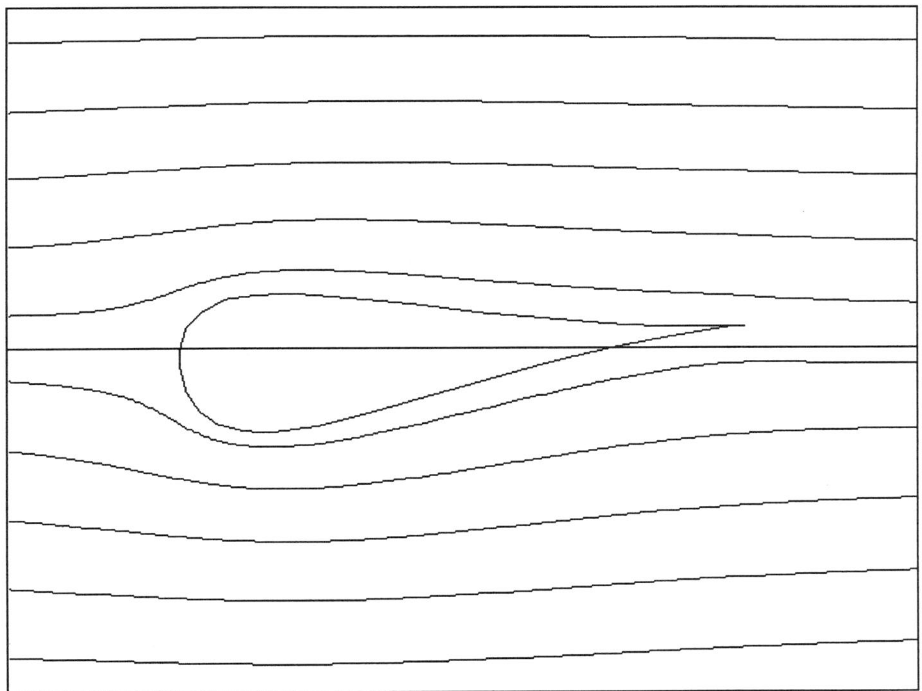

Figure 1. Stream lines picture for flow around a symmetric Zhukovsky airfoil under angle of attack 6°. The Zhukovsky point is situated on the airfoil chord.

4.3 CONTOUR WITH CORNER POINTS

4.3.1. *Sharp trailing edge*
Three approaches are possible in this case:

Figure 2. Stream lines picture near the trailing edge for the same flow as in Figure 1. Circulation is equal 0.0052.

a) The velocity on the upper surface near the trailing edge is equal to the velocity on the lower surface:

$$\left(\frac{\partial \overline{\Psi}}{\partial n}\right)_j + n_{y_j} = \left(\frac{\partial \overline{\Psi}}{\partial n}\right)_{j+1} + n_{y_{j+1}} \tag{15}$$

b) The Zhukovsky point is chosen after the trailing edge:

$$C - y_c = \sum_{k=1}^{N}\left(\frac{\partial \overline{\Psi}}{\partial n}\right)_k \int_{\Gamma k} P(x_c, y_c)dS - C\int_{\Gamma}\frac{\partial P}{\partial n}(x_c, y_c)dS + \int_{\Gamma} y\frac{\partial P}{\partial n}(x_c, y_c)dS \tag{16}$$

4.3.2. *Corner point in general case*

The velocity in this point must be equal to zero, otherwise there is a singularity at this point. The usual zero order boundary element method is not applicable in this case and it is necessary to use first or higher order algorithms:

$$\left(\frac{\partial\overline{\Psi}}{\partial n}\right)_j + n_{y_j} = 0, \qquad \left(\frac{\partial\overline{\Psi}}{\partial n}\right)_{j+1} + n_{y_{j+1}} = 0 \tag{17}$$

Thus $\dfrac{\partial\overline{\Psi}}{\partial n}$ is a discontinuous function because of the n_y discontinuity.

5. Vortical flow

5.1. NUMERICAL ALGORITHM

Discrete vortices algorithm for vorticity approximation inside the flow domain was used. It is a quite usual approach, as described in publications [1-4, 14].

Vorticity will be approximated by a discrete vortices system or by a vortex sheet. Let

$$\omega(x,y) = \sum_{i=1}^{M} \omega_i \delta(x - x_i, y - y_i) \tag{18}$$

where M - number of free vortices, ω_i - their intensities, δ - Dirac delta function. Thus, by (6):

$$\chi(x_0,y_0)\overline{\Psi}(x_0,y_0) = \oint_\Gamma P(x,y,x_0,y_0)\frac{\partial\overline{\Psi}}{\partial n}dS$$
$$-\oint_\Gamma \overline{\Psi}\frac{\partial P}{\partial n}(x,y,x_0,y_0)dS + \sum_{i=1}^{M}\omega_i P(x_i,y_i,x_0,y_0) \tag{19}$$

It has been assumed that the vortex does not influence the velocity at a point where it is situated. The velocity at any point of the flow domain can be calculated as follows:

$$V_x = 1 + \frac{\partial\overline{\Psi}}{\partial y_0} = 1 + \oint_\Gamma \frac{\partial P}{\partial y_0}(x,y,x_0,y_0)\frac{\partial\overline{\Psi}}{\partial n}dS$$

$$-\oint_\Gamma \overline{\Psi}\frac{\partial^2 P}{\partial n\partial y_0}(x,y,x_0,y_0)dS + \sum_{i=1}^{M}\omega_i\frac{\partial P}{\partial y_0}(x_i,y_i,x_0,y_0),$$

$$\tag{20}$$

$$V_y = -\frac{\partial\overline{\Psi}}{\partial x_0} = -\oint_\Gamma \frac{\partial P}{\partial x_0}(x,y,x_0,y_0)\frac{\partial\overline{\Psi}}{\partial n}dS$$

$$+\oint_\Gamma \overline{\Psi}\frac{\partial^2 P}{\partial n\partial x_0}(x,y,x_0,y_0)dS - \sum_{i=1}^{M}\omega_i\frac{\partial P}{\partial x_0}(x_i,y_i,x_0,y_0)$$

and the vortices motion is described by the system of equations:

$$\dot{x}_k = V_x(x_k, y_k) \qquad \dot{y}_k = V_y(x_k, y_k) \qquad (k = 1 \ldots M), \qquad (21)$$

where V_x, V_y are determined from (20).

5.2. VORTICITY DIFFUSION AND ALTERNATIVE VORTEX MODELS

Generally speaking, singularity is only a mathematical model, it is impossible in real flow, and for points situated near the singularity it must be replaced by another model, for example vorticity diffusion in viscous flow. There is no mechanism of vortices attraction or repulsion in ideal fluid flow, but if we consider a vortices pair, situated one near the other, at points (x_1, y_1) and (x_2, y_2), the stream function of that flow is

$$\Psi(x, y) = \omega_1 P(x, y, x_1, y_1) + \omega_2 P(x, y, x_2, y_2)$$

with function P determined by relation (7). Now, replace the stream function by its expansion in Taylor series, and neglect the second order terms to obtain

$$\Psi(x, y) = (\omega_1 + \omega_2) P(x, y, x_0, y_0) + (\omega_1 h_1 - \omega_2 h_2) \frac{\partial P}{\partial h} + \cdots$$

where the point (x_0, y_0) is the position of the new structure, approximating the vortices pair, h_1 is the distance between the points (x_1, y_1) and (x_0, y_0), h_2 - between points (x_2, y_2) and (x_0, y_0). But $\dfrac{\partial P}{\partial h}$ is a formula of a doublet, thus the approximation of vortices pair by a doublet (of value $\omega_1 h_1 - \omega_2 h_2$) and a vortex with value ($\omega_1 + \omega_2$). Since h_1 and h_2 are small, the far stream function and velocity fields are determined by the vortex, but the doublet determines the near field.

If we neglect convective effects, we obtain the vorticity distribution:

$$\omega(x, y, t) = \frac{\text{Re}}{4 \cdot \pi \cdot t} \exp\left[-\text{Re}\frac{(x - x_0)^2 + (y - y_0)^2}{4 \cdot t} \right] \qquad (22)$$

where (x_0, y_0) is the centre of the vortex location, corresponding to discrete vortex position, and t is the time after the start of the vorticity diffusion process.

The right-hand side of (22) decreases fast so that (19) can be put in the form

$$\chi(x_0, y_0)\overline{\Psi}(x_0, y_0) = \oint_\Gamma P(x, y, x_0, y_0)\frac{\partial \overline{\Psi}}{\partial n} dS - \oint_\Gamma \overline{\Psi}\frac{\partial P}{\partial n}(x, y, x_0, y_0) dS$$

$$+ \sum_{i=1}^{M} \omega_i \int_{\Omega_j} \frac{\text{Re}}{4 \cdot \pi \cdot t} \exp\left[-\text{Re}\frac{(x - x_i)^2 + (y - y_i)^2}{4 \cdot t} \right] P(x, y, x_0, y_0) dxdy \qquad (23)$$

where Ω_j is a small domain, within which diffusion effects are essential.

If a discrete vortex was generated by vortex sheet disintegration, two models can be proposed. The first one: vorticity and doublets are distributed along the boundary of vortical structure and (19) can be written as:

$$\chi(x_0,y_0)\overline{\Psi}(x_0,y_0)=\oint_\Gamma P(x,y,x_0,y_0)\frac{\partial\overline{\Psi}}{\partial n}\,dS-\oint_\Gamma\overline{\Psi}\frac{\partial P}{\partial n}(x,y,x_0,y_0)\,dS$$

$$+\sum_{i=1}^{M}\int_{\partial\Omega_i}[\omega_i P(x,y,x_0,y_0)+d_i(\frac{(x-x_0)a_x+(y-y_0)a_y}{((x-x_0)^2+(y-y_0)^2)})]ds \qquad (24)$$

where $\partial\Omega_i$ is the boundary of the i-th vortical structure, d_i is the doublet intensity, and a_x, a_y are the doublet axis direction cosines.

The second one is a region with constant vorticity:

$$\chi(x_0,y_0)\overline{\Psi}(x_0,y_0)=\oint_\Gamma P(x,y,x_0,y_0)\frac{\partial\overline{\Psi}}{\partial n}\,dS$$

$$-\oint_\Gamma\overline{\Psi}\frac{\partial P}{\partial n}(x,y,x_0,y_0)\,dS+\sum_{i=1}^{M}\omega_i\int_{\Omega_i}P(x,y,x_0,y_0)\,dxdy \qquad (25)$$

6. Examples of calculations

Several ideal fluid flow problems were solved to illustrate the proposed algorithms.

6.1. HYDRODYNAMICS INTERACTION

The problem of two circular contours' interaction was considered as an example. The case of cylinder's attraction is shown in Figure 3. The case of the cylinder's repulsion is shown in Figure 4.

A variety of different pictures can be obtained, changing the circulations along the cylinders, as shown in Figure 5.

Consider now an interaction of several discrete vortices with circular contours. There may be a situation, when discrete vortex or discrete vortices are attracted to the contours or are repulsed from them. An example of such a situation is shown in Figure 6.

Consider the problem of vortex-vortex interaction in real flow using, as an example, disintegration of vortex sheet, separated from the trailing edge of a symmetric Zhukovsky airfoil with a small angle of attack. A spiral structure, created by discrete vortices only, is seen in Figure 7. Combined discrete vortices-doublets created a closed structure, which is seen in Figure 8. It is rather difficult to choose the best algorithm, because any approach, excluding an infinite velocity near the discrete vortex, gives similar results in the general case, but, for example, using vortex-doublet combination is convenient for calculation of the vortex sheet after a thin body.

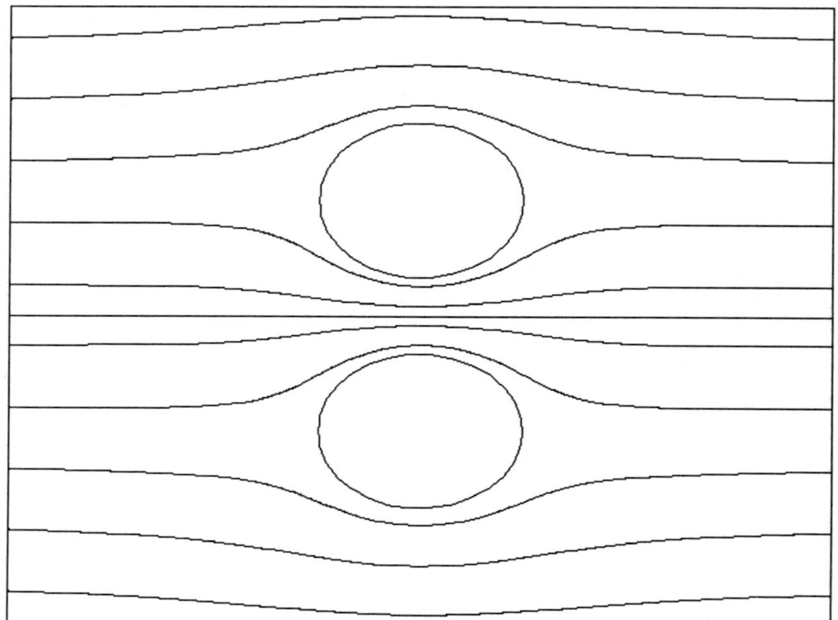

Figure 3. Stream lines picture of flow about two circular cylinders, with unit radii. The distance between centres is 3. An attraction of cylinders takes place. Circulations are $\gamma_1=2.44$, $\gamma_2=-2.44$. Interaction forces are $F_{x1}=-0.26\times10^{-5}$, $F_{x2}=0.26\times10^{-5}$, $F_{y1}=-8.65$, $F_{y2}=8.65$.

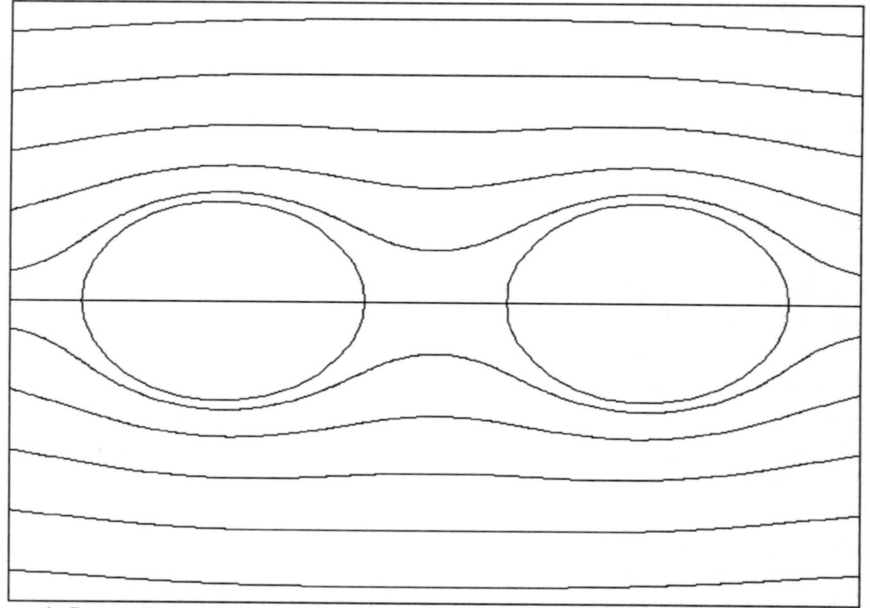

Figure 4. Stream lines picture of flow about two circular cylinders, radii are 1, the distance between centres is 3. A repulsion of cylinders takes place. Circulations are $\gamma_1=0.13\times10^{-5}$, $\gamma_2=-0.22\times10^{-5}$. Interaction forces are $F_{x1}=-0.68$, $F_{x2}=0.680.26\times10^{-5}$, $F_{y1}=-0.39\times10^{-6}$, $F_{y2}=0.45\times10^{-5}$.

Figure 5. Stream lines picture of flow around two circular cylinders, of unit radii. The distance between centres is 3. Circulations are $\gamma_1=3.89$, $\gamma_2=-3.89$. Interaction forces are $F_{x1}=1.46$, $F_{x2}=-1.46$, $F_{y1}=-5.79$, $F_{y2}=5.79$.

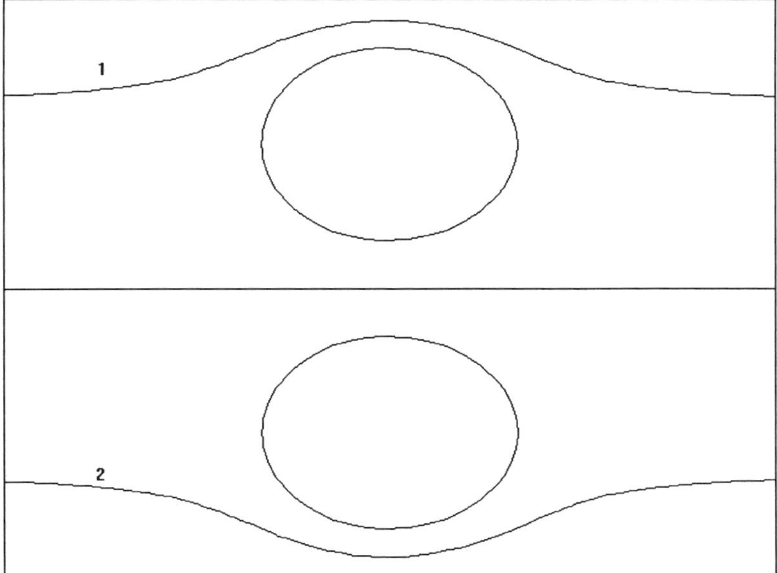

Figure 6. Trajectories of two vortices with opposing signs in flow around two circular cylinders. The cylinders' radii are 1, and the distance between centres is 4. Vortices are $\omega_1=-1.0$, $\omega_2=1.0$.

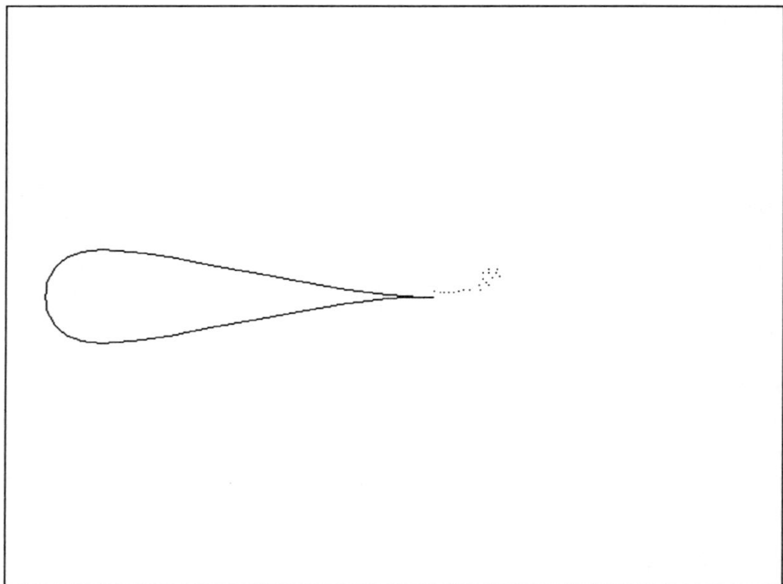

Figure 7. Calculation of starting stage of a vortex sheet after trailing edge of a symmetric Zhukovsky airfoil with small angle of attack. Pure discrete vortices model is used in calculation. The dots correspond to discrete vortices locations.

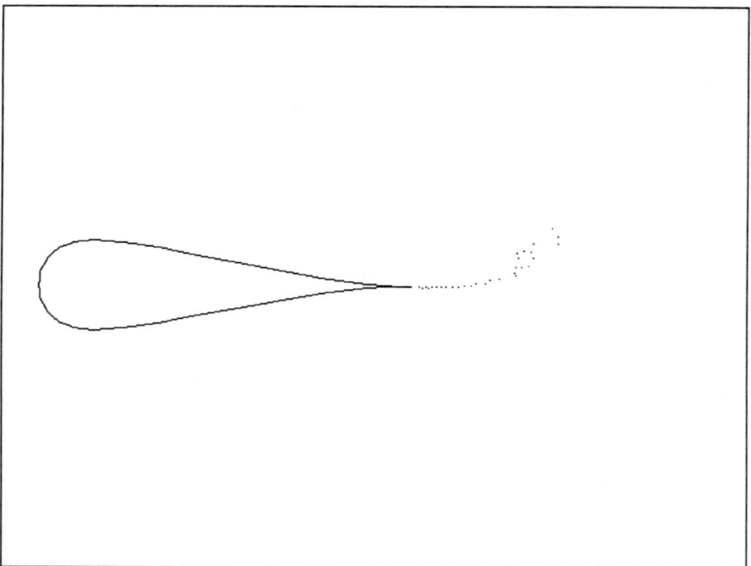

Figure 8. Calculation of starting stage of a vortex sheet after trailing edge of a symmetric Zhukovsky airfoil with small angle of attack. Proposed combination of discrete vortices doublets is used in calculation. The dots correspond to discrete vortices locations.

7. Conclusions

The proposed algorithms, based on boundary element and discrete vortices method combination, can successfully compete with other numerical methods for external incompressible fluid flow calculations. Moreover, since the proposed algorithms are based on representation of singularities, they are more convenient for physical singularity modelling than other numerical methods. The proposed algorithm cannot improve the full viscous flow calculations, but it can be useful as a first approximation. The consideration of physical singularities, which was used in alternative vorticity distribution analysis in the present work, can be considered only as a first approximation too, because convective terms were not taken into account. The proposed algorithm has better accuracy than the discrete vortices method, and is more convenient for vortical flow calculation than the traditional boundary element method. The analysis in the present work was restricted to the simplest examples, but it is enough to prove the algorithm effectiveness.

Further development of the proposed approach is connected with the search for stable vortical structures and using them instead of discrete vortices. It is desirable, that the new algorithms accurately describe the flow field and not only the integral values.

8. References

1. Belotserkovsky, S.M., Lifanov, I.K. (1993) *Method of Discrete Vortices*, CRC Press.
2. Belotserkovsky, S.M., Lifanov, I.K. (1985) *Numerical methods in singular integral equations*, Nauka, Moscow (in Russian).
3. Belotserkovsky, S.M., Kotovsky, V.N., Nisht, M.I., Fedorov, R.M. (1988) *Mathematical modelling of plane-parallel separated flows around bodies*, Nauka, Moscow (in Russian).
4. Lifanov, I.K. (1995) *Singular integral equations method and numerical experiment*, Yanus, Moscow (in Russian).
5. Leonard, A. (1987) Vortex Method for Flow Simulation, *J. Computational Physics*, 13, 183-202.
6. Leonard, A. (1985) Computing Three-Dimensional Incompressible Flows with Vortex Elements, *Ann. Rev. Fluids Mechanics*, 17, 523-559.
7. Chorin, A. J. (1973) Numerical Study of Slightly Viscous Flow, *J. Fluid Mechanics*, 57, 785-796.
8. Chorin, A.J., Bernard, P.S. (1973) Discretization of Vortex Sheet with an Example of Roll-up, J. Computational. Physics, 13, 423-429.
9. Chorin, A.J. (1994) *Vorticity and Turbulence*, Springer-Verlag, Berlin, New York.
10. Banerjee, P.K., Butterfield, R. (1981) *Boundary element methods in engineering science*, McGraw-Hill, London, New York.
11. Brebbia, C.A., Telles, J.C.F., Wrobel, L.C. (1984) *Boundary element techniques*, Springer-Verlag, Berlin, New York.
12. Wu, J.C., Thompson, J.F. (1973) Numerical Solutions of Time-Dependent Incompressible Navier-Stokes Equations using an Integro-Differential Formulation, *Computers and Fluids*, 1, 197-215.
13. Wu, J.C. (1976) Numerical Boundary Conditions for Viscous Flow Problems, *AIAA Jnl*, 14, 1042-1049.
14. Wu, J.C. (1982) Problems of General Viscous Flow, in P.K. Banerjee and R. Show (eds.), *Development in Boundary Element Method, Vol. 2*, Applied Science Publishers, London, pp. 69-111.
15. Patterson, C., Sheikh, M.A. (1983) On the Use of Higher Order Weighting Functions in the Boundary Integral Method for Fluid Flow, in S.A. Taylor (ed), *Numerical Methods in Laminar and Turbulent Flow*, Pineridge Press, Swansea, pp. 1163-1173.

16. Belyaev, N.M., Yevdokymov, D.V. (1986) Calculation of irrotational plane ideal fluid flows by boundary element method, in N.M. Belyaev (ed.), *Mathematical methods of fluid and gas mechanics*, Dniepropetrovsk University, Dniepropetrovsk, pp. 83-86 (in Russian).

17. Belyaev, N.M., Yevdokymov, D.V. (1989) Modelling of vortical plane ideal fluid flows by boundary element method, in N.M. Belyaev (ed.), *Calculation of fluid and gas flows*, Dniepropetrovsk University, Dniepropetrovsk, pp. 4-11 (in Russian).

18. Yevdokymov, D.V. (1988) Boundary element method for potential ideal fluid flow calculations, in N.M. Belyaev (ed.), *Numerical solutions of fluid and gas mechanics problems*, Dniepropetrovsk University, Dniepropetrovsk, pp. 113-116 (in Russian).

19. Yevdokymov, D.V. (1997) Combination of boundary element and discrete vortices method, in V.K. Khrutch (ed.), *Mathematical methods of fluid and gas mechanics*, Dniepropetrovsk University, Dniepropetrovsk, (in press) (in Russian).

METHOD OF NUMERICAL ANALYSIS OF STRESS SINGULARITY AT SINGULAR POINTS IN TWO- AND THREE-DIMENSIONAL BODIES

V.P. MATVEYENKO, S.M. BORZENKOV, S.G. MINAKOVA
Institute of Continuous Media Mechanics, Ural Department
of Russian Academy of Science, 1 Korolyov Str., Perm,
614061, Russia

1. Introduction

The analysis of stress singularity in the vicinity of singular points is basically made by two approaches. The first approach involves construction of solutions satisfying homogeneous equations and homogeneous boundary conditions for the regions incorporating singular points [1-5]. The second approach uses Mellin's transformation and the residue theory [6-9]. These approaches were applied to examine singularity in the vicinity of singular points practically in all situations arising in two-dimensional regions of isotropic materials. They were also effective in studying the stress singularity at the wedge edges in three-dimensional problems, which can be generally reduced to the analysis of the plane-strained state and the analysis of the antiplane strain [3,9].

To our knowledge, the two-dimensional problems of stress singularity in the vicinity of singular points in isotropic bodies have been well-studied, whereas an estimation of stress singularity in anisotropic bodies, as a more complicated problem, is yet far from complete understanding. Some of the pioneering studies on this subject based on the analytical methods are reported in [9,10]. However, in three-dimensional regions the analysis of stress singularity at singular points, for example at the vertices of a multi-edged wedge cannot be reduced to a two-dimensional problem. This probably explains the fact that, until recently, there have been no publications in literature presenting numerical estimations of stress singularity at such singular points.

To advance in this direction one needs to develop an effective numerical-analytical method allowing estimation of stress singularity at a vertex of a plane composite wedge made from the anisotropic materials and at a vertex of a three-dimensional wedge. In this paper, we consider one of the possible methods based on numerical development of eigensolutions for plane and spatial wedge problems.

2. Statement of eigensolution problem for a plane and composite wedge

We consider a semi-infinite composite wedge in the system of polar co-ordinates r,φ (Fig.1). Let S_1 and S_2 be the parts of the composite wedge which are assumed to satisfy

231

D. Durban and J.R.A. Pearson (eds.),
IUTAM Symposium on Non-Linear Singularities in Deformation and Flow, 231-242.
© 1999 *Kluwer Academic Publishers. Printed in the Netherlands.*

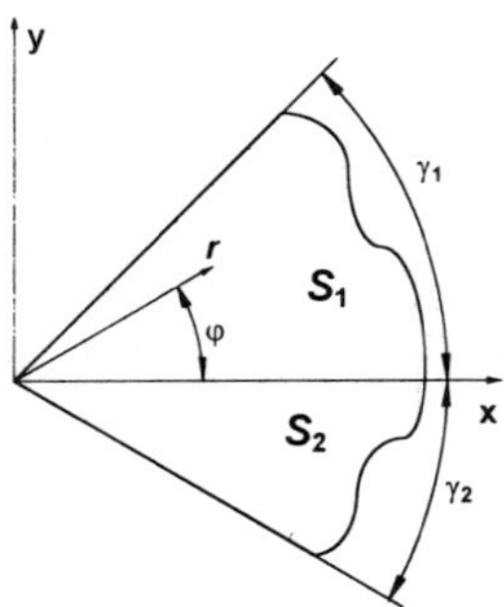

Figure 1.

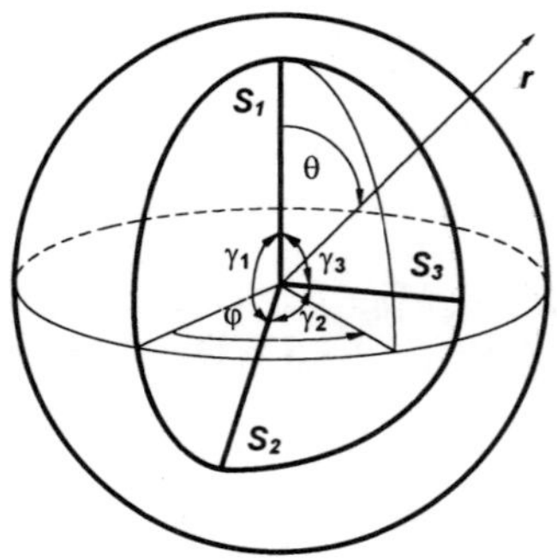

Figure 2.

one of the versions of the boundary conditions

$$u_r^1 = u_r^2 , \qquad u_\varphi^1 = u_\varphi^2 , \qquad \sigma_{\varphi\varphi}^1 = \sigma_{\varphi\varphi}^2 , \qquad \sigma_{r\varphi}^1 = \sigma_{r\varphi}^2 , \tag{1.a}$$

$$u_\varphi^1 = u_\varphi^2 , \qquad \sigma_{\varphi\varphi}^1 = \sigma_{\varphi\varphi}^2 , \qquad \sigma_{r\varphi}^1 = 0 , \qquad \sigma_{r\varphi}^2 = 0 \tag{1.b}$$

at the contact surface of the wedge components, and

$$u_r = 0 , \qquad u_\varphi = 0 , \tag{2.a}$$

$$\sigma_{\varphi\varphi} = 0 , \qquad \sigma_{r\varphi} = 0 , \tag{2.b}$$

$$u_r = 0 , \qquad \sigma_{r\varphi} = 0 \tag{2.c}$$

at the lateral facets. Thus, for anisotropic elastic body the physical relations in the system of polar co-ordinates are

$$\sigma_{rr} = C_{11}\varepsilon_r + C_{12}\varepsilon_\varphi + C_{13}\gamma_{r\varphi} ,$$

$$\sigma_\varphi = C_{21}\varepsilon_r + C_{22}\varepsilon_\varphi + C_{23}\gamma_{r\varphi} , \tag{3}$$

$$\sigma_{r\varphi} = C_{31}\varepsilon_r + C_{32}\varepsilon_\varphi + C_{33}\gamma_{r\varphi}$$

in which the elastic constants $C_{ij} = C_{ji}$ are the function of φ-co-ordinate. The eigensolutions to a wedge problem are found in the following form

$$u_i = r^\lambda \xi_i(\varphi) , \qquad (i = 1, 2) , \tag{4}$$

where $u_1 = u_r$, $u_2 = u_\varphi$. A substitution of solution (4) in the equilibrium equation and cancellation of $r^{\lambda-2}$ lead to a system of ordinary differential equations, where the function $\xi_i(\varphi)$ and parameter λ have variable coefficients.

$$L_1(\lambda, \xi_1, \xi_2) = 0, \qquad L_2(\lambda, \xi_1, \xi_2) = 0. \tag{5}$$

These equations written for each component of the composite wedge are to be supplemented with the boundary conditions for the function $\xi_i(\varphi)$.

Thus, we arrive at the eigenvalue problem, in which the values of the parameter λ provide the information on the character of the stress singularity.

3. Formulation of eigensolution problem for three-dimensional wedge

Consider a semi-infinite trihedral wedge (Fig.2), whose vertex lies at the centre of spherical co-ordinates, one edge coincides with the axis $\theta=0$, and one of the facets adjacent to the edge $\theta=0$ lies in the plane $\varphi=0$. Consideration of a trihedral wedge instead of polyhedral one does not infringe upon the general line of our reasoning. To analyse the character of stress singularity we need to develop eigensolutions, such that, they can satisfy in the examined domain the equilibrium equations

$$\frac{1}{1-2v}\,\text{grad div }\mathbf{u} + \nabla^2\mathbf{u} = 0 \tag{6}$$

and the mixed uniform boundary conditions on the lateral facets, for which we can prescribe either zero displacements

$$\mathbf{u} = 0, \tag{7}$$

or zero stresses

$$\frac{v}{1-2v}\,\mathbf{n}\,\text{div }\mathbf{u} + \mathbf{n} \bullet \nabla\mathbf{u} + \frac{1}{2}\mathbf{n}\times\text{rot }\mathbf{u} = 0, \tag{8}$$

or combination of zero values of the displacement vector $\mathbf{u}$ and the stress tensor. Here v is the Poisson ratio, $\mathbf{n}$ is the unit vector of the outward normal, the dot denotes a scalar product , and the cross denotes a vector product.

In view of [11,12], the eigensolutions allowing for the stress singularities at the edges of the three-dimensional wedge can be defined by

$$u_k(r,\theta,\varphi) = r^{\eta}\rho_1^{\alpha_1}\rho_2^{\alpha_2}\rho_3^{\alpha_3}\xi_k(\theta,\varphi), \qquad (k{=}1,2,3), \tag{9}$$

where $u_1 = u_r$, $u_2 = u_\theta$, $u_3 = u_\varphi$, $\xi_k(\theta,\varphi)$ - are sufficiently smooth functions, $\rho_i(r,\theta,\varphi)$, ($i{=}1,2,3$) is the equivalent of the distance to the i-th edge , α_i are the eigenvalues for a plane wedge formed by cross-sectioning of the plane perpendicular to the i-th edge of the wedge and the facets of the spatial wedge.

After determining functions $\rho_i(r,\theta,\varphi)$ in terms of a distance to the i-th edge we can write them in the following form:

$$\rho_i(r,\theta,\varphi) = r\beta^{1/2}, \tag{10}$$

$$\beta_i = 1 - \left[\cos\theta\cos\theta_i + \sin\theta\sin\theta_i\cos\left(\varphi - \varphi_i\right)\right]^2,$$

where the straight line $\theta = \theta_i$, $\varphi=\varphi_i$ specifies the i-th edge of the wedge. Then, relation (9) becomes

$$u_k(r,\theta,\varphi) = r^{\lambda}\sqrt{\beta_1^{\alpha_1}(\theta,\varphi)\,\beta_2^{\alpha_2}(\theta,\varphi)\,\beta_3^{\alpha_3}(\theta,\varphi)}\;\xi_k(\theta,\varphi), \tag{11}$$

$$\lambda = \eta + \alpha_1 + \alpha_2 + \alpha_3.$$

Substituting (11) into (6) gives the system of differential equations in terms of partial derivatives of the functions $\xi_k(\theta,\varphi)$ and parameter λ:

$$L_k(\lambda,\xi_1,\xi_2,\xi_3) = 0, \qquad (k{=}1,2,3). \tag{12}$$

In view of (11), boundary conditions (7) and (8) are transformed to

$$\xi_i = 0, \qquad M_k\left(\lambda, \xi_1, \xi_2, \xi_3\right) = 0. \tag{13}$$

Hence, the formulated problem is reduced to an eigenvalue problem for partial differential equations and the character of stress singularity can be determined from the values of parameters λ.

4. Algorithms for numerical solutions of eigenvalue problems for a plane and spatial wedge

Incorporation of variable coefficients in the system of equations (5) makes impossible the application of traditional approaches based on searching an analytical solution for ξ_j. Here we develop another computational algorithm for these equations. First, we write the differential equations for each wedge component in a weak form [13]. To this end, it is necessary to multiply these equations by the test functions $\psi_i(\varphi)$ and to integrate over the arc φ determining the value of the wedge angle

$$\int_0^\gamma \left(L_1 \psi_1 + L_2 \psi_2\right) d\varphi = 0. \tag{14}$$

These equations with boundary conditions are written for each component of the wedge. The problem defined in such a way is solved by the Galerkin method , which is an explicit discretization of a weak form of equations (14). The choice of the functions approximating the solution $\xi_i(\varphi)$ and the test functions $\psi_i(\varphi)$ is made using the finite-element method. With consideration for the order of the derivative functions ξ_i, involved in L_1 and L_2, the finite element used for solution approximation must ensure the continuity of the functions $\xi_i(\varphi)$, as well as the continuity of their first derivatives. For approximation of the test functions we employ the same finite elements as for the functions $\xi_i(\varphi)$. As a finite element we choose a two-node one-dimensional element with Hermite interpolation of the function by a third-degree polynomial [13]. A combined application of the Galerkin procedure and the finite-element method renders the problem under consideration reducible to an algebraic problem of determining eigenvalues λ and eigenvectors, which are the discrete analogs of the eigenfunctions $\xi_i(\varphi)$.

By making use of the Gauss-Ostrogradsky transformation we can reduce the order of derivatives involved in L_1 and L_2. Representation of differential equation in a weak form (5) requires that during approximation of solutions and test functions in the element the functions be continuous only in the region of their definition. In this case, for finite elements we have used one-dimensional elements with Lagrangian interpolation of the function by a linear polynomial.

In this work, we have develop a numerical algorithm for seeking eigensolutions to the spatial wedge problem. According to this algorithm, differential equations (12) are written in a weak form, suggesting their multiplication by the test functions $\psi(\theta, \varphi)$ and integrating over the domain S generated on the sphere by a trihedral angle

$$\int_S \left[\sum_{j=1}^{3} L_j(\lambda,\xi_j)\psi_j\right] dS = 0. \tag{15}$$

After making identical transformations to reduce the order of the derivative functions in equation (15) and taking into account conditions (13), we obtain the final equation

$$\int_\varphi \int_\theta \left[\beta_1^{\alpha_1-2}\beta_2^{\alpha_2-2}\beta_3^{\alpha_3-2}\left\{\left[2(1-v)(\lambda^2+\lambda-2)\sin\theta\xi_1+(\lambda-3+4v)(\cos\theta\xi_2+\right.\right.\right.$$

$$+\sin\theta\frac{\partial\xi_2}{\partial\theta}+\frac{\partial\xi_3}{\partial\varphi}\,\left.\right)\right]\psi_1-(1-2v)\left(\sin\theta\frac{\partial\xi_1}{\partial\theta}\frac{\partial\psi_1}{\partial\theta}+\frac{1}{\sin\theta}\frac{\partial\xi_1}{\partial\varphi}\frac{\partial\psi_1}{\partial\varphi}\right)+$$

$$+\left[(\lambda-3+4v)(\sin\theta A\xi_2+B\xi_3)+(1-2v)\right[\left(F\sin\theta+\frac{D}{\sin\theta}\right)\xi_1+\sin\theta A\frac{\partial\xi_1}{\partial\theta}+$$

$$+\frac{B}{\sin\theta}\frac{\partial\xi_1}{\partial\varphi}\,\left]\psi_1+\left[(1-2v)\lambda(\lambda+1)\sin\theta\xi_2-\frac{2(1-v)}{\sin\theta}\xi_2+(4+\lambda-4v)\sin\theta\frac{\partial\xi_1}{\partial\theta}-\right.\right.$$

$$-(3-4v)\,\mathrm{ctg}\,\theta\frac{\partial\xi_3}{\partial\varphi}\,\left]\psi_2-2(1-v)\frac{\partial\xi_2}{\partial\theta}\frac{\partial\psi_2}{\partial\theta}\sin\theta-\right.$$

$$-(1-2v)\left(\frac{1}{\sin\theta}\frac{\partial\xi_2}{\partial\varphi}\frac{\partial\psi_2}{\partial\varphi}+\frac{\partial\xi_3}{\partial\theta}\frac{\partial\psi_2}{\partial\varphi}\right)-2v\frac{\partial\xi_3}{\partial\varphi}\frac{\partial\psi_2}{\partial\theta}+\left[(4+\lambda-4v)\sin\theta A\xi_1+\right.$$

$$+2(1-v)F\sin\theta\xi_2+\frac{(1-2v)}{\sin\theta}D\xi_2+2(1-v)\sin\theta A\frac{\partial\xi_2}{\partial\theta}-\mathrm{ctg}\,\theta(3-4v)B\xi_3+$$

$$+E\xi_3+2vB\frac{\partial\xi_3}{\partial\theta}+(1-2v)\left(A\frac{\partial\xi_3}{\partial\varphi}+\frac{B}{\sin\theta}\frac{\partial\xi_2}{\partial\varphi}\right)\,\left]\psi_2+\left[(1-2v)(\lambda^2\sin\theta+\right.\right.$$

$$+\lambda\sin\theta-\frac{1}{\sin\theta}\,\left)\xi_3+(4+\lambda-4v)\frac{\partial\xi_1}{\partial\varphi}+(3-4v)\mathrm{ctg}\,\theta\frac{\partial\xi_2}{\partial\varphi}\,\right]\psi_3-$$ \hfill (16)

$$-(1-2v)\left(\sin\theta\frac{\partial\xi_3}{\partial\theta}\frac{\partial\psi_3}{\partial\theta}+\frac{\partial\xi_2}{\partial\varphi}\frac{\partial\psi_3}{\partial\theta}\right)-\frac{2(1-v)}{\sin\theta}\frac{\partial\xi_3}{\partial\varphi}\frac{\partial\psi_3}{\partial\varphi}-2v\frac{\partial\xi_2}{\partial\theta}\frac{\partial\psi_3}{\partial\varphi}+$$

$$+\left[(4+\lambda-4v)B\xi_1+\xi_2\big(\mathrm{ctg}\,\theta(3-4v)B+E\big)+\frac{2(1-v)}{\sin\theta}\left(D\xi_3+B\frac{\partial\xi_3}{\partial\varphi}\right)+\right.$$

$$+(1-2v)\left(F\sin\theta\xi_3+B\frac{\partial\xi_2}{\partial\theta}+A\sin\theta\frac{\partial\xi_3}{\partial\theta}\right)\psi_3\right\}d\theta d\varphi+$$

$$+\int_l\left[\beta_1^{\alpha_1-2}\beta_2^{\alpha_2-2}\beta_3^{\alpha_3-2}\left\{(1-2v)\left[n_2\big(\xi_2(1-\lambda)-A\xi_1\big)+n_3\big(\xi_3(1-\lambda)-\right.\right.\right.$$

$$-\frac{B}{\sin\theta}\xi_1\,\left)\right]\psi_1-2n_2\left[(1+v\lambda)\xi_1+\big(v\mathrm{ctg}\,\theta+(1-v)A\big)\xi_2+\frac{vB}{\sin\theta}\xi_3\,\right]\psi_2-$$

$$-(1-2v)n_3\left[\xi_3(A-\mathrm{ctg}\,\theta)+\frac{B}{\sin\theta}\xi_2\right]\psi_2-2n_3\left[(1+v\lambda)\xi_1+\big(\mathrm{ctg}\,\theta(1-v)+vA\big)\xi_2+\right.$$

$$+\frac{(1-v)}{\sin\theta}B\xi_3\right]\psi_3-(1-2v)n_2\left[(A-\mathrm{ctg}\,\theta)\xi_3+\frac{B}{\sin\theta}\xi_2\right]\psi_3\,\right\}\right]dl=0,$$

where l is the boundary of the surface S subjected to the prescribed stresses, n_1, n_2, n_3 are the vector components of the normal $\mathbf{n}$,

$$A = \frac{1}{2}\sum_{i=1}^{3}\frac{\partial\beta_i}{\partial\theta}\frac{\alpha_i}{\beta_i}, \qquad B = \frac{1}{2}\sum_{i=1}^{3}\frac{\partial\beta_i}{\partial\varphi}\frac{\alpha_i}{\beta_i}, \qquad C = \frac{1}{2}\sum_{i=1}^{3}\left[\frac{\partial^2\beta_i}{\partial\theta^2} - \left(\frac{\partial\beta_i}{\partial\theta}\right)^2\bigg/\beta_i\right]\frac{\alpha_i}{\beta_i} + A^2,$$

$$D = \frac{1}{2}\sum_{i=1}^{3}\left[\frac{\partial^2\beta_i}{\partial\varphi^2} - \left(\frac{\partial\beta_i}{\partial\varphi}\right)^2\bigg/\beta_i\right]\frac{\alpha_i}{\beta_i} + B^2, \qquad E = \frac{1}{2}\sum_{i=1}^{3}\left[\frac{\partial^2\beta_i}{\partial\varphi\partial\theta} - \frac{\partial\beta_i}{\partial\theta}\frac{\partial\beta_i}{\partial\varphi}\frac{1}{\beta_i}\right]\frac{\alpha_i}{\beta_i} + AB,$$

$$F = \operatorname{ctg}\theta A + C.$$

The functions $\xi_k(\theta,\varphi)$, $\psi_k(\theta,\varphi)$, incorporated in equation (15) have been chosen using the finite element method. The finite element implementation of equation (15) is a rather complicated procedure, because it requires a usage of two-dimensional elements providing continuity of the functions ξ_k, ψ_k as well as continuity of their first derivatives. Without going into details we only note, that until recently, all attempts to effectively solve this problem by the finite element method have been a failure. A reduction of the derivative orders allows one to use such finite elements which assure continuity only of the functions ξ_k, ψ_k. For finite elements we use triangular elements with Lagrangian approximation of the functions ξ_k, ψ_k by linear polynomials. Fig. 3 illustrates a finite - element discretization of the domain S generated by a trihedral wedge on a sphere.

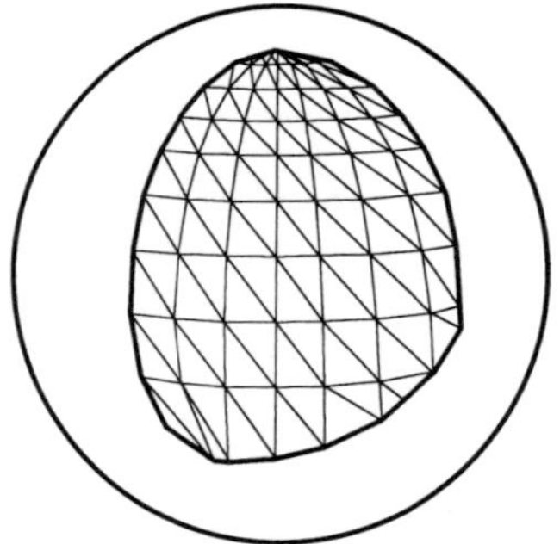

Figure 3.

It should be noted, however, that this particular formulation of the elasticity theory problem cannot be applied to the case when Poison's ratio is equal to 0.5. Furthermore, it is a well established fact that for weakly compressible materials an application of the finite element method to numerical treatment of some solid mechanics problems traditionally formulated in terms of displacements leads to essential computational errors. In this paper, an alternative approach is proposed to analyse the stress singularities in the bodies made of incompressible or weakly compressible materials. The basic idea of the method is to develop resolving equations in terms of the elasticity theory problem, which is valid for the whole range of Poisson's ratios [15]. In this problem, the desired quantities are the displacement vector and the function of the mean pressure $H = 3\sigma\big/\left[2\mu(1+\nu)\right]$, where σ is the mean stress, and μ is the shear module.

A scheme for constructing eigensolutions is similar to the scheme outlined above. In the finite element realisation of this problem the displacements in a triangular element are approximated by the linear polynomials, and H is assumed constant. After choosing the type of the finite elements we develop a resolving system of equations using the algorithms based on the standard finite element method procedures. The Galerkin procedure combined with the finite element method reduces the formulated problem to seeking eigenvalues λ and eigenvectors for an algebraic nonsymmetric matrix with band structure.

The complex eigenvalues of nonsymmetric band matrix of high order are found with the help of algorithms generated on the basis of Mueller's method [14].

5. Calculation of stress singularity power

Computational experiments were performed to demonstrate the capabilities of the proposed numerical-analytical method in calculating stress singularities at the vertices of homogeneous and composite anisotropic wedges.

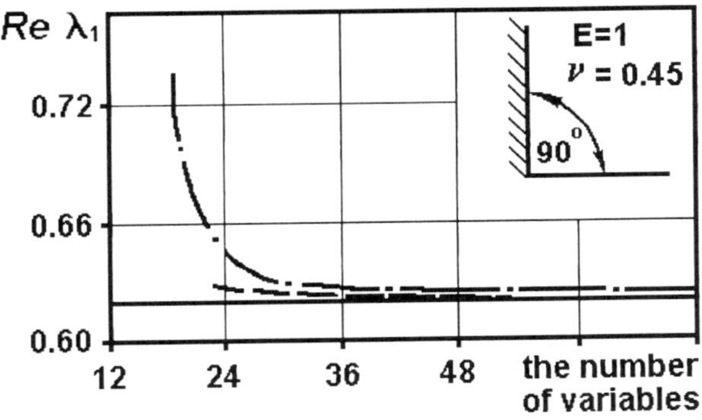

Figure 4.

The eigenvalues λ were determined for a homogeneous isotropic wedge with the mixed boundary conditions in a plane strain state. Fig. 4 shows the calculated minimum values of λ_1 for different numbers of nodal variables. Here the solid line denotes an exact value λ_1 determined by the known analytical methods , and the dashed line denotes values calculated with numerical-analytical method using in a finite element realisation of equation (14) the element with Hermite approximation of the function by a cubic polynomial. The dashed- dotted line represents the results of calculations mad for the lower order derivatives in equation (14) using the finite element with the Lagrangian interpolation of the function by a linear polynomial. The results in the figure show that the convergence of the numerical- analytical method depends on the degree of discretization. It is evident that the element with cubic polynomial provides a more rapid convergence.

We calculated eigenvalues λ for a composite wedge with rigidly fastened components and free facets. In Fig.5 the results from the numerical-analytical method are compared with the results reported in [4]. Variations of the first real eigenvalues and the real parts of the first complex eigenvalues are denoted by solid and dashed lines, respectively; the results from the numerical -analytical method are denoted by dots.

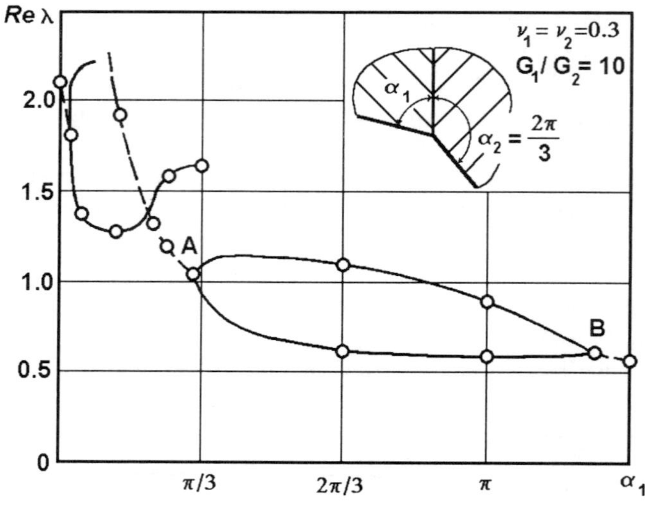

Figure 5.

We also estimated the eigenvalues λ for a composite wedge with free facets composed of rigidly bonded carbon plastic (angle α) and aluminium (angle β) components. The characteristic parameters of the carbon plastic are the following: $E_L = 1.962\cdot10^5$ MPa is the Young module in the direction of reinforcement coinciding with the wedge edge, $E_1 = E_Z = 0.20601\cdot10^5$ MPa is the transverse Young's module, $G_{LT} = G_{TZ} = G_{LZ} = 8,3385\cdot10^3$ MPa are the shear modules, and $\nu_{LT} = \nu_{LZ} = \nu_{TZ} = 0.21$ are the Poisson's ratio. The aluminium is characterised by $E = 0.70632\cdot10^5$ MPa, $\nu = 0.3$. Table 1 gives the real parts of the first eigenvalues λ obtained in the work [9] (column A) and by the numerical-analytical method (column B). The results are presented for several angles defining the composite wedge. A good agreement of numerical results demonstrates the efficiency and reliability of the proposed method.

TABLE 1.

α	β	λ_1		λ_2	
		A	B	A	B
80	90	0.75	0.752	0.55	0.5
90	180	0.97	0.978	0.67	0.65
45	180	0.84	0.840	-	-
112.5	180	0.85	0.852	0.61	0.6
112.5	90	0.79	0.788	-	-
157.5	90	0.82	0.827	0.57	0.562
225	90	0.55	0.562	0.47	0.455

To analyse the influence of anisotropy on the character of stress singularity we performed calculations for a wedge made of transversal-isotropic material with mixed boundary conditions on the lateral facets. Fig.6 illustrates different versions of the plane wedge region orientation with respect to the plane of isotropy X_1OX_2. The shaded facet is fixed. Fig. 7 gives the first eigenvalue as a function of elastic material constants for the second version at $\gamma=90°$. The plots in the figure illustrate the effect of anisotropy on the character of the stress singularity. A series of computations were performed to verify the reliability and efficiency of the proposed method for spatial wedges.

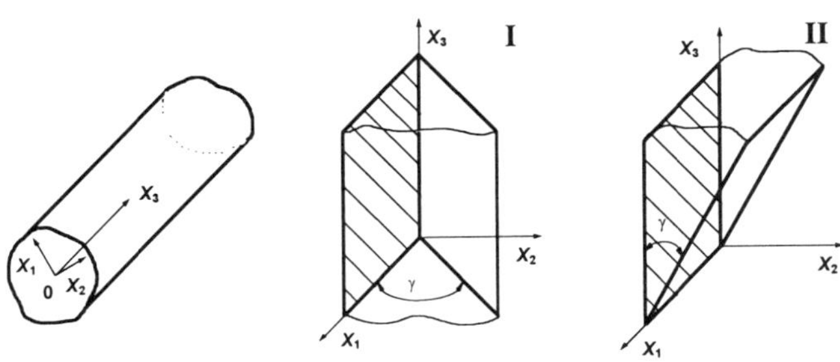

Figure 6. Schematic representation of wedge orientation with respect to isotropy plane for mixed boundary conditions.

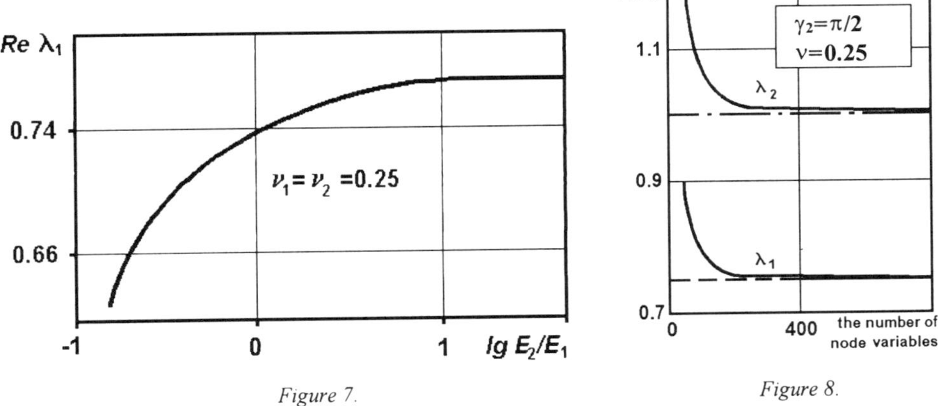

Figure 7.

Figure 8.

Here, we consider a particular case of three-dimensional wedge, for which we can compare the calculated results with the results from the plane wedge solutions. If the angles γ_1 and γ_2 are directed toward π a trihedral wedge degenerates to a dihedral one.

Fig. 8 presents the first two minimum eigenvalues λ versus the number of nodal displacements N for the domain (φ,θ) with a uniform finite-element mesh. The calculations are made for the angles $\gamma_1 = \gamma_3 =\pi$, $\gamma_2 =\pi/2$ under the requirement that the facet. S_1 is free of stress and the facet, S_3 is fixed, the Poisson's ratio being 0.25. In this example, the first eigenvalue for a dihedral angle coincides with the first eigenvalue for a plane wedge in a plain strain state and the second eigenvalue coincides with the first eigenvalue for a plane wedge in an antiplane strain state. Dashed lines correspond to the analytical solutions.

Similar numerical experiments were carried out for different variants of the boundary conditions (both facets are free of stress or fixed) and for different values of Poisson's ratio including $\nu=0.5$. Like in the problem defined in Fig.3, the obtained results agree fairly well with the results from the corresponding two-dimensional problem solution.

The next numerical experiment supporting the validity of the proposed approach is a numerical solution of a harmonic problem. Numerical estimation of solution singularities in this problem have been reported in [11,12]. The harmonic problem defined as $\Delta u = 0$ was considered in the framework of the proposed method. The solution to this problem was searched in the form of $u\left(r,\theta,\varphi\right) = r^{\lambda}\,\beta_1^{\alpha_1/2}\,\beta_2^{\alpha_2/2}\,\beta_3^{\alpha_3/2}\xi\left(\theta,\varphi\right)$, using the same notation as in (11). For the finite elements we used triangular elements with Lagrangian interpolation of the functions $\xi(\theta,\varphi)$ by a linear polynomial. Consideration is given to a case, when a computational domain extends over the whole half-space $-\pi < \varphi < 0$ and one part of the plane (a wedge with a cone angle 2α) is under the constraint of zero value for the function u and the other part is under the constraint of zero value for its normal derivative $\partial u/\partial n$ (Fig.9). Table 2 gives the least values Re $\lambda_1 > 0$ as a function of the cone angle obtained in [11] (row 1) and in the present work (row 2).

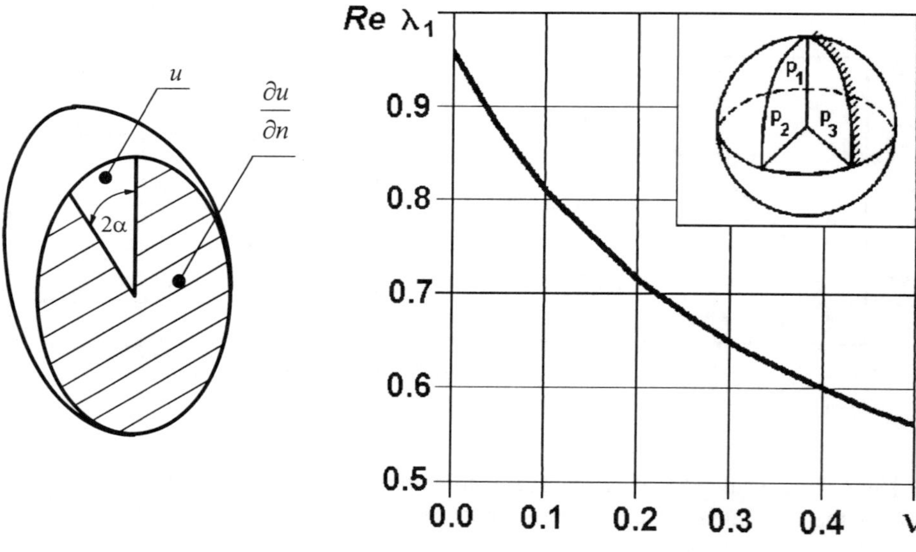

Figure 9.   Figure 10.

TABLE 2.

2α	0	$\pi/2$	π	$3\pi/2$	2π
1	0	0.296	0.5	0.816	1
2	0.001	0.301	0.503	0.823	1

Using the developed method we analysed the character of stress singularity at the vertex of trihedral wedge for different values of the cone angle $\gamma_1 = \gamma_2 = \gamma_3 = \pi/2$ and under different boundary conditions on the lateral wedge facets.

In Fig. 10, the least eigen value Re $\lambda_1 > 0$ is plotted against Poisson's ratio for a wedge, in which one facet is fixed, and the other two facets are stress-free. These results were obtained by a uniform discretization of the domain S into 722 elements. From the analysis of the convergence response to a number of variables used $\nu = 0.3$ it can be concluded that this number of elements suffices to perform calculations within the 1% error. Figures 11 and 12 show the results of calculation for a trihedral wedge with three fixed facets and for a wedge with two fixed facets and one stress-free facet.

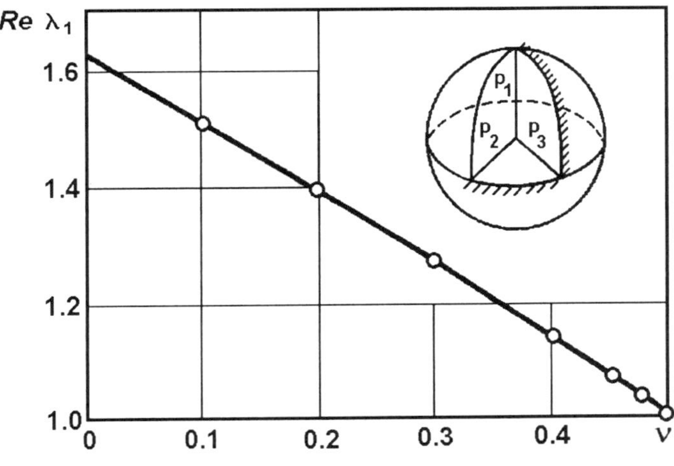

Figure 11.

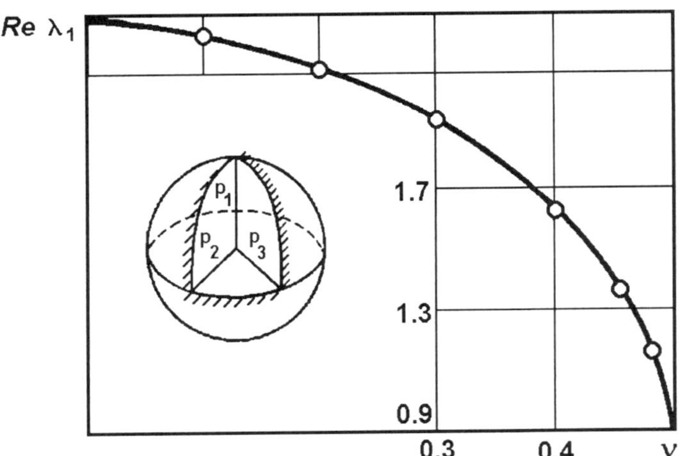

Figure 12.

The proposed method allows one to consider a variety of values for the angles γ_1, γ_2, γ_3 and as mentioned earlier, can be effectively applied to polyhedral wedge problems.

6. Conclusion

In the present work, a version of numerical algorithm for calculating the stress singularity power at the vertices of plane and spatial wedges is introduced. The results of numerical experiments demonstrate a reliability and efficiency of the proposed method. The computational capabilities of the method have been tested by calculating the singularity power at a vertex of a spatial wedge under different versions of the boundary conditions. This work was supported by RFFI under the grant N 96-01-00471.

7. References

1. Williams, M.L.: Stress singularities resulting from various boundary conditions in angular corners of plates in extension, *J. Appl. Mech. ASME* **19** (1952), 526-528.
2. Zak, A.R., Williams, M.L.: Crack point stress singularities at a bimaterial interface, *J. Appl. Mech. ASME* **30** (1963), 142-143.
3. Aksentyan, O.K.: Stress-strain singularities of a plate in the vicinity of edge, *J. Applied Mathematics and Mechanics* **31**, 1 (1967), 178-186 (published in Russia).
4. Lushik, O.N.: On response of radicals in equation determining stress singularities at the vertex of composite wedge, *Proceedings of the Russian Academy of Sciences, Solid Mechanics* 5 (1979), 82-92 (published in Russia).
5. Aleksanyan, R.K., Chobanyan, K.S.: Stress behaviour at the contact surface boundary in anisotropic composite rod under torsion, *J. Applied Mechanics* **13**, 6 (1977), 90-96 (published in Russia).
6. Bogy, D.B.: Edge-bonded dissimilar orthogonal elastic wedges under normal and shear loading, *J. Appl. Mech. ASME* **35** (1968), 460-466.
7. Bogy, D.B.: On the problem of edge-bonded elastic quarter-planes loaded at the boundary, *Int. J. Solids and Struct.* **6**, 9 (1970), 1287-1313.
8. Bogy, D.B.: Two edge-bonded elastic wedges of different materials and wedge angles under surface tractions, *J. Appl. Mech. ASME* **38** (1971), 377-386.
9. Mikhailov, S.E.: Stress singularity in the vicinity of a rib in a compound nonhomogeneous anisotropic body and some applications to composites, *Proceedings of the Russian Academy of Sciences, Solid Mechanics* 5 (1979), 103-110 (published in Russia).
10. Delale, F.: Stress singularities in bonded anisotropic materials, *Int. J. Solids and Struct.* **20**, 1 (1984), 31-40.
11. Bazant, Z.P.: Three-dimensional harmonic functions near termination or intersection of gradient singularity lines: a general method, *Int. J. Eng. Sci.* **12**, 3 (1974), 221-243.
12. Parton, V.Z., Perlin, P.I.: *Methods of mathematical theory of elasticity*, Nauka, Moscow, 1981 (published in Russia).
13. Strang, G., Fix, G.J.: *An analysis of the finite element method*, Prentice-Hall Inc., Englewood Cliffs, New Jersey, 1973.
14. Matveyenko, V.P.: On finite element based-algorithm for solving natural vibration problem for elastic bodies, *Boundary-value problems of elasticity theory*, USC of AS of USSR, Sverdlovsk, 1980, 20-24 (published in Russia).
15. Herrmann, L.R., Toms, R.M.: A reformulation of the elastic field equations, in terms of displacements, valid for all admissible values of Poisson's ratio, *J. Appl. Mech. ASME* **31** (1964), 148-149.

CAPILLARY-ELASTIC INSTABILITIES WITH AN OSCILLATORY FORCING FUNCTION

DAVID HALPERN
Department of Mathematics
University of Alabama
Tuscaloosa, AL 35487

J.A. MORIARTY
Baxter Healthcare
RLP-30
Rte 120 and Wilson Rds.
Round Lake IL 60073-0490

AND

JAMES B. GROTBERG
Departments of Biomedical Engineering and Anesthesia
Northwestern University
Evanston, IL 602081

1. Introduction

The lung is comprised of a network of bifurcating airway tubes which are coated with a thin viscous film. Often times, especially in the case of disease, the liquid film can form a meniscus which blocks the tube, thus obstructing airflow. The formation of the liquid meniscus is due to capillary driven instabilities which can arise in the lining, causing the lining to close up. In addition, airflow can also be obstructed if the airway tube collapses in on itself. This occurs when the elastic forces of the tube are not large enough to sustain the negative fluid pressures caused by the surface-tension of the air-liquid interface. Premature babies, whose lungs haven't developed sufficient surfactant to maintain the surface tension of the lung at a sufficiently low level for healthy functioning, are especially predisposed to problems caused by airway closure. In such cases, the patients are sometimes put on high frequency ventilation machines to improve gas exchange (Heldt *et al.* 1992, Patel 1995, Paulson it el al. 1996). The frequency of the breathing cycle, as

243

D. Durban and J.R.A. Pearson (eds.),
IUTAM Symposium on Non-Linear Singularities in Deformation and Flow, 243-255.
© 1999 *Kluwer Academic Publishers. Printed in the Netherlands.*

well as the tidal volume, are two control parameters at the disposal of the clinician. Little is known, however, of the effect that these two parameters can have on the occurrence of airway closure. Indeed, if airways can be kept open by using some given machine frequency, then this would be of vital importance in the treatment of such patients. The present work examines and quantifies these particular issues.

Hammond (1983) developed a model to study the stability of a thin film lining a rigid tube and solved a nonlinear evolution equation for the interface position. He found that waves initially amplified but eventually saturated into almost disconnected liquid collars. Gauglitz and Radke (1988) extended Hammond's model by including a more accurate representation for the curvature of the air-liquid interface. They computed a critical film thickness $\epsilon_c = 0.12$ necessary for closure. Johnson *et al.* (1991) analyzed the stability of a viscous annular film, incorporating the full interfacial curvature term and including the effects of inertia. Their results showed that the initial growth rate agreed with Goren's linear results (Goren, 1962), and their model was able to predict closure for sufficiently thick films. Halpern and Grotberg (1992) extended the approach used by Gauglitz and Radke (1988) to demonstrate how wall compliance can enhance the growth of the interfacial instability.

Time periodic perturbations to two-phase flows have been shown to be both stabilizing and destabilizing. Coward and Papageorgiou (1994) analyzed two-phase Couette flow between two plates, where one of the plates moves with a velocity that consists of a steady part combined with an oscillatory component. The interface between the two layers is unstable if the thinner of the two layers contains the more viscous fluid as in the case of steady Couette flow (Yih, 1967). They showed, using Floquet theory, that this system can be stabilized if the amplitude of the oscillatory component is sufficiently large. Coward *et al.* (1995) studied the stability of core-annular flows, where the pressure gradient driving the core flow is a constant modulated by an oscillatory component. Using a weakly nonlinear analysis, they obtained a Kuramoto-Sivashinsky equation with time-periodic coefficients, which revealed a spectrum of solutions, such as steady states, time-periodic and chaotic solutions.

Imposed external oscillations have also been investigated in annular extrusion methods for manufacturing of small hollow fibers (Wong *et al.* 1990, Isayev *et al.* 1990). In the manufacturing of optical fibers, an inner plug core of liquid and an outer annulus of a different liquid are co-extruded simultaneously, one around the other. An important problem is to have a smooth interface between the materials so that the desired mechanical and optical properties are achieved. Instabilities of this interface during concentric co-extrusion have been examined by Chen (1991).

In the present work, we shall consider an individual airway tube coated with a thin viscous film, and examine this simplified system for instabilities. This will require solving two partial differential equations for the tube wall and the air-liquid interface. In order to simulate breathing, the airway tube is assumed to oscillate, and there is an oscillatory shear stress at the air-liquid interface. This paper will only discuss these two types of oscillatory forcing separately. The model we develop is an extension of that developed by Halpern and Grotberg (1992, 1993) and is described in section 2 of the paper. In section 3 we present a linear analysis of the problem, in which the disturbance is linearized about the unperturbed state. Since the oscillatory shear stress introduces oscillatory terms into the governing equations, Floquet theory is used to study the system. The results determine which wave number disturbances are the most unstable. In section 4 the most dangerous wave number disturbances are studied in a full nonlinear analysis for the case where the unperturbed airway radius is periodically forced.

2. Basic Model and Scaling

The model is formulated in accordance with the analysis developed by Halpern and Grotberg (1992). However the present model incorporates the effects of breathing or imposed mechanical ventilation (Moriarty *et al.*, 1995). In this case, the tube radius is oscillating to simulate breathing, and there is an oscillatory shear stress on the air-liquid interface. The airway tube is assumed to be axisymmetric, and elastic. The axisymmetric assumption is valid until the onset of compliant collapse, at which time it must be discarded. Deflections are assumed to be small so that linear elasticity theory can be used to model the motion of the tube. The film is thin so that lubrication theory can be used to simplify the Navier Stokes equations.

The wall interface is at a radius $r^* = a(t^*) + \eta^*(z^*, t^*)$, and the air-liquid interface is at a radius $r^* = b(t^*) + h^*(z^*, t^*)$. The base state radii $a(t^*)$, $b(t^*)$ are time dependent to simulate the effects of breathing. The functional form of $a(t^*)$ and $b(t^*)$ are initially assumed to be sinusoidal. Although expiration is generally a faster process than inspiration, this is a reasonable model which contains all of the relevant aspects of the breathing cycle. The model can readily be extended to incorporate other functional forms of breathing. Hence $a(t^*) = a_0(1 - \delta \cos \omega^* t^*)$ where $a_0 \delta$ is the amplitude of the wall oscillations (dependent on the tidal volume of air inhaled each breathing cycle) and ω^* is the oscillatory frequency. Conservation of mass in the fluid, then demands that $b(t^*) = a_0((1 - \epsilon - \delta/(1 - \epsilon)) \cos \omega^* t^* + O(\epsilon \delta^2))$ where the unperturbed radius of the air-liquid interface is $a_0(1 - \epsilon)$, with $\epsilon \ll 1$.

2.1. SHEAR STRESS AT THE AIR-LIQUID INTERFACE.

Since the core viscosity is so much smaller than the film viscosity, we neglect the air flow details during breathing. Instead, we replace its effects with an oscillatory shear stress $\tau(t^*)$ at the air-liquid interface, derived from an uncoupled problem. We assume that the core air-flow rate, Q^*, is purely periodic, with constant frequency ω^* and amplitude A^*:

$$Q^* = A^* \sin \omega^* t^* \tag{1}$$

The amplitude can be expressed in terms of the stroke length, l_T^*, so that $A^* = \frac{1}{2}\omega^* l_T^* \pi a^2$, where a is the tube radius. Hence the mean axial core velocity, w_c^*, is

$$w_c^* = \frac{1}{2}\omega^* l_T^* \sin(\omega^* t^*) \tag{2}$$

Adjacent to the wall we assume a Stokes boundary layer of thickness $\delta_s = (\nu_a/\omega^*)^{1/2}$ for the basic unperturbed flow. The velocity field is approximately linear if the film is thin or the frequency is small. The linear axial velocity profile is given by

$$w_s^* = \frac{w_c^*}{\delta_s}(a - y^*) \tag{3}$$

where ν_a is the core viscosity and y^* is a coordinate perpendicular to the axial flow. Hence the shear stress is given by

$$\tau^* = -\frac{\mu_a}{2}\left(\frac{\omega^*}{\nu_a}\right)^{1/2}\omega^* l_T^* \sin(\omega^* t^*) \tag{4}$$

2.2. GOVERNING EQUATIONS IN VISCOUS FILM

In the viscous film we scale the radial and axial coordinates with respect to a_0 $(r, z \sim a_0)$, the air-liquid and wall-liquid perturbations $h^*, \eta^* \sim \epsilon a_0$, the radial and axial velocity components $u, w \sim \epsilon W, W$ where $W = (\sigma \epsilon^3/\mu)$, and the pressure $p \sim (\sigma \epsilon/a_0)$, where μ is the fluid viscosity and σ is the surface- tension (Halpern, 1992). Changing to a co-ordinate system moving with the liquid-wall interface, (y, z, t) such that $y = (a(t) + \epsilon \eta(z, t) - r)/\epsilon$, the continuity, and radial and axial momentum equations reduce to

$$\frac{\epsilon u}{a + \epsilon(\eta - y)} - \frac{\partial u}{\partial y} + \frac{\partial w}{\partial z} + \frac{\partial \eta}{\partial z}\frac{\partial w}{\partial y} = 0, \tag{5}$$

$$\frac{\partial p}{\partial y} = 0 + O(\epsilon^2), \tag{6}$$

and

$$w_{yy} = p_z + O(\epsilon^2) \tag{7}$$

At the liquid-wall interface, $y = 0$, we impose no slip

$$w(0, z, t) = 0 \tag{8}$$

and match the vertical fluid and wall components of velocity:

$$u(0, z, t) = \eta_t + \frac{\dot{a}}{\epsilon} \tag{9}$$

At the liquid/air interface $y = Y = (a(t) - b(t))/\epsilon + \eta(z, t) - h(z, t)$, the tangential and normal stress boundary conditions are respectively,

$$\frac{\partial w}{\partial y} = -T_s \sin(\Omega t + \psi) \tag{10}$$

and

$$p - p_i = -\frac{1}{\epsilon b(b + \epsilon h)} + h_{zz} \tag{11}$$

where T_s is the magnitude of the dimensionless applied shear stress (scaled with $\sigma \epsilon^2 / a_0$)

$$T_s = \frac{1}{2}\Omega^{3/2}\gamma\epsilon\left(\frac{\sigma\epsilon^3\rho}{\mu\mu_a a_0}\right)^{1/2}l_T^* , \quad \Omega = \frac{\mu a_0}{\sigma\epsilon^3}\omega^* \tag{12}$$

with $\gamma = \mu_a/\mu$, and $p_i = 1/(\epsilon b)$ is the pressure in the inner core. The dimensionless frequency, Ω, is the ratio of the fluid time scale (a_0/W) to the oscillation time scale $(1/\omega^*)$, and ψ is the phase lag between the shear stress and breathing cycles. Integrating equation (7), using boundary conditions (8) and (10) gives

$$w = \frac{1}{2}p_z(y^2 - 2yY) - T_s \sin(\Omega t + \psi)y \tag{13}$$

Applying the kinematic condition on the air-liquid interface at y=Y, gives,

$$h_t = \frac{1}{a + \epsilon(\eta - Y)}((a + \epsilon\eta)(\frac{\dot{a}}{\epsilon} + \eta_t) - \frac{a}{3}\frac{\partial}{\partial z}(p_z Y^3 + \frac{3}{2}T_s \sin(\Omega t + \psi)Y^2)) - \frac{\dot{b}}{\epsilon} \tag{14}$$

which agrees with the Halpern-Grotberg model when $\delta = 0$, $T_s = 0$.

2.3. EQUATION OF MOTION OF THE TUBE.

The equation of motion for the tube is obtained by assuming the tube is tethered at each end, so that deflections are only in the radial direction.

Thus, by considering the forces acting on the tube, the equation of motion of the tube is given by Atabek (1966) and Halpern & Grotberg (1992)

$$\phi\eta_t = \frac{h}{b(b+\epsilon h)} + h_{zz} + T_l\eta_{zz} + \frac{T_0\eta}{a(a+\epsilon\eta)} - \frac{1}{\Gamma}\frac{\eta}{(a+\epsilon\eta)^2} \tag{15}$$

where ϕ is the wall damping parameter, T_l is the longitudinal tension of the tube, T_0 is the initial tension, and Γ is a ratio of surface tension forces to elastic forces. Thus, $\Gamma = 0$ is the rigid tube case, and $\Gamma = O(1)$ indicates the tube is very compliant, and hence more predisposed to compliant collapse.

3. Floquet theory for the oscillatory shear stress forcing

We will first study the effects of the oscillatory shear stress only, and set $\delta = 0$, so that the tube radius does not oscillate around some base state. We consider the linearized system, so that the evolution equations for the air-film interface, (14), and the wall displacement, (15), are given respectively by

$$h_t = \eta_t - \frac{1}{3}(h_{zzzz} + h_{zz}) - T_s\sin(\Omega t + \psi)(\eta_z - h_z) \tag{16}$$

$$\phi\eta_t = h + h_{zz} + T_l\eta_{zz} - \frac{\eta}{\Gamma} \tag{17}$$

Since the equations are linear, we seek solutions of the form

$$h = a_1(t)e^{ikz} \text{ and } \eta = b_1(t)e^{ikz} \tag{18}$$

where k is the wavelength of the initial disturbance. Substituting these forms into (16) and (17) yields an ODE for $b_1(t)$ such that

$$\frac{d^2b_1}{dt^2} + (A - T_sik\sin(\Omega t + \psi))\frac{db_1}{dt} + (C - T_sik\sin(\Omega t + \psi)D)b_1 = 0 \tag{19}$$

where

$$A = \frac{T_lk^2 + 1/\Gamma}{\phi} + (k^2 - 1)(\frac{1}{\phi} + \frac{k^2}{3}) \tag{20}$$

$$C = \frac{(T_lk^2 + 1/\Gamma)k^2(k^2 - 1)}{3\phi} \tag{21}$$

$$D = (\frac{(1 - k^2)}{\phi})(1 - \frac{T_lk^2 + 1/\Gamma}{1 - k^2}) \tag{22}$$

Equation (19) is a nonlinear spring equation, with $A - T_sik\sin(\Omega t + \psi)$ being the damping coefficient. Note that this can become negative for various values of k. The range of k over which it can become negative increases with increasing Γ, but is always negative for $k > 1$. This negative damping is reflected in the capillary instability of the system. The capillary instability does not arise in the planar geometry because there is no out of plane curvature to destabilize the system. In this case the damping coefficient is always positive. Since equation (19) is a linear periodic differential equation for $b_1(t)$, Floquet theory is used to study the system (Yih, 1968, Coward & Papageorgiou, 1994). We therefore seek solutions to (19) in the form

$$b_1(t) = e^{\alpha t} \sum B_K e^{iK\Omega t} \tag{23}$$

where α is the Floquet exponent, analogous to the growth rate when $\Omega = 0$. When substituted into (19) it gives the recurrence relation for B_K;

$$(\alpha + iK\Omega)^2 B_K + \sum_{j=-1}^{1} \theta_j B_{K-j} = 0 \tag{24}$$

for $-N \le K \le N$, where

$$\theta_0 = (\alpha + iK\Omega)^2 + A(\alpha + iK\Omega) + C \tag{25}$$

$$\theta_{-1} = \frac{T_s k}{2}(\alpha + i(K+1)\Omega + \frac{1}{\phi}(k^2 - 1 + T_l k^2 + 1/\Gamma)) \tag{26}$$

$$\theta_1 = \frac{-T_s k}{2}(\alpha + i(K-1)\Omega + \frac{1}{\phi}(k^2 - 1 + T_l k^2 + 1/\Gamma)) \tag{27}$$

and $\theta_j = 0$, $\forall |j| > 1$. Since we require nontrivial values of B_K, the determinant of the system (24) must be zero. This gives an infinite order polynomial for the Floquet exponent α. We truncate the system at some value N (ensuring all off diagonal elements are zero, and diagonal elements $\to 1$ as $N \to \infty$) and solve the system $\Delta(\alpha) = 0$ where Δ refers to the determinant. Solutions to this system are obtained numerically by iterating on one of the diagonal elements of the matrix. A Newton-Raphson complex algebra technique is used to do this.

In the limit as $\Omega \to 0$, the results of Halpern & Grotberg (Halpern, 1992) are recovered. For $\Omega = T_s = 0$, $\Delta(\alpha) = 0$ reduces to a quadratic equation for α. One solution is always unstable for $0 < k < 1$ and stable for $k > 1$. The other solution is stable for all values of k provided $\Gamma < 1$, and becomes neutrally stable in the long wavelength limit, $k = 0$, for $\Gamma \ge 1$. For finite values of Ω, again there is an unstable solution if $0 \le k \le 1$, and stable solutions if $k > 1$. The oscillatory forcing function does not stabilize the capillary instability, however the growth rates are modified so that breathing can reduce the rate of capillary destabilization.   For a rigid tube, $\Gamma = 0$, linear stability analysis shows that frequency has no effect on the real part of the Floquet multiplier, $\Re(\alpha)$. Fig. 1(a) shows the

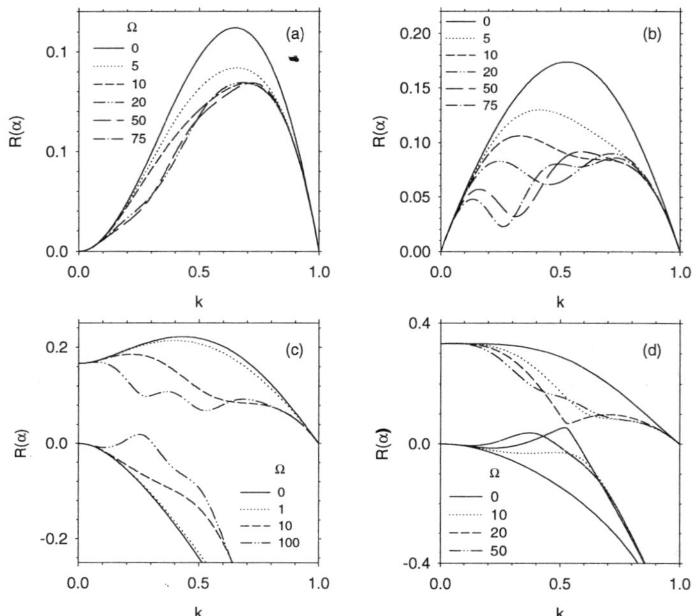

Figure 1. The real part of the Floquet multiplier versus wavenumber for different frequencies, Ω and compliance, Γ. (a) $\Gamma = 0.5$, (b) $\Gamma = 1$. (c) $\Gamma = 1.2$, (d) $\Gamma = 1.5$.

effect of Ω on $\Re(\alpha)$ for a mildly compliant tube, $\Gamma = 0.5$, and with $\phi = 1$. The amplitude of the shear stress is chosen to be $T_s = \Omega^{3/2}$, which is the appropriate value for the small airways of infants. As Ω increases from 0 to 75, the maximum value of $\Re(\alpha)$, $\Re(\alpha_{max})$, decreases by approximately 20%. $\Re(\alpha)$ does not always decrease monotonically with increasing Ω for fixed k. For example, as Ω increases from 50 to 75, there are small ranges of k for which $\Re(\alpha)$ is bigger for the larger value of Ω.

For a more compliant wall, $\Gamma = 1$, $\Re(\alpha_{max})$ still decreases with increasing Ω, and the value of k at which $\Re(\alpha_{max})$ occurs also decreases with increasing Ω (Fig. 1(b)). As Ω increases above 10, a local minimum develops at some $k \in (0, 1)$, and at $\Omega \approx 20$, $\Re(\alpha_{max})$ occurs at two distinct wavenumbers. For $\Omega \geq 20$, the local maximum that occurs at small k continues to decrease in magnitude, and a new local minimum develops for larger k.

For $\Gamma = 1.2$, the growth rate becomes positive at $k = 0$, but $\Re(\alpha_{max})$ still occurs at some $k \in (0, 1)$ (Fig. 1(c)). As for smaller values of Γ, increasing Ω causes $\Re(\alpha_{max})$ to decrease, and local minima to develop. However, the rate of decrease is not as significant since the growth rate is relatively

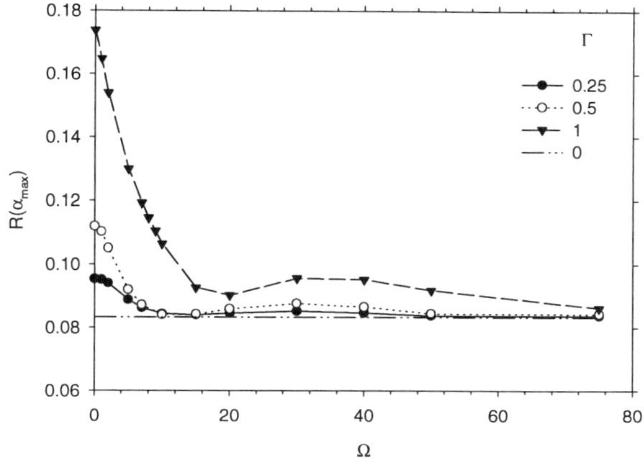

Figure 2. Maximum growth rate versus frequency for different Γ

insensitive to Ω for $k \ll 1$, and $\Re(\alpha) \neq 0$ at $k = 0$. The secondary root, which is always stable for $\Gamma < 1$, can now become unstable for a small range of k at sufficiently large Ω. For $\Omega = 1.5$, $\phi = 1$, $\Re(\alpha_{max})$ occurs at $k = 0$, independent of Ω (Fig. 1(d)). For $k \in (0, 1)$, increasing Ω does in general cause $\Re(\alpha)$ to decrease. At large Ω, the second root can become positive and in some instances is the dominant, unstable root.

The dependency of $\Re(\alpha_{max})$ on Ω is shown in Fig. 2 for different values of Γ. For $\Omega \leq 20$, a moderately compliant wall, $\Re(\alpha_{max})$ decreases quite rapidly with Ω, reaching the rigid tube value of $1/12$, and becomes less dependent on Γ for $\Omega > 6$. Above $\Omega = 20$, the maximum growth rate increases slightly, before leveling off to the rigid tube result.

4. Nonlinear Stability for radially imposed oscillations

We consider next the effect of radially imposed oscillations only, and neglect the oscillatory shear stress at the air-liquid interface, $T_s = 0$. In this case, linear stability yields the same results as the unforced system, and so we turn immediately to the nonlinear stability problem for this system. We have used a finite difference time marching scheme to solve for the wall and air-liquid positions, Eqs. (14) and (15), as they evolve in time. Periodic boundary conditions have been imposed. Initial conditions are chosen $h = h_0 \cos(\lambda z)$, $\eta = \eta_0 \cos(\lambda z)$, where λ is the wavelength of the disturbance to the profiles and h_0 and η_0 are the initial amplitudes. Tracking the evolution of the profiles with time will allow us to examine the stability of the system to any given initial disturbance.

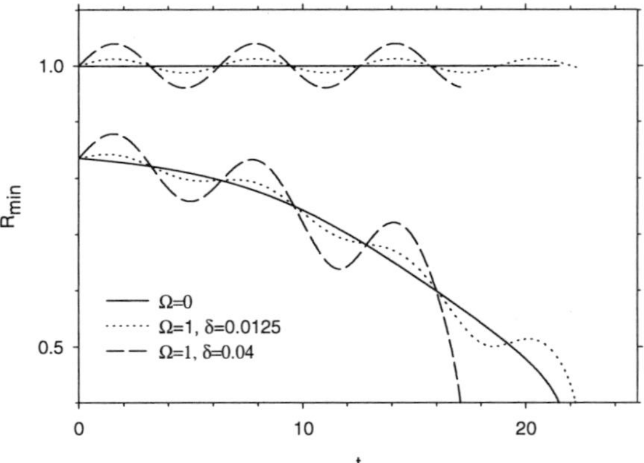

Figure 3.   Minimum core radius versus time for $\epsilon = 0.15$.

Closure occurs when the minimum radius of the fluid-air interface, R_{min}, reaches zero at some point in time. For a rigid, non-oscillating tube, closure will occur if $\epsilon > 0.12$ and λ is set to $2^{3/2}\pi$, which is the most dangerous wavelength based on linear stability theory. Fig. 3 shows R_{min} and the wall position where the air-liquid interface is a minimum, versus time for the forced and unforced case. R_{min} is an important measure of the system, since when $R_{min} = 0$, the system is closed. We stop the computations when $R_{min} \approx 0.4$, since the model assumptions break down (Halpern & Grotberg 1992). However, since R_{min} decreases quite rapidly at this point, the closure time, T_c, can be accurately estimated. Fig. 3 shows that closure can be delayed slightly for small amplitudes, but the instability can be enhanced for sufficiently large amplitudes. Note that for a uniform film, mass conservation dictates that the film has to be thicker in the part of the cycle when the wall is moving inwards. This will therefore enhance the destabilizing effect of the transverse component of curvature.   In Fig. 4(a), T_c is plotted versus the amplitude of the wall oscillation, δ. For $\epsilon = 0.15$, $\Omega = 1$, T_c initially increases with δ, where $\delta \in [0, 0.021]$, but for $\delta > 0.021$, T_c first decreases very rapidly and then levels off for $\delta > 0.03$. Increasing the frequency to $\Omega = 5$ causes T_c to decrease monotonically with increasing δ. There is a small range of δ, $0.028 < \delta < 0.045$, where the higher frequency case is slightly more stable. Closure time increases if the film thickness is reduced to $\epsilon = 0.14$. For small values of δ, we find that T_c decreases slightly with δ, before decreasing rapidly as in the $\epsilon = 0.15$ case, suggesting that the small increase in T_c observed for $\epsilon = 0.15$ may be an artifact of the initial condition and the initial phase of the oscillatory cycle.

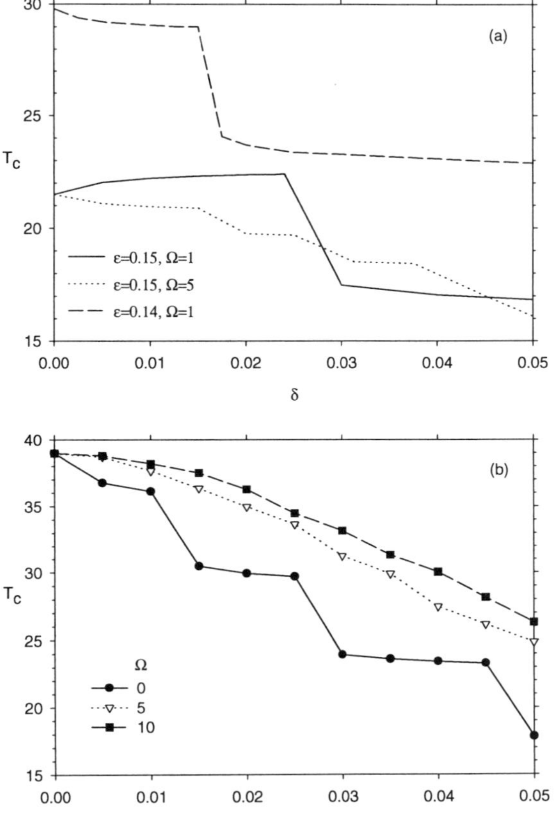

Figure 4. (a) Closure time versus δ for the rigid tube case; (b) Closure time versus δ for $\epsilon = 0.15$, $\Gamma = 0.2$, $\phi = 1$.

Fig. 4(b) shows how closure time varies with δ for a compliant tube, $\Gamma = 0.2$. Because the film is thin, $\epsilon = 0.1$, the effects of the initial conditions on the closure time are not so significant as in the previous case. T_c decreases with δ for a range of frequencies. But for this compliant tube, increasing the frequency delays the onset of closure.

5. Conclusions

The linear stability results indicate that a periodic shear stress at the air-liquid interface due to a core flow can significantly dampen the fastest growing mode. This is especially the case for compliant airway walls: for $\Gamma = 1$, $\Re(\alpha_{max})$ is more than halved as Ω is increased from 0 to 20, which is within the range used in high frequency ventilators for infants. Above $\Omega = 20$, the maximum growth rate approaches the rigid tube value.

Externally imposed wall oscillations both enhance and dampen the capillary instability. In general, the period of the cycle that is most destabilizing is when the wall is moving inwards, and increasing the amplitude of the tube radius increases the transverse component of curvature. However, there are small windows in parameter space where the nonlinear growth rate can be diminished. Airway closure can also be delayed by increasing the frequency of oscillation, keeping the amplitude fixed.

These results support the use of high frequency ventilator machines on premature neonates to prevent airway closure.

Acknowledgements

This work was supported by NSF grant CTS9412523 and NASA grant NAG3-1959.

References

Atabek, H.B. & Lew, H.S. (1966) Wave propagation through a viscous incompressible fluid contained in an initially stressed elastic tube, *Biophys. J.* **6, pp. 481-503**

Chen, K. (1991) Interfacial instability due to elastic stratification in concentric co-extrusion of two viscoelastic fluids, *J. Non-Newtonian Fluid Mech.* **40, pp. 155-175**

Coward, A.V. & Papageorgiou, D.T. (1994) Stability of oscillatory two-phase Couette flow, *IMA Journal of Applied Mathematics* **55, pp. 75-93**

Coward, A.V., Papageorgiou, D.T. & Smyrlis, Y.S. (1995) Nonlinear stability of oscillatory core-annular flow: A generalized Kuramoto-Sivashinsky equation with time period coefficients, *Z. angew Math. Phys.* **46, pp. 1-39**

Gauglitz, P.A. & Radke, C.J. (1988) An extended evolution equation for liquid film breakup in cylindrical capillaries,*Chem. Eng. Sci.* **43, pp. 1457-1465**

Goren, S. L. (1962) The instability of an annular thread of fluid, *J. Fluid Mech.* **12, pp. 309-319**

Halpern, D. & Grotberg, J.B. (1992) Fluid-elastic instabilities of liquid lined flexible tubes: Airway closure, *J. Fluid Mech.* **244, pp. 615-632**

Halpern, D. & Grotberg, J.B. (1993) Surfactant effects on fluid elastic instabilities of liquid lined flexible tubes: a model of airway closure,*J. Biomech. Eng.* **115, pp. 271-277**

Hammond, P. S. (1983)Nonlinear adjustment of a thin annular film of viscous fluid surrounding a thread of another within a circular pipe,*J. Fluid Mech.* **137, pp. 363-384**

Heldt, G.P., Merritt, T.A., Golembeski, D., Gilliard, N., Bloor, C. & Spragg, R. (1992) Distribution of surfactant, lung compliance, and aeration of preterm rabbit lungs after surfactant therapy and conventional and high-frequency oscillatory ventilation, *Pediatric Research* **31(3), pp. 270-275**

Isayev, A.I., Wong, C.M. & X. Zeng (1990) Flow of thermoplastics in an annular die under orthogonal oscillations, *J. Non-Newtonian Fluid Mech.* **34, pp. 375-397**

Johnson, M., Kamm, R., Ho, L.W., Shapiro, A. & Pedley, T.J. (1991) The nonlinear growth of surface-tension driven instabilities of a thin annular film,*J. Fluid Mech.* **233, pp.141-156**

Moriarty, J.A., D. Halpern & J.B. Grotberg (1995) Capillary-elastic instabilities in liquid-lined flexible tubes with an external forcing function, *Ann. Biomed. Eng.* **23, S8**

Patel, C. A. & Klein, J. M. (1995) Outcome of infants with birth weights less than 1000 g with respiratory distress syndrome treated with high-frequency ventilation and surfactant replacement therapy, *Archives of Pediatrics & Adolescent Medicine* **149(3), pp. 317-21**

Paulson, T. E., Spear, R. M., Silva, P. D., Peterson, B. M. (1996) High-frequency pressure-control ventilation with high positive end-expiratory pressure in children with acute respiratory distress syndrome, *Journal of Pediatrics* **4, pp. 566-73**

Wong, C.M., Chen, C.H. & A.I. Isayev (1990) Flow of thermoplastics in an annular die under parallel oscillations, *Polymer Engineering Science* **30, pp. 1574-1584**

Yih, C.-S. (1967) Instability due to viscosity stratification, *J. Fluid Mech.* **27, pp. 337-52**

Yih, C.-S. (1968) Instability of unsteady flows or configurations. Part 1. Instability of a horizontal liquid layer on an oscillating plate, *J. Fluid Mech.* **31, pp. 737-51**

SINGULARITIES AND SIMILARITY SOLUTIONS IN CAPILLARY BREAKUP

JOHN R. LISTER
Institute of Theoretical Geophysics, University of Cambridge
DAMTP, Silver St, Cambridge CB3 9EW, UK

MICHAEL P. BRENNER
Department of Mathematics, M.I.T.

RICHARD F. DAY AND E. JOHN HINCH
DAMTP, University of Cambridge

AND

HOWARD A. STONE
Division of Applied Sciences, Harvard University

Abstract. A variety of dynamical regimes for capillary breakup gives rise to a zoo of asymptotic similarity solutions for the flow immediately before the breakup singularity. These solutions, some new and some previously published, are summarized and the temporal order of transitions between them determined by scaling. Among the new results are that the perfectly inviscid solution is an equidimensional potential flow rather than near-unidirectional and that the asymptotic viscous solution includes the effects of the external viscosity. Breakup due to van-der-Waals forces is discussed briefly.

1. Introduction and 1D models

The capillary breakup of a long liquid thread into droplets is a well-known fluid-mechanical instability first analysed over a hundred years ago by Lord Rayleigh. Linear stability theory (Rayleigh 1892) shows that a sinusoidal perturbation to the interface of a cylindrical thread will grow if the wavelength is greater than the circumference of the thread, with the growth driven by interfacial tension and the decrease in surface area caused by such a perturbation. The growth rate and most unstable wavelength de-

257

D. Durban and J.R.A. Pearson (eds.),
IUTAM Symposium on Non-Linear Singularities in Deformation and Flow, 257-269.
© 1999 *Kluwer Academic Publishers. Printed in the Netherlands.*

pend on the ratio of the internal and external viscosities and we note for later reference that the most unstable wavelength is infinite when the external viscosity is zero.

The nonlinear growth of the capillary instability to large amplitude is observed experimentally to lead to fragmentation of the thread into droplets, often with the production of many small satellite droplets by secondary capillary instabilities on the filaments of fluid left between the primary drops (e.g. Kowalewski 1996; Henderson et al. 1997). Motivated by the importance of satellite-drop formation in industrial applications such as ink-jet printing, many studies have been made of the spatial and temporal development of capillary instabilities on a jet (e.g. Bogy 1979; Chaudhary & Redekopp 1980; Tjahjadi et al. 1992), usually based on experiments, numerical calculations or weakly nonlinear analyses.

Recently, there has been much interest in the formation and structure of singularities in nonlinear partial differential equations, such as those occurring in fluid mechanical problems due to topology transitions or cusp formation in flows with interfaces (e.g. Goldstein et al. 1993; Tanveer 1993; Pugh & Shelley 1997). Capillary-driven breakup of a thread into drops provides a natural example of a topological transition and there have been a number of studies of the self-similar structure in the vicinity of the singularity at breakup in various dynamical regimes (e.g. Ting & Keller 1990; Eggers 1993, 1995; Papageorgiou 1995; Brenner et al. 1996). The expectation of a self-similar structure is based on the assumption that, when a physical quantity such as the velocity or pressure diverges to infinity at a point in time and space, very close to the singularity the flow is determined locally and does not depend on distant boundary or initial conditions.

A comprehensive review of the literature on capillary breakup has been given recently by Eggers (1997) and we make no attempt to repeat that here. Instead we provide a brief survey of some published and novel similarity solutions for capillary breakup just *before* rupture, giving particular emphasis to the variety of assumed physical balances in the solutions and commenting on the relationship between them. In particular, we describe in §6 the temporal sequence in which these solutions might be observed owing to transitions from one dynamical regime to another as a neglected effect comes into play.

Many, though not all, of the studies of breakup are based on a long-wavelength approximation to the Navier-Stokes equations in which the breaking thread is axisymmetric with radius $h(z, t)$ and the ratio ϵ of the scale of h to the lengthscale in the axial direction z is assumed to be much less than 1 so that, for example, $|\partial h/\partial z| \ll 1$. To leading order in $\epsilon \ll 1$, the axial velocity $v(z, t)$ and pressure $p(z, t)$ are uniform across any cross-section of the thread. The long-wavelength approximations can then be discussed in the context of the following one-dimensional equations for mass, momentum and pressure:

$$(h^2)_t + (h^2 v)_z = 0 \tag{1}$$

$$\rho(v_t + v v_z) = -p_z + \frac{3\mu}{h^2}(h^2 v_z)_z - \frac{C\mu_{\text{ext}} v}{h^2} \tag{2}$$

$$p = \frac{\gamma}{h} + \frac{A}{h^3}, \tag{3}$$

where ρ and μ are the density and viscosity of the thread, μ_{ext} is the viscosity of the surrounding fluid, γ and A are the coefficients of surface tension and van-der-Waals forces between the fluids and subscripts denote partial derivatives. The two viscous terms in (2) correspond, respectively, to extensional stresses within the thread and a simple heuristic approximation to shear stresses from the external fluid owing to the axial motion of the thread. If the external fluid is in Stokes flow then the drag coefficient C is given by slender-body theory as $2/|\ln \epsilon|$ (cf. the 'sliding-rod' regime of Lister & Stone 1996). An axial curvature term $-\epsilon^2 \gamma h_{zz}$, though formally of higher order in the long-wavelength approximation, is often included in (3) to provide a short-wavelength cut-off for the secondary Rayleigh instabilities which inevitably occur in all long-wave models (Brenner et al. 1996).

The solutions to be discussed can now be variously divided into those driven by surface tension (§§2–4) and those driven by van-der-Waals forces (§5), those with an external viscous fluid (§4) and those without (§§2–3), and those in which the Reynolds number $Re = \rho v z/\mu$ of the internal extensional flow is negligible (§2.2), order 1 (§2.1) or infinite (§3). In each case it will be useful to define dimensionless variables

$$\tilde{z} = \frac{z - z^*}{\ell}, \qquad \tilde{h} = \frac{h}{\ell}, \qquad \tilde{t} = \frac{t^* - t}{\hat{t}}, \qquad \tilde{v} = \frac{v\hat{t}}{\ell} \tag{4}$$

where ℓ and $\hat{t}$ are length and timescales that depend on the problem and (z^*, t^*) is the location of the singularity; $\tilde{t}$ is positive *before* breakup.

2. Viscous solutions without an external fluid

Throughout this section we neglect the effects of the external fluid viscosity ($\mu_{\text{ext}} = 0$) and van-der-Waals forces ($A = 0$) and work with the correspondingly simplified versions of (1)–(3).

2.1. VISCOUS THREAD WITH INERTIA: $Re = O(1)$

Eggers (1993) was the first to realize that a unique similarity scaling emerges out of the assumption that the inertial, viscous and capillary terms in (2) are comparable in the vicinity of the singularity. Estimates of the terms in (1)–(3) suggest the scalings

$$v \sim \frac{z}{t}, \qquad \frac{\rho v}{t} \sim \frac{\gamma}{hz} \sim \frac{\mu v}{z^2} \qquad \Longrightarrow \qquad \tilde{z} \sim \tilde{t}^{1/2}, \quad \tilde{h} \sim \tilde{t}, \quad \tilde{v} \sim \tilde{t}^{-1/2}, \qquad (5)$$

where $\ell = \mu^2/(\gamma\rho)$ and $\hat{t} = \mu^3/(\gamma^2\rho)$.

Substitution of the similarity forms $\tilde{h}(\tilde{z}, \tilde{t}) = H(\eta)\tilde{t}$ and $\tilde{v}(\tilde{z}, \tilde{t}) = V(\eta)\tilde{t}^{-1/2}$, where $\eta = \tilde{z}/\tilde{t}^{1/2}$, into (1)–(3) then gives rise to the third-order system of ordinary differential equations

$$(V + \tfrac{1}{2}\eta)H' + (\tfrac{1}{2}V' - 1)H = 0 \tag{6}$$

$$\tfrac{1}{2}H^2(\eta V + V^2)' = H' + 3(H^2 V')'. \tag{7}$$

It turns out that there a unique point η_0, which we term the stagnation point, where $2V(\eta) + \eta = 0$ and (6) is singular. The boundary conditions on (6) and (7) can be written as

$$V'(\eta_0) = 2, \qquad V \to 0 \text{ as } \eta \to \pm\infty, \tag{8}$$

which express conditions of regularity of H at $\eta = \eta_0$ and of no forcing of the singularity from infinity. Eggers (1993, 1995) found one numerical solution to (6)–(8) by guessing values for η_0 and $H_0 = H(\eta_0)$, using series expansions to move away from the stagnation point and then shooting to hit the boundary conditions at plus and minus infinity.

Brenner et al. (1996) examined the recurrence relationships for the coefficients in the series expansions $H = \sum H_i(\eta - \eta_0)^i$, $V = \sum V_i(\eta - \eta_0)^i$ about the stagnation point,

$$3H_0^2 i(i+1)V_{i+1} + (12H_0 + 1)iH_i = \alpha_{i-1} \tag{9}$$

$$H_0(i+1)V_{i+1} + 5iH_i = \beta_i \qquad i = 1, 2, \ldots \tag{10}$$

where α_{i-1} and β_i depend on lower-order coefficients, and noticed that they are singular (linearly dependent) when $i = n$ if $H_0 = (15n - 12)^{-1}$. Motivated by the observation that Eggers' value $H_0 = 0.0304\ldots$ is very close to the third singular value $\frac{1}{33}$, they made a more extensive study of (6)–(8) and found an infinite family of solutions, one near the 3rd, 7th, 11th, 15th etc. singular values (Figure 1). While it is clear from Figure 1 that the singular values of H_0 have a strong influence on the family of similarity solutions, it is unclear how to interpret physically this apparent 'resonance' between the continuity and momentum equations. All the similarity solutions are highly asymmetric (Figure 2) and have the same time exponents (e.g. $\tilde{z} \sim \tilde{t}^{1/2}$). The lowest-order solution ($n = 3$) is the one

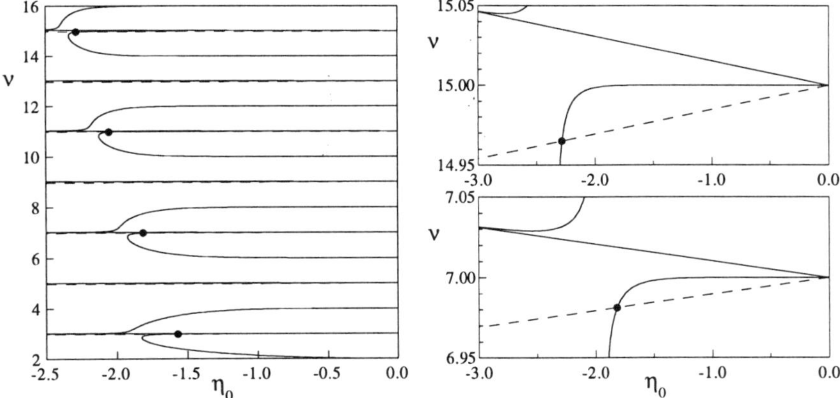

Figure 1. The results of numerical integration of (6)–(8a) from $\eta = \eta_0$, $H = H_0$, where $\nu(H_0)$ is defined by $H_0 = 1/(15\nu - 12)$. The solid and dashed curves give solutions satisfying the boundary conditions (8b) at $\pm\infty$ respectively and their intersections (solid dots) correspond to similarity solutions for a viscous thread with inertia.

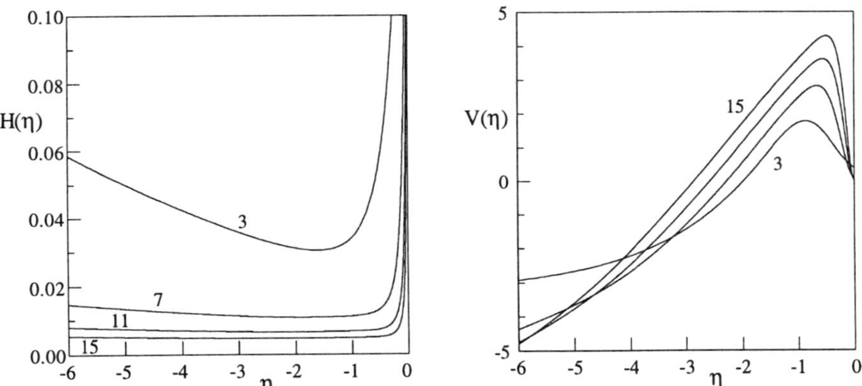

Figure 2. The first four solutions of (6)–(8), which occur near the 3rd, 7th, 11th and 15th singular values of H_0 in (9) and (10)

selected by the dynamics in calculations and experiments since it is least prone to finite-amplitude secondary Rayleigh instabilities (Brenner et al. 1996).

2.2. VISCOUS THREAD WITHOUT INERTIA: $Re = 0$

Papageorgiou (1995) considered the case in which inertia is also negligible in (2) ($\rho = 0$). The scaling estimates are

$$v \sim \frac{z}{t}, \qquad \frac{\gamma}{hz} \sim \frac{\mu v}{z^2} \quad \Longrightarrow \quad \tilde{z} \sim \tilde{t}^{\beta}, \quad \tilde{h} \sim \tilde{t}, \quad \tilde{v} \sim \tilde{t}^{\beta-1}, \qquad (11)$$

where ℓ is now an arbitrary lengthscale, perhaps best understood as the lengthscale of the initial conditions, $\hat{t} = \mu\ell/\gamma$ and β is an as yet undetermined exponent. The fact that there are insufficient terms in the equations to determine unique exponents by scaling is sometimes referred to as incomplete self-similarity or similarity of the second kind (Barenblatt 1979) and, in this case, the exponent β must be found as a nonlinear eigenvalue of the similarity equations.

Substitution of the similarity forms $\tilde{h}(\tilde{z}, \tilde{t}) = H(\eta)\tilde{t}$ and $\tilde{v}(\tilde{z}, \tilde{t}) = V(\eta)\tilde{t}^{\beta-1}$, where $\eta = \tilde{z}/\tilde{t}^{\beta}$, into (1)–(3) yields the ordinary differential equations

$$(V + \beta\eta)H' + (\tfrac{1}{2}V' - 1)H = 0 \tag{12}$$

$$H' + 3(H^2V')' = 0 \tag{13}$$

subject to the boundary conditions (8), where the stagnation point η_0 is now where $V(\eta) + \beta\eta = 0$.

For general values of H_0, solution of the corresponding recurrence relationships

$$3H_0^2 i(i+1)V_{i+1} + (12H_0 + 1)iH_i = \alpha_{i-1} \tag{14}$$

$$H_0(i+1)V_{i+1} + 2(2+\beta)iH_i = \beta_i, \qquad i = 1, 2, \dots \tag{15}$$

leads only to the exact solution $H = H_0$, $V = 2(\eta - \eta_0)$, which does not satisfy the boundary conditions at infinity. Thus, in contrast to solutions of (9)–(10), we actually need $H_0 = [6(2+\beta)n - 12]^{-1}$ so that (14)–(15) is singular at $i = n$ and (H_n, V_{n+1}) can be set equal to a multiple of the null eigenvector. Brenner et al. (1996) showed that the boundary conditions at infinity can then be satisfied if n is even, $\eta_0 = 0$ and β takes a particular (numerically determined) value depending on n. The case $n = 2$ was first calculated by Papageorgiou (1995). There is thus again an infinite family of similarity solutions, but all are left-right symmetric and each has a different time-exponent β. As $\tilde{t} \to 0$ all solutions satisfy the long-wavelength assumption since $\beta(n) < 1$ but fail to satisfy neglect of inertia since $\beta(n) < \tfrac{1}{2}$. The lowest-order solution ($n = 2$) is again the one observed in simulations as it is the least prone to instability.

3. Inviscid solutions

In this section we neglect the effects of viscosity ($\mu = \mu_{\text{ext}} = 0$). We also assume that $\rho \gg \rho_{\text{ext}}$ in order to neglect any dynamic pressure gradients in the external fluid.

3.1. INVISCID THREAD: $Re = \infty$

Inviscid dynamics gives the scalings

$$v \sim \frac{z}{t}, \qquad \frac{\rho v}{t} \sim \frac{\gamma}{hz} \quad \Longrightarrow \quad \tilde{z} \sim \tilde{t}^{\beta}, \quad \tilde{h} \sim \tilde{t}^{2-2\beta}, \quad \tilde{v} \sim \tilde{t}^{\beta-1}, \qquad (16)$$

where ℓ is an arbitrary lengthscale, $\hat{t} = (\rho\ell^3/\gamma)^{1/2}$ and β is again an undetermined exponent. The long-wavelength approximation requires $\beta > \frac{2}{3}$.

The scalings (16) correspond to the second-order system of similarity equations

$$(V + \beta\eta)H' + (\tfrac{1}{2}V' + 2\beta - 2)H = 0 \qquad (17)$$

$$(1 - \beta)V + (V + \beta\eta)V' = H'/H^2 \qquad (18)$$

The solution is specified by the location η_0 and radius $H(\eta_0)$ of the stagnation point, and the recurrence relationship for the local series solution is never singular. A rescaling symmetry of the equations allows $H(\eta_0)$ to be set equal to 1 and leaves η_0 and β as parameters with which to try to satisfy boundary conditions at infinity.

Letting $V = \Phi'$ allows (18) to be integrated once (equivalent to the Bernoulli integral) and reproduces the equations studied by Ting & Keller (1990), who focused on $\eta_0 = 0$ and certain values of β. While (17) and (18) are thus only partially explored, they prove to be of limited interest in describing capillary breakup since numerical simulations of the inviscid version of (1)–(3) always form a shock (or 'turnover' singularity) before breakup occurs (Papageorgiou 1996; Brenner et al. 1997), thus violating the long-wavelength assumption. The remedy is to return to full equidimensional Euler flow as described next.

3.2. IRROTATIONAL EULER FLOW

Looking for an equidimensional inertia–capillary balance,

$$v \sim \frac{z}{t}, \qquad \frac{\rho v}{t} \sim \frac{\gamma}{hz}, \qquad h \sim z, \qquad (19)$$

is equivalent to setting $\beta = \frac{2}{3}$ in (16) and leads to

$$h \sim z \sim \left(\frac{\gamma t^2}{\rho}\right)^{1/3}, \qquad |\mathbf{u}| \sim \left(\frac{\gamma}{\rho t}\right)^{1/3}, \qquad \phi \sim \left(\frac{\gamma^2 t}{\rho^2}\right)^{1/3}, \qquad (20)$$

where ϕ is the velocity potential and $\mathbf{u} = \nabla\phi$. Neglect of the viscous term is eventually invalid as $t \to 0$.

The axisymmetric time-dependent irrotational flow is given by

$$\nabla^2\phi = 0, \qquad \rho\frac{D\phi}{Dt} - \tfrac{1}{2}\rho|\nabla\phi|^2 + \gamma\left(\frac{(1 + h_z^2)/h - h_{zz}}{(1 + h_z^2)^{3/2}}\right) = 0, \qquad (21)$$

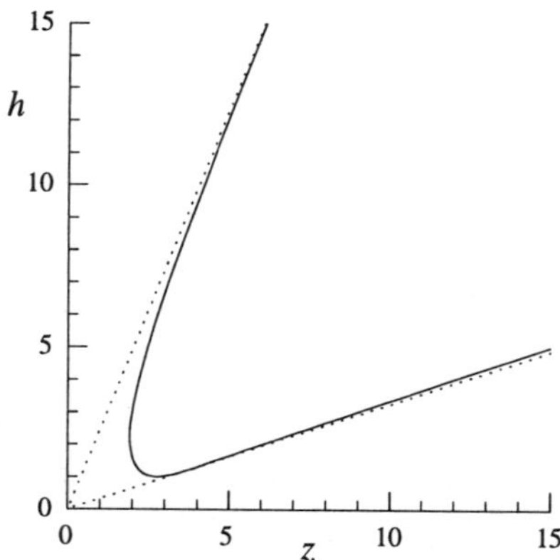

Figure 3. The self-similar shape of breakup in equidimensional Euler (potential) flow scaled so that $h_{\min} = 1$. The internal half-angles of the asymptotic fluid double cone (dotted) are about 18.1° and 112.8°. Flow is directed from the higher-curvature narrow cone towards the lower-curvature reverse cone.

where (21b) is Bernoulli's equation on the free surface with the full curvature in the capillary pressure. Numerical solutions of (21) from a variety of initial conditions by Day et al. (1997) confirm the scalings in (20) as $h_{\min} \to 0$ and show that the asymptotic self-similar interfacial shape is a blunted double cone with approximate half-angles 18.1° and 112.8° (Figure 3). There are two free parameters in the far-field expansion of the similarity solution in each cone and four matching conditions between the cones, which confirms that the solution is discrete and probably unique. Since the full irrotational Euler problem gives an asymptotic similarity shape that bends back on itself, it is clear that attempts at long-wavelength (one-dimensional) approximations were bound to fail. It should, however, be noted that flow through a bent-back shape might separate in a fluid with small, but nonzero viscosity, which would then modify the pressure distribution and the observed shape.

4. Solutions with an external viscous fluid

The possible balances considered so far, which ignore the external fluid, suggest evolution to an asymptotic scaling (5) in which both viscosity and inertia play a role. However, this scaling, with $\tilde{h}/\tilde{z} \to 0$, is inconsistent with the neglect of the external drag in (2) even if $\mu_{\text{ext}} \ll \mu$. (The same considerations explain why the most unstable linear perturbation in the viscous Rayleigh problem only has infinite wavelength when $\mu_{\text{ext}} \equiv 0$.)

4.1. SLENDER-BODY THEORY

When external drag is reintroduced, inertia becomes negligible. The relevant estimates are

$$v \sim \frac{z}{t}, \qquad \frac{\gamma}{hz} \sim \frac{\mu v}{z^2} \sim \frac{C\mu_{\text{ext}}v}{h^2} \quad \Longrightarrow \quad \tilde{z} \sim \delta^{-1}\tilde{t}, \quad \tilde{h} \sim \tilde{t}, \quad \tilde{v} \sim \delta^{-1}, \tag{22}$$

where $\delta = (C\mu_{\text{ext}}/\mu)^{1/2}$, ℓ is arbitrary and $\hat{t} = \mu\ell/\gamma$. Since $\tilde{z}$ and $\tilde{h}$ have the same time-dependence, use of a long-wavelength approximation relies here on $\delta \ll 1$ or $\mu_{\text{ext}} \ll \mu$ rather than asymptotic behaviour in time.

Detailed calculations by Lister et al. (1997) show that the asymptotic behaviour is not quite as straightforward as suggested by (22). First, the similarity equations

$$(V + \eta)H' + (\tfrac{1}{2}V' - 1)H = 0 \tag{23}$$

$$H' + 3(H^2V')' = V \tag{24}$$

obtained by the substitutions $\tilde{h}(\tilde{z}, \tilde{t}) = H(\eta)\tilde{t}$ and $\tilde{v}(\tilde{z}, \tilde{t}) = \delta^{-1}V(\eta)$, where $\eta = \delta\tilde{z}/\tilde{t}$, do not appear numerically to have any solutions satisfying the boundary conditions at the stagnation point and plus and minus infinity. Second, numerical integrations of (1) and

$$\gamma h_z + 3\mu(h^2 v_z)_z = C\mu_{\text{ext}}v \tag{25}$$

over three decades in $\tilde{t}$ show that $v_{\max}$ increases slowly and $dh_{\min}/dt$ decreases slowly rather than tending to a constant (Figure 4). These calculations suggest that the naive scalings (22) have the right power-law dependence on $\tilde{t}$ but need to be modified by a slow dependence on $|\ln(\tilde{t})|$.

4.2. STOKES FLOW

In order to understand whether the logarithmic modifications above might be an artifact of the slender-body approximation, Lister et al. (1997) studied the case of Stokes flow with $\mu = \mu_{\text{ext}}$. The relevant scalings are

$$v \sim \frac{z}{t}, \qquad \frac{\gamma}{hz} \sim \frac{\mu v}{z^2} \sim \frac{C\mu v}{h^2} \quad \Longrightarrow \quad \tilde{z} \sim \tilde{t}, \quad \tilde{h} \sim \tilde{t}, \quad \tilde{v} \sim 1 \tag{26}$$

and the time-dependent problem was solved numerically by integrating the boundary-integral representation

$$\mathbf{u}(\mathbf{x}) = \frac{\gamma}{8\pi\mu} \int_{S(t)} \left(\frac{\mathbf{I}}{r} + \frac{\mathbf{rr}}{r^3}\right) \cdot \mathbf{n}\kappa \, dS(\mathbf{y}) \tag{27}$$

where $\mathbf{r} = \mathbf{y} - \mathbf{x}$ and κ is the full curvature (cf. 21b).

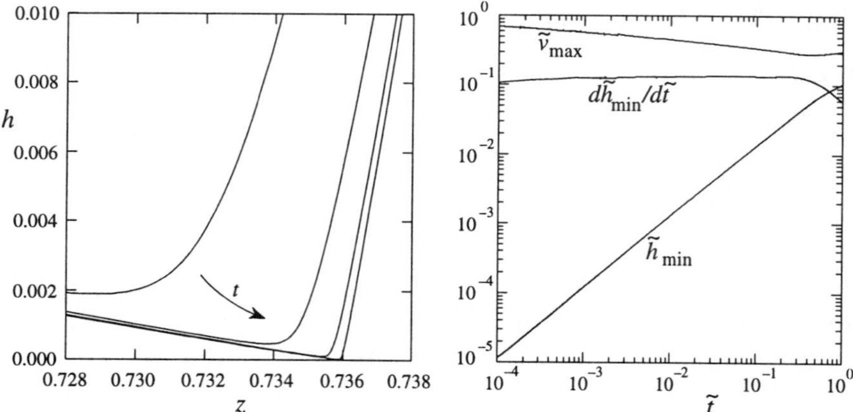

Figure 4. The numerical evolution of a thread according to the slender-body equations (1) and (25). The results suggest that the naive scalings (22) need a slowly varying correction.

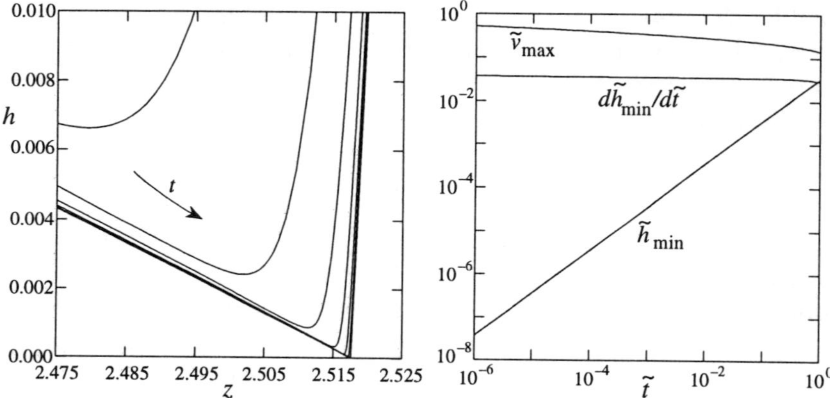

Figure 5. The numerical evolution of capillary-driven Stokes flow (27) with $\mu = \mu_{\text{ext}}$. The results suggest that the naive scalings (26) need a slowly varying correction.

Numerical results (Figure 5) suggest that the naive scalings (26) are correct for $h_{\min}$ but again need modification by a slowly varying logarithmic correction for $v_{\max}$. Some understanding of the origin of this factor is obtained by observing, first, that the far-field shape of a similarity solution with the naive scaling would be conical and, second, that the analytic solution for the Stokes flow driven by surface tension acting on a perfect cone is logarithmically infinite at the cone's vertex. It is thus to be expected that logarithms will enter the scaling when the far-fields on either side of the pinching region are matched to the neck of fluid that joins them.

5. Breakup due to van-der-Waals forces

It is clear from (3) that when $h_{\min} = O(A/\gamma)^{1/2}$, van-der-Waals forces must come into play. Though this lengthscale is comparable to molecular dimensions, it is of mathematical interest to see how a thread breaks up under van-der-Waals forces. The possible balances are analogous to those considered in previous sections but important differences emerge in the scalings and ordering of exponents.

5.1. VISCOUS THREAD WITH INERTIA: $Re = O(1)$

Setting $\gamma = 0$ and $\mu_{\text{ext}} = 0$ in (1)–(3) gives

$$v \sim \frac{z}{t}, \quad \frac{\rho v}{t} \sim \frac{A}{h^3 z} \sim \frac{\mu v}{z^2} \quad \Longrightarrow \quad \tilde{z} \sim \tilde{t}^{1/2}, \quad \tilde{h} \sim \tilde{t}^{1/3}, \quad \tilde{v} \sim \tilde{t}^{-1/2}, \quad (28)$$

where $\ell = A\rho/\mu^2$ and $\hat{t} = A^2\rho^3/\mu^5$. The same scalings apply to two-dimensional rupture of a sheet, as studied numerically by Erneaux & Davis (1993) and Ida & Miksis (1996), and the corresponding similarity equations have been solved by Vaynblat et al. (1997). Intriguingly, the plot analogous to Figure 1 has much less regularity, despite a similar equation structure, and all solutions are left-right symmetric. It should be noted, however, that $\tilde{h} \gg \tilde{z}$ as $\tilde{t} \to 0$ and so the long-wavelength assumption is violated. This suggests examination of alternative scalings in which $\tilde{h} \sim \tilde{z}$.

5.2. EQUIDIMENSIONAL SCALINGS FOR $Re = 0$ AND $Re = \infty$

To obtain $\tilde{h} \sim \tilde{z}$, we must either use the inertia-less balance

$$v \sim \frac{z}{t}, \quad \frac{A}{h^3 z} \sim \frac{\mu v}{z^2}, \quad h \sim z \quad \Longrightarrow \quad \tilde{z} \sim \tilde{h} \sim \tilde{t}^{1/3}, \quad \tilde{v} \sim \tilde{t}^{-2/3} \quad (29)$$

or the inviscid balance

$$v \sim \frac{z}{t}, \quad \frac{\rho v}{t} \sim \frac{A}{h^3 z}, \quad h \sim z \quad \Longrightarrow \quad \tilde{z} \sim \tilde{h} \sim \tilde{t}^{2/5}, \quad \tilde{v} \sim \tilde{t}^{-3/5} \quad (30)$$

Evaluation of the neglected terms shows that the correct balance is inviscid, in contrast to the inertia-less result of §4.2. Since $h \sim z$, detailed solutions will require evaluation of the van-der-Waals potential by pairwise integration over the intermolecular dipole interactions for a general interfacial configuration.

6. Discussion and temporal transitions

It is satisfying to see that the classical problem of capillary breakup, studied in the 19th century by Lord Rayleigh and others, is still producing surprises

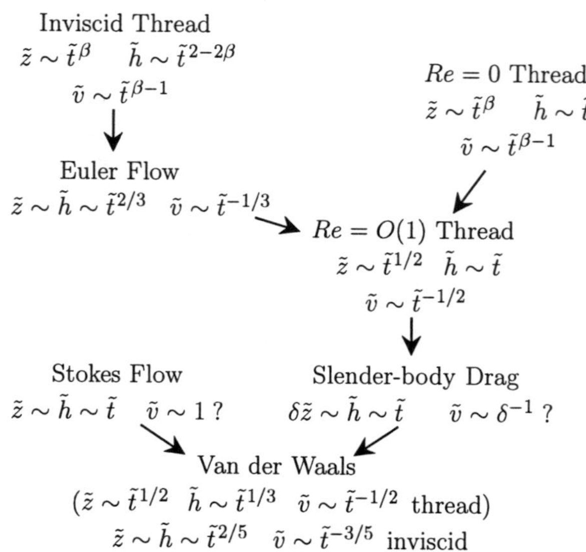

Figure 6. The temporal transitions between the various dynamical regimes (see discussion). The Stokes and slender-body scalings probably need logarithmic modifications.

and a wealth of analyses at the end of the 20th. A zoo of singularity structures has been uncovered by considering various dynamical regimes. Many interesting questions remain, such as the significance of 'resonance' at the stagnation point and why some solutions are left-right symmetric and others asymmetric.

We close by observing that the dynamical regimes can be ordered into a temporal cascade as $t \to 0$ (Figure 6), though experimental limitations will prevent the full cascade being observed in practice. The inviscid-thread approximation (§3.1) produces overturning before breakup and hence gives way to an equidimensional Euler flow (§3.2). The neglect of viscous stresses in the Euler flow or of inertia in a highly viscous thread (§2.2) both become invalid so that these regimes give way to a thread with internal Reynolds number $O(1)$ (§2.1). Even with a small external viscosity, neglect of the external shear stresses becomes invalid giving rise to the viscous slender-body motion (§4.1) or, if the external viscosity were comparable to the internal, there will be an equidimensional Stokes flow (§4.2). When the thread has thinned sufficiently, surface tension gives way to van-der-Waals forces, for which the most likely dynamical balance is inviscid. As an illustration of the usefulness of scaling and the richness of fluid dynamics, this familiar phenomenon of droplet formation has much to commend it.

Acknowledgement. We are very grateful to Jens Eggers for stimulating discussion and thoughtful comments on a draft of this manuscript.

References

Barenblatt, G.I. 1979 *Similarity, Self-Similarity, and Intermediate Asymptotics*. Plenum, New York.

Bogy, D.B. 1979 Drop formation in a circular liquid jet. *Ann. Rev. Fluid Mech.* **11**, 207–228.

Brenner, M.P., Lister, J.R. & Stone, H.A. 1996 Pinching threads, singularities and the number 0.0304... *Phys. Fluids A*, **8** 2827–2836.

Brenner, M.P., Eggers, J., Joseph, K., Nagel, S.R. & Shi, X.D. 1997 Breakdown of scaling in droplet fission at high Reynolds number. *Phys. Fluids A*, **9** 1573–1590.

Chaudhary, K.C. & Redekopp, L.G. 1980 The nonlinear capillary instability of a liquid jet. Part I. Theory. *J. Fluid Mech.*, **96**, 257–274.

Day R.F., Hinch, E.J. & Lister, J.R. 1997 Self-similar pinch-off of an inviscid fluid. *Phys. Rev. Lett.* (submitted)

Eggers, J. 1993 Universal pinching of 3D axisymmetric free-surface flow. *Phys. Rev. Lett.* **71**, 3458–3460.

Eggers, J. 1995 Theory of drop formation. *Phys. Fluids* **7**, 941–953.

Eggers, J. 1997 Nonlinear dynamics and breakup of free-surface flows. *Rev. Modern Phys.* **3**, 865–929.

Erneux T. & Davis, S.H. 1993 Nonlinear rupture of free films. *Phys. Fluids A*, **5** 1117–1122.

Goldstein, R., Pesci, A.I. & Shelley, M.J. 1993 Topology transitions and viscous singularities in viscous flows. *Phys. Rev. Lett.* **70**, 3043–3046.

Henderson D.M., Pritchard, W.G. & Smolka, L.B. 1997 On the pinch-off of a pendant drop of viscous fluid. *Phys. Fluids A* **9**, 1–13 (in press).

Ida, M.P. & Miksis, M.J. 1996 Thin Film Rupture. *App. Math. Lett.*, **9**, 35–40.

Kowalewski, T.A. 1996 On the separation of droplets from a liquid jet. *Fluid Dyn. Res.* **17**, 121–145.

Lister, J.R. & Stone, H.A. 1996 Time-dependent viscous deformation of a drop in a rapidly rotating denser fluid. *J. Fluid Mech.*, **317**, 275–299.

Lister, J.R., Brenner, M.P. & Stone, H.A. 1997 Capillary breakup of a viscous thread surrounded by another viscous liquid. (in preparation)

Papageorgiou, D. 1995 On the breakup of viscous liquid threads. *Phys. Fluids* **7**, 1529–1544.

Papageorgiou, D. 1996 Analytical description of the breakup of liquid jets. *J. Fluid Mech.*, **301**, 109–132.

Rayleigh, Lord 1892 On the stability of a cylinder of viscous fluid under a capillary force. *Phil. Mag.* **34**, 145.

Pugh, M. & Shelley, M.J. 1997 Singularity formation in thin jets with surface tension. *Comm. Pure App. Math.* (in press).

Tanveer, S. 1993 Evolution of Hele-Shaw interface for small surface tension. *Phil. Trans. R. Soc. Lond.* **A343**, 155–204.

Ting, L. & Keller, J.B. 1990 Slender jets and thin sheets with surface tension. *SIAM J. Appl. Math.* **50**, 1533–1546.

Tjahjadi, M., Stone, H.A. & Ottino, J.M. 1992 Satellite and subsatellite formation in capillary breakup. *J. Fluid Mech.*, **243**, 297–317.

Vaynblat, D., Brenner, M.P., Lister, J.R. & Witelski, T.P. 1997 Scaling theory of thin film rupture. (in preparation).

THE LINEAR STABILITY OF A TWO-PHASE COMPOUND JET

Temporal, spatial and absolute stability of viscous and inviscid compound jets

A. Chauhan[1], C. Maldarelli[1], D. Papageorgiou[2] and D. Rumschitzki[1*]

[1]Department of Chemical Engineering and the Levich Institute
City College of the City University of New York
Convent Avenue at 140th Street
New York, New York 10031 USA
[2]Dept. of Mathematics, Center for Applied Mathematics & Statistics
New Jersey Institute of Technology
University Heights, Newark, New Jersey 07102 USA
*Author for correspondence

1. Introduction

Axisymmetric circumferential capillary forces destabilize a liquid thread. Rayleigh (1879) considered the temporal linear stability of a static, doubly infinite, inviscid liquid thread of tension σ, radius a and density ρ by imposing an interfacial disturbance of wave numbers k (real) and n in the θ and z directions, respectively, at $t = 0$ and examining its time evolution using normal modes $e^{i(kz+n\theta)+s(k,n)t}$, s complex (s_r = temporal growth rate, s_i=wave speed). The balance between the destabilizing longitudinal curvature and the stabilizing axial curvature is such that axisymmetric (n=0) waves longer than the undisturbed jet circumference grow ($s_r>0$) without traveling ($s_i=0$), those shorter and non-axisymmetric waves decay and there exists a wavelength of maximal instability. Since the instability is capillary-driven, consideration of fluid viscosity (Chandrasekhar, 1961) and of an infinite surrounding fluid (Tomitika, 1935) do not change the qualitative picture, with viscosity only lowering growth rates and elongating the fastest growing wave.

Thread stability impacts liquid bridge collapse in microgravity and the setting of thin-walled annular polymer molds, but its main application is to the breakup of jets emerging from a nozzle tip. Examples of the latter occur in ink jet printing, fuel injection, particle sorting and polymer fiber spinning. In jets, disturbances generally arise at the nozzle tip, and, for a uniformly convecting base state, one would expect them to simply convect with the jet velocity V, i.e., $s_i(k)=-kV$. However, since these disturbances arise at the nozzle tip (z=0) continuously, not just at t=0, Keller *et al.* (1973) suggested that a spatial analysis $s=\pm i\varpi$ imaginary, $k=k(s(\varpi))$ complex (k_r=wave length, k_i=spatial growth rate), $e^{ik(\omega)z+i\omega t}$, was more appropriate. They found that for Weber numbers ($W:=\rho V^2 a/\sigma$) W>3.2 and non-dimensional frequencies $\omega a/V$ between 0 and a cutoff depending on W, but close to 1, disturbances grow axially. If the linear theory holds to breakup, the k with the largest negative imaginary part can predict drop size and the distance z to breakup. For

271

D. Durban and J.R.A. Pearson (eds.),
IUTAM Symposium on Non-Linear Singularities in Deformation and Flow, 271-282.
© 1999 *Kluwer Academic Publishers. Printed in the Netherlands.*

asymptotically large W, disturbances do in fact simply convect with the base velocity and grow with the temporal growth rate. For W<3.2, the jet becomes absolutely unstable and grows both in space and in time due to a pinch singularity in the dispersion equation (Lieb and Goldstein, 1986a, b). Including viscosity and density (Li and Lian, 1989) lowers the critical W with decreasing Reynolds number $Re := \rho Va / \mu$.

A large body of experiments (see review by Yarin, 1993), most at high enough W to be in the asymptotically large W regime, follow the inviscid results predictions almost to breakup and show that non-linear coupling of harmonics other than those imposed also only enter near breakup. Breakup events such as satellite formation and finite time singularities are clearly non-linear, and a series of weakly non-linear, fully non-linear one-dimensional and numerical approaches have captured some of them.

This paper studies the linear stability of a compound thread comprised of two immiscible fluids, fluid 1 in a circular core and fluid 2 in a surrounding concentric annulus. Fluid properties and other parameters are subscripted with the fluid number, e.g., core (outer) radius R_1 (R_2), inner (outer) interfacial tensions $\sigma_1 (\sigma_2)$. Applications of compound jets are similar to those of single jets, but having the second fluid and the second destabilizing interface provide additional physical properties that one can manipulate to control size and breakup length. It is also of interest for the formation of compound fibers and particles. Looked at in another light, compound jets are a marriage between liquid film and liquid jet problems and allow study of their combination.

Sanz and Masseguer (1985) and Radev and Shkadov (1985) performed one dimensional (pressure and velocity only a function of axial position z) temporal analyses of the inviscid compound jet and found two growing modes. One, with the larger growth rate, is a stretching mode (both interfaces move in phase) driven by the capillary forces at the inner interface and the other is a squeezing mode (interfaces exactly out of phase) driven by the capillary forces at the outer interface. The two types of modes are reminiscent of planar film instabilities (Felderhof, 1968, Taylor, 1959). Inclusion of viscosity and radial motions (Radev and Tchavdarov, 1988, Shkadov and Sisoev, 1996, Chauhan et al., 1997a) do not qualitatively change these features, but allow study of the effects of the outer-to-inner ratios d, m, γ of density, viscosity and surface tensions on the growth rate s_r, its maximum s_m and corresponding wave number k_m, and on the amplitude ratio amp_m, the latter useful in determining whether the core or the film breaks first.

In a series of papers (Chauhan et al., 1996, 1997 a-c), we have for each of the inviscid and the viscous compound jets examined its temporal, spatial and absolute instabilities using the unified Fourier-Laplace approach which treats the temporal stability as an initial value problem. In addition to characterizing the types of instabilities possible, the corresponding parameter regimes and the effects of the physical property ratios on linear drop size and breakup length, we examine both the long wave ($k \rightarrow 0$) and the thin film ($\varepsilon := (R_2 - R_1)/R_1 << 1$) asymptotics. These asymptotics shed light on the physical processes that drive the particular instabilities. Here we review the main results of these analyses and contrast the predictions and the physics of the inviscid and viscous analyses. We note how the various analyses compare with the Herz and Hermanrud's experiments (1983) aimed at developing a method for ink-jet printing using compound jets.

2. Formulation

2.1 GOVERNING EQUATIONS AND BOUNDARY CONDITIONS

Consider a compound jet as described above with a cylindrical coordinate system (r, θ, z), whose z axis is coincident with the fluid cylinder's axis. The base state compound jet is moving in a vacuum with a uniform velocity V in the positive z direction. The undeflected inner interface is $r=R_1$ and the outer interface is $r=R_2$.

The equations of motion and continuity in dimensionless form are

$$\nabla . v_i = 0 \tag{1}$$

$$d_i \frac{Dv_i}{Dt} = -\nabla P_i + \frac{m_i}{J^{1/2}} \nabla^2 v_i + F_i(r,t)\delta(z - z_0), \qquad (i = 1,2) \tag{2}$$

where in fluid i, P_i is the pressure, v_i is the velocity vector with components u_i and w_i in the r and z directions, respectively, and F_i is a source of a spatial disturbance at z_0. A temporal disturbance can enter into the boundary conditions. In the governing equations and the boundary conditions, we have non-dimensionalized using inviscid scalings, i.e., lengths by $[R_1]$, time by $[(\sigma_1/\rho_1 R_1^3)^{-1/2}]$, pressure by the surface tension pressure $[\sigma_1/R_1]$ and velocity by $[(\sigma_1/\rho_1 R_1)^{1/2}]$, to facilitate viscous and inviscid comparisons. The dimensionless parameters are $a=R_2/R_1$, $\gamma=\sigma_2/\sigma_1$, $d_2=d=\rho_2/\rho_1$ and $d_1=1$. For the viscous problem, $m_2=m=\mu_2/\mu_1$, $m_1=1$ and $J=\rho_1\sigma_1 R_1/\mu_1^2$, whereas for the inviscid problem, $m_i=m_i/\sqrt{J}=0$, $i=1,2$. Henceforth V will be non-dimensional, i.e., $V=W^{1/2}$. The interfaces are $r=h_i(z,t)$, $i=1,2$. Define the mean curvature H_i of interface i ($i=1,2$) by,

$$2 H_i = -\frac{1}{h_i(1 + h_{i'}^2)^{\frac{1}{2}}} + \frac{h_i''}{(1 + h_{i'}^2)^{\frac{3}{2}}} \tag{3}$$

Aside from the centerline condition (v_1,P_1 finite at $r=0$), the dimensionless jump in normal stress, the continuity of normal velocity and kinematic conditions are:

$$-[[P - \frac{2m}{\sqrt{J}}\frac{\partial u}{\partial r} - (-P + \frac{2m}{\sqrt{J}}\frac{\partial w}{\partial z})h'^2 + \frac{2m}{\sqrt{J}}(\frac{\partial u}{\partial z} + \frac{\partial w}{\partial r})h']]_i = 2H_i\gamma_i , \quad r=h_{1,2}(z,t)$$

$$[[n \bullet v]]_i = 0 \quad \text{at } r=h_1(z,t) \tag{4},(5)$$

$$[\frac{\partial}{\partial t} + v_i \bullet \nabla](r - h_i) = 0 \qquad \text{at } r=h_i(z,t) \quad i=1,2 \tag{6}$$

where $[[\ ]]_i$ denotes the difference between inside and outside of interface i and n_i and t_i are the vectors normal and tangential (in the r-z "plane") to interface i. For the viscous problem, these are augmented by the continuity of tangential velocity and shear stresses:

$$[[t \bullet v]]_i = 0 \qquad \text{at } r=h_1(z,t) \tag{7}$$

$$[[\frac{m}{\sqrt{J}}(\frac{\partial u}{\partial z} + \frac{\partial w}{\partial r})(1 - h'^2) + \frac{2m}{\sqrt{J}}\frac{\partial u}{\partial r}h' - \frac{2m}{\sqrt{J}}\frac{\partial w}{\partial z}h']]_i = 0 \quad \text{at } r=h_1(z,t) \tag{8}$$

$$[\frac{m_2}{\sqrt{J}}(\frac{\partial u_2}{\partial z}+\frac{\partial w2}{\partial r})(1-h_2'^2)+\frac{2m_2}{\sqrt{J}}\frac{\partial u_2}{\partial r}h'_2-\frac{2m_2}{\sqrt{J}}\frac{\partial w_2}{\partial z}h'_2]=0 \quad \text{at } r=h_2(z,t)$$

$$(9)$$

In the base state there is no bulk forcing disturbance and the interfaces are undeflected. For the inviscid case, any solution of the form, u=0, w=Φ(r), where Φ is differentiable, but otherwise arbitrary, satisfies the base state equations exactly. Since we are interested in capillary driven instabilities, we restrict the analysis to the solutions w=V= constant in both the core and annulus, which is also the unique viscous base state solution. This base state has no slip at the interface to prevent Kelvin Helmoltz instabilities. The base state (super 0) pressure is uniform in each fluid and equal to $P_2^0=\gamma$ /a, $P_1^0=1+\gamma$ /a.

If a small ($\sim \lambda$), axisymmetric disturbance perturbs the jet, the disturbances to the velocities and the interface from the base state are also of order λ. For i=1,2, we have

$$P_i(r,z,t) = P_i^0 + P_{i,1}(r,z,t)\lambda + O(\lambda^2)$$

$$(10)$$

$$v_i(r,z,t) = (0,V) + v_{i,1}(r,z,t)\lambda + O(\lambda^2)$$

$$(11)$$

$$h_1(z,t) = 1 + \zeta_1(z,t)\lambda + O(\lambda^2)$$

$$(12)$$

$$h_2(z,t) = a + \zeta_2(z,t)\lambda + O(\lambda^2)$$

$$(13)$$

Substituting these in the governing equations gives the leading order perturbation equations which, if we delete the first order subscript 1 for clarity, are identical to (1) and (2) except that Dv_i/Dt becomes $V\partial v_i/\partial z$. We choose the origin to be at the source; thus $z_0=0$.

2.2 TEMPORAL INSTABILITY

In the temporal instability, $F_i=0$, i.e., there is no spatial source. The initial conditions set off the instability. Choose a reference frame moving in the z direction with velocity V in which the jet is stationary ($V_{\text{new frame}}=0$). Choose the initial conditions $v_i(r,z,0)=0$, $\zeta_i(z,0)=\zeta_{i0}(z)$, i=1,2. We now take the Fourier-Laplace transform, defined by

$$f((k,s) = \int_0^\infty (\int_{-\infty}^\infty f(z,t)e^{-ikz}dz)e^{-st}dt,$$

$$(14)$$

of the governing equations. (We omit any hats or overbars to denote the transforms to keep the notation simple.) The general solution to these equations are sums of modified Bessel functions, each multiplied by a constant of integration. There is a total of four constants for the inviscid and eight for the viscous problem. Applying the centerline conditions, inserting these solutions into the remaining transformed, linearized boundary conditions gives an equation A(k,s)x(k,s)=b(k), where A(k,s) is a 5x5 (inviscid) or an 8x8 (viscous) matrix, x is the vector of constants of integration plus the Fourier-Laplace transforms of the two interfacial deflection functions and the inhomogeneity b(k) is the vector whose only two non-zero components are the Fourier transforms of the two initial interfacial disturbance functions. The equations corresponding to these two components are the two interfaces' normal stress balances. If A(k,s) is non-singular, the solution for the time evolution of the interfacial positions is $x(k,s)=A^{-1}(k,s)b(k)$.

To get the solution in the time and space domain, one must invert the solution from the Fourier-Laplace domain, i.e.,

$$\zeta_i(z,t) = \frac{1}{4\pi^2 i} \int_{-\infty}^{\infty} e^{ikz} \left[\int_{c(k)-i\infty}^{c(k)+i\infty} \zeta_i(k,s) e^{st} ds \right] \qquad (15)$$

where $c(k)$ is a real number that lies to the right of all singularities of $\zeta_i(k,s)$ in the s plane for each k. See Chauhan *et al.* (1996, 1997a) for details of the inversion. The Laplace inversion can be done by converting the line integral to a Bromwich contour integral and the Fourier by completing the contour with the semi-circle in the upper (lower) half plane for z>0 (<0) and using the residue theorem to evaluate the integrals. The poles contributing to instability arise from the roots of the system's dispersion equation $f(k,s)=\det A=0$. For each k, the s plane poles $\{s_n(k)\}$ are the corresponding roots of the dispersion equation.

In both the viscous and inviscid cases there are at most two roots of the dispersion equation with positive real parts for each k, corresponding to stretching (amplitude ratio $\zeta_2(k,s)/\zeta_1(k,s)$ real and positive) and squeezing (amplitude ratio real and negative) modes as in Sanz and Masseguer. For both the viscous and inviscid problems, s for each of these modes is real and non-negative for k from 0 to 1/a for the stretching (0 to 1 for the squeezing) mode and is purely imaginary (inviscid) or complex (viscous) beyond. The only poles of $\zeta_i(k,t)$ in the k plane turn out to be the poles k_0 of the initial conditions $\zeta_{i0}(k)$. If $\zeta_{i0}(k)$ derives from a disturbance that is smooth and decays at $\pm\infty$, it has no poles on the real axis. The method of stationary phases gives the long-time linear behavior of the interface as $\zeta_I(z,t) \sim \sum_n \text{coef}(k_m,s_n(k_m)) e^{ik_m z + s_n(k_m)t} /t^{1/2}$ as $t \to \infty$, where k_m is the point(s) on the real k axis for which ds/dk=0, i.e., the mode of maximum instability. If $\zeta_{i0}(k)$ derives from a sum of periodic disturbances, its singularities consist solely of a sum of delta distributions on the real k axis (at $\{k_0\}$). The k inversion then is explicit, giving, $\zeta_I(z,t) \sim \sum_{k_0} \sum_n \text{coef}(k_0,s_n(k_0)) e^{ik_0 z + s_n(k_0)t}$, where the s plane poles that contribute to a motion in time are the roots of the dispersion equation for k $=k_0$.

2.3 SPATIAL AND ABSOLUTE INSTABILITIES

The spatial instability derives from spatial disturbances that are localized in space. They can either grow or decay in place or they can convect with the jet. These disturbances can either arise in the bulk of the fluid (as assumed) or at either of the interfaces. In the latter case, the disturbance would enter the normal stress condition instead of the governing equations and would only change the form of the inhomogeneity, leaving the results on convective and absolute instabilities unchanged.

We solve the spatial problem in the stationary coordinate (lab) frame, the natural choice for experimentation, because the location of the onset of absolute instability depends on the observer. Thus the base state velocity no longer transforms out of the problem. If the source $F_i(z,t)$ is separable, i.e., $F_i(r,t)=H_i(r)T(t)$, one can still find a general solution to the Fourier-Laplace transformed equations. The homogeneous part is the same as in the temporal problem, but with s replaced by s+ikV, the base flow contribution, and with an inhomogeneous part deriving from $F_i(r,t)$. As in the temporal problem, insertion into the boundary conditions yields the matrix equation $A_s(k,s)x(k,s)=b(k)$, where $A_s(k,s)=A(k,s+ikV)$, $x(k,s)$

276   A. CHAUHAN ET AL.

again are the integration constants and the transformed disturbances. $\mathbf{b}(k)$ involves the inhomogeneity and, if also present, an initial distributed disturbance; we presume its absence. If $A_s(k,s)$ is non-singular, $\mathbf{x}(k,s)=A^{-1}(k,s)\mathbf{b}(k)$ and one only needs to invert to the z-t domain by complex plane integration.

If D_{mn} is the cofactor of $A_{s,nm}$, the cofactor expansion of $A_s^{-1}(k,s)$ gives,

$$x(z,t) = \frac{1}{4\pi^2} \int_{c-i\infty}^{c+i\infty} [\int_{-\infty}^{\infty} A_s^{-1}(k,s)b(k,s)e^{ikz}\,dk]e^{st}\,ds =$$

$$\frac{1}{2\pi} \int_{c-i\infty}^{c+i\infty} T(s)[\int_{-\infty}^{\infty} \frac{D(k,s)c(k,s)}{(s+ikv)|A_s(k,s)|} e^{ikz}\,dk]e^{st}\,ds = \frac{1}{2\pi} \int_{c-i\infty}^{c+i\infty} T(s)\phi(s,z)e^{st}\,ds$$

$$(16)$$

where c is chosen to be larger than the singularities of $T(s)\phi$ and T and (s+ikV) enter via the initial condition and the substantial derivative, respectively. For each s along the Laplace (Bromwich) contour, the poles of the function in the k plane will arise from the roots of the dispersion equation, from the poles of $D(k,s)c(k,s)$ and from k such that s+ikv=0. The poles of $D(k,s)c(k,s)$ occur when any of these matrices' components becomes singular. This happens in the k plane for k=0. As $k\rightarrow 0$, the pressure gradients, i.e., $\partial P/\partial z$, $\partial P/\partial r$ vanish; there is no driving force to accelerate the fluid and the growth rate vanishes. We close the k-contour with the semicircle in the upper (lower) half k-plane for z>0 (<0).

The singularities of the s integral can arise either from T(s) or from $\phi(s,z)$. If the source is periodic in time, T(s) has its poles on the imaginary s axis. For example, $T(t)=\sin(\omega t)$, $T(s)=1/(s^2+\omega^2)$ and the poles are $s=\pm i\omega$. If ϕ has no poles to the right of the imaginary s axis, its poles can only lead to decay in time, cf., e^{ikz+st}, and so the system oscillates with the frequency (frequencies) of the disturbance. To search for $\phi(s,z)$'s pole (Briggs, 1964), we note that, by choice of c,$T(s)\phi$ has no singularities along the Bromwich contour, i.e., the k-integrand has no pole on the real-k axis. Thus the k-integrand for any $s=c+is_i$ is the sum of the residues of poles in the upper half k-plane. Assuming that the k-plane residues and the poles at which they are evaluated are continuous functions of s, none of the k-plane residues crosses the real k axis (that is, the sum of this well-defined subset of k-plane residues defines the s-integrand) and the k-integral, i.e., the s-integrand, is a continuous function of s_i for s along $s=c+is_i$. As s_r decreases through $0<s_r<c$, k-plane residues may cross from the upper to the lower half k-plane or vice-versa. The former, part of the s-integrand for z>0, have negative k_i at $s_r=0$ and contribute a growth in space at the disturbance frequency, cf., e^{ikz+st}, $-k_i z>0$, for z>0; the latter do the same for z<0 ($-k_i z>0$) but are evanescent, i.e., are not part of the s-integrand, for z>0. On the other hand, if a pole below and a pole above the real k axis for $s=c+is_i$, $s_i \in \Re$ merge at some s with $0<s_r<c$, $T(s)\phi(s,z)$ has a singularity (in fact, a branch point) at the s of coalescence. Since this occurs for positive s_r, it contributes growth in both space and time: the system is absolutely unstable. This is a "pinch" singularity since the pinching/merging of roots from opposite halves of the k-plane causes it. Being a double or higher root in k of the dispersion equation, $f_s(k_0,s_0)=0$ and $[\partial f_s(k(s),s)/\partial s]|_{k_0,s_0}=0$, where $f_s(k,s)=\det(A_s(k,s))$. If $\partial f_s(k,s(k))/\partial s \neq 0$ at k_0,s_0, then a necessary condition for absolute instability is $ds/dk=-(\partial f_s/\partial k)/(\partial f_s/\partial s)=0$ at $(k,s)=(k_0,s_0)$, which is particularly useful when, as in the inviscid case, an explicit solution of the dispersion equation for the growth rate exists and if the poles in the s plane are simple.

3. Results and Discussion

3.1 TEMPORAL INSTABILITY

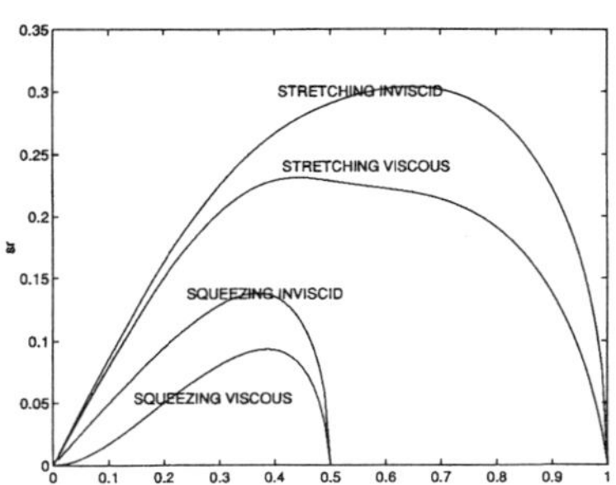

Fig. 1. Growth rate s_r of (top down) stretching mode, inviscid and viscous; squeezing mode, inviscid and viscous, as a function of wave number. $d=1, \gamma=2, a=2$.

The inviscid dispersion equation is quadratic in s^2 and thus allows explicit solutions for $s(k)$. There are 4 roots, but two of them are the negative of the other two. Figure 1 shows its two unstable roots with their long wave ($k\rightarrow 0$) limits. All the modes are linear in k as $k\rightarrow 0$, i.e., $s \sim s_1 k$, where s_1 satisfies $4s_1^4 - 2s_1^2[1 + \gamma /(ad) + (d-1) \gamma /(da^3)] + \gamma /(da) - \gamma /(da^3)=0$. As shown in Chauhan $et\ al.$ (1996), both s^2 and the amplitude ratio is purely real for both unstable modes. Thus the growth rate s is either purely real or purely imaginary. The interfaces in the stretching mode grow in phase ($0<k<1$) and oscillate out of phase ($k>1$) without damping, while the squeezing mode grows in phase ($0<k<1/a$) and oscillates out of phase ($k>1/a$). The lack of damping is a result of the absence of viscosity. Even though the stretching mode has a higher growth rate than the squeezing mode, proper choice of initial disturbance can favor either mode and, although clearly beyond the reach of the linear theory, may influence whether the core or film breaks first.

The viscous dispersion equation is transcendental and can, in principle, have an infinite number of roots s_n for each k. However, numerically, one finds three interfacial roots (where $s_r(k\rightarrow 0)\rightarrow 0$), two of which - stretching and squeezing as per their purely real amplitude ratios - also have purely real growth rates (of lower amplitude) for the same unstable k bands as in the inviscid case (Fig. 1). Unlike the inviscid fluids, the growth rates for each mode become complex beyond the unstable k bands. As $k\rightarrow 0$, the stretching mode is again first order in k $(s=\pm [(1+a\gamma)/(2(a^2d-d+1))]^{1/2}k+O(k^2))$ with coefficient independent of fluid viscosities; the squeezing mode is quadratic in k $(s=[\gamma J^{1/2}/(16ma^3(1+\gamma a)][4a^4\ln(a)+(m-3)a^4+2(2-m)a^2+m-1]k^2+O(k^3))$, with a leading coefficient independent of the density ratio. Since the instabilities occur at long waves, the physics of the analytic long wave limit, where the axial scales are much longer than both fluids' radial scales, can explain the similarity in the stretching and the difference in the squeezing modes between the viscous and inviscid cases. In all cases (viscous and inviscid; stretching and squeezing) the outer interfacial disturbance creates a order k^0 perturbed pressure in the annulus via the leading order normal stress balance. The order k^0 r-component of the equation of motion determines this pressure to be uniform across the film. From the kinematic and continuity equations, u and s are of the same order and one order in k higher than w. w can have a leading order (k^0) contribution and, for the inviscid case, the resulting scalings imply that the film's z-pressure gradient accelerates the annular fluid uniformly, i.e., $w \neq w(r)$, for both modes. In the viscous case, however, a leading order axial velocity means that the axial equation of motion has an order k^0 contribution from the circular radial derivative of the viscous shear stress set to zero.

Since the outer interface is stress-free, w is uniform across the film, and (continuity) u proportional to r. This is consistent only with a positive amplitude ratio, i.e., the stretching mode. Thus, both the viscous and inviscid stretching modes, w is independent of r; u, by continuity, is linear in r in the annulus; and the motion is simply a biaxial extension to leading order that is independent of the viscosities.

In the squeezing mode, the interfaces move out of phase and thus u can no longer be proportional to r. In the inviscid case, this means that the leading order annular u must have an additional term proportional to 1/r with coefficient of opposite sign. (Such a term is allowed despite the fact that the core u cannot have such a term because the lack of viscosities allows the axial velocities in the core and film to slip.) Otherwise the inviscid scalings remain unchanged. In contrast, in the viscous squeezing mode, u not being proportional to r disallows an order k^0 axial velocity. As a result, w~k and u and s~k^2. These scales make the axial acceleration of higher order in k than the perturbation pressure. Thus the viscous dissipation balances the perturbation pressure to leading order without accelerations. That is why the coefficient is independent of the densities.

Most applications aim to control the breakup size and time. The system parameters are manipulatable: the most important are, γ, d and a. As a $\rightarrow$ 1, the film becomes thin. The region where the squeezing mode becomes unstable shrinks and disappears, while the equations for the stretching mode reduce, in both viscous and inviscid problems, to those for a single jet limit with $\sigma = \sigma_1 + \sigma_2$. Since the instability is tension-driven, the effect of γ should be large and basically viscosity-independent. An increase in γ, σ_2 in our scalings, increases the driving force and increases the growth rate for all modes. Increasing σ_2 from σ_1 by a factor of 5 can double the wavelength of maximum instability. Increasing the viscosity or density ratios (here the annular properties) makes the annular fluid harder to move around and thus decreases the growth rates slightly. For curves and discussion, see Chauhan *et al.* (1996, 1997a).

3.2 SPATIAL INSTABILITY

We investigate the response of the system to a disturbance (s-plane poles at s=±iω) that is sinusoidal in time and of known frequency for a given velocity and jet parameters. If the system is not absolutely unstable the response is sinusoidal in time with the frequency of the source and spatial growth or decay depending on the k plane poles corresponding to s=±iω.

The non-zero poles in the k plane arise either from the dispersion equation or when (s+ikV)=0. The latter poles are always in the upper k plane for s_r>0 and touch the axis for s_r=0. Thus, absent pinching with a root of the dispersion equation, these roots are neutral. For such pinching to lead to growth , the root of |A_s(k,s)|=0, (s+ikV)=0 would need to have k_i>0.

Figure 2 shows two of the poles "1" and "2" of the dispersion equation in the k plane for the inviscid equations and d=1, γ=2, α=2, V=2. Each curve is a mapping of the line s_r=constant onto the k-plane through the dispersion relation, with arrows indicating increasing ω. For large s_r, root 1 lies in the

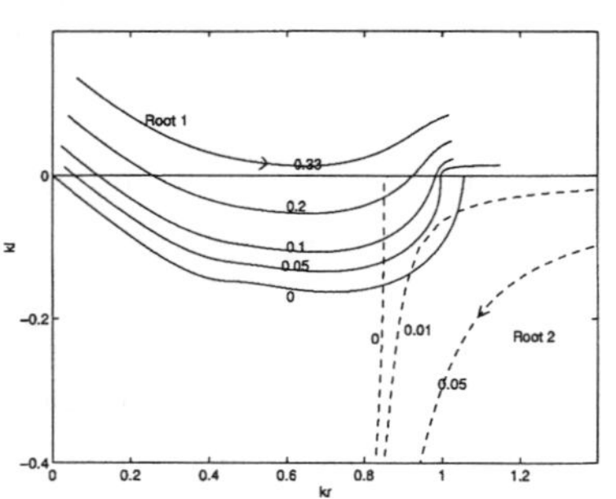

Fig. 2. Inviscid. Mapping of constant s_r lines in k-plane for the spatial roots 1 and 2. s=s_r-iω. Arrows show direction of increasing ω. Numbers denote s_r values. d=1,γ=2,a=2,V=2.

upper and root 2 lies in the lower half k-plane. Thus the residue at pole 1, but not that at pole 2, contributes to the integrand of the s-integration for z>0. (For z<0, the opposite holds.) On reducing s_r, root 1 moves down and root 2 up. For V=2, the roots do not merge. They intersect in the k plane but the value of s_i at those intersections is different for the two roots; so, these intersections do not represent a double root. They arise because s(k) is a multi-valued function. At s_r=0, a part of root 1 lies in the lower half k plane and this represents a spatially unstable ($-k_i z$>0) wave for z>0. Root 2 is evanescent (i.e., not part of the s-plane integrand) for z>0 and stable ($-k_i z$<0) for z<0. For these parameters another pair of roots, 3 and 4, behave similarly (curve not shown). Both lie in the upper half k plane for large s_r. On reducing s_r, root 3 moves down and crosses into the lower k plane and hence represents an spatially unstable mode for z>0. Root 4 moves down too, but it just touches the real k axis at s_r=0. Thus it is a neutrally stable mode for z>0. Root 3 does not merge with any evanescent root and thus the jet has no absolute instability. No other roots cross the real k axis for these parameters.

Thus, roots 1 and 3 are unstable for z>0. The cut-off Strouhal number ($|s_i|/V$) above which the modes are stable are about 1 and 1/a for modes 1 and 3, respectively. The exact values depend on the jet velocity. Leib and Goldstein (1986) found similar results for a single inviscid jet. The growth rate of root 1 is larger; henceforth we call it the primary and root 3 the secondary mode. Each mode has a maximum s_m in its spatial growth rate at $k_{i,max}$. In response to a broad band disturbance, the system will at long times (if it is still in the linear regime) select the frequency which gives the highest growth rate.

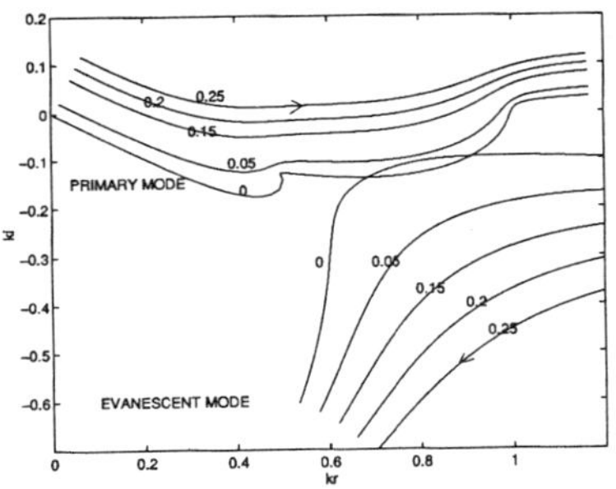

Fig. 3. Viscous. Mapping of constant s_r lines in the k-plane for the primary and evanescent spatial roots. $s=s_r-i\omega$. Arrows show increasing ω. Numbers denote s_r. d=1,γ=2,a=2,V=2.

The viscous jet is quite similar, having two spatially growing modes for sufficiently high velocities. Figure 3 illustrates the real k-axis crossing of the primary mode, the one with the higher growth rate and the critical k_r close to 1, for V=1.65. At this velocity, the growing and evanescent modes do not merge; the system is convectively, but not absolutely unstable.

The spatial growth rates k_i are strong functions of the jet velocity and increase as V decreases. The wave number k_r increases almost linearly with the driving frequency $s_i=\pm\omega$ for both unstable modes. At high velocities, for each mode, $k_r=-s_i/V$ for $|s_i|<V$ (primary) or $|s_i|<V/\alpha$ (secondary mode). For s_i above this cutoff, there is a sudden change in the slope. For high velocities, $k_r=-s_i/V$ implies the disturbances just convect in the +z direction. It is reasonable to suspect that they at a rate that depends on their wavelengths, as in temporal instability and Chauhan et al. 1996) show this. Thus, although the spatial analysis is more pertinent to free flowing jets, the temporal analysis gives the same predictions as the spatial at sufficiently high jet velocities. Since the single jet is a limiting case of the compound jet, this result is true for the single jet as well. This and the surprising reach of the linear theory in general are the reasons why predictions on drop size based on the temporal model agree well with experiments. Also, in experiments the jet is actually semi-infinite. But since the doubly infinite jet is stable in the negative z direction, it is reasonable to model the semi-infinite jet as a doubly infinite jet.

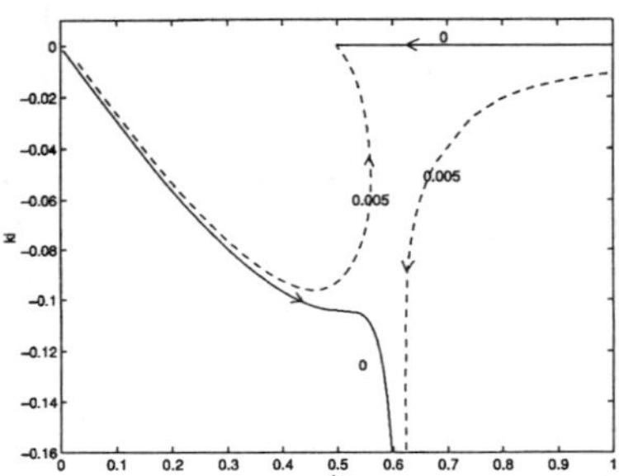

Figure 4. Merging of secondary and evanescent inviscid roots causing absolute instability. Numbers on curves denote s_r values. $d=1, \gamma=2, a=2, V=1.7$.

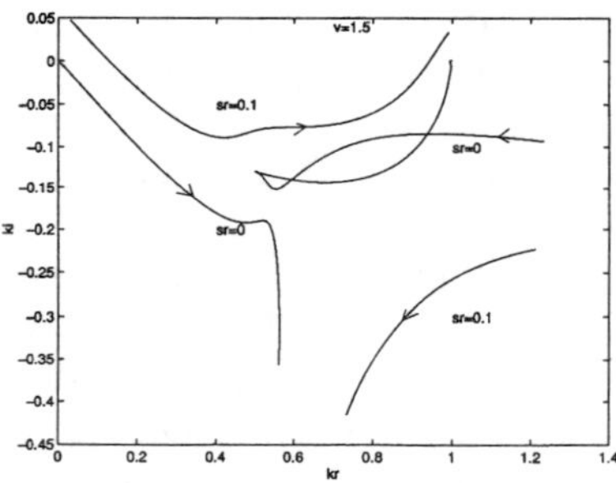

Figure 5. Merging of viscous roots causing absolute instability. Numbers on curves denote s_r. $d=1, \gamma=2, a=2, V=1.6, m=1, J=1000$.

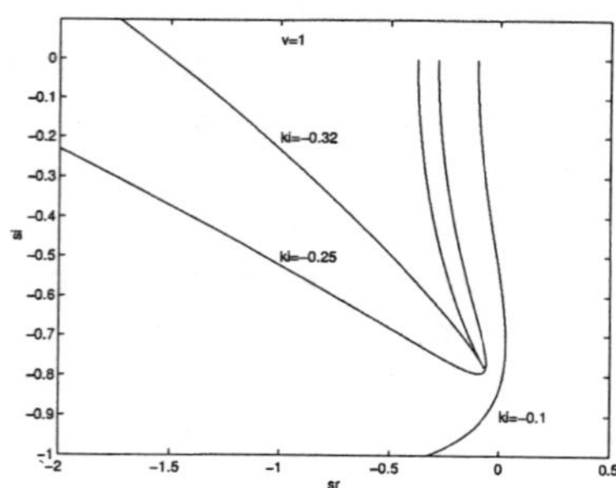

Figure 6. Cusp formation in s-plane implying $ds/dk=0$ - viscous case, thin annular limit. Curves are lines of constant k_i in s plane. Numbers are k_i. $d=1, \gamma=2, V=1, m=1, J=1000$.

3.3 ABSOLUTE INSTABILITY

If the jet velocity is below a critical value, the jet becomes absolutely unstable. Fig 4 shows the movement of the inviscid evanescent root (root 2), and the growing secondary root on reducing s_r for $V=1.7$. For $s_r=0.005$, the roots are not merged. At some s_r between 0.005 and 0, the curves merge and then split to give a topologically different configuration at $s_r=0$ than at $V=2$ (Fig 3). This results in an absolute instability. (Figure 5 shows a similar story for the viscous case at $V=1.6$, with merging of the secondary and an evanescent mode between $s_r=.1$ and 0). At $V=1.7$, the primary mode does not yet merge with an evanescent root in the lower k plane and thus is only convectively unstable.

On further reducing the jet velocity, the primary mode also undergoes such a merge to contribute another root which grows in time. But, the critical velocity for the primary mode is lower than that of the secondary mode. In an experiment it would be desirable to stay above the (higher) critical velocity of the secondary mode to maintain a jet that is only convectively unstable. At $V=0$, the merger for either mode takes place on the real k axis at $k= k_r$ and $s=g(k_r)$. The k_r for the primary and the secondary modes correspond to the k at which the temporal growth rate ($s=g(k)$) is maximum for the stretching and the squeezing mode, respectively. This happens because at $V=0$, the spatial dispersion equation, i.e., $f_s(s,k)=f(s+ikV,k)$ reduces to the temporal form. Also, the required condition for absolute instability, $ds/dk=0$, is same as the condition for the maxima in $s=s(k)$.

The critical velocity for the jet is a function of all the system parameters. For the inviscid case, an explicit solution of $s=g(k)-ikV$ exists. A necessary condition for the onset of an absolute instability is $ds/dk=\partial s_r/\partial k_r+i\partial s_i/\partial k_r=0$. The root can be found numerically or graphically by plotting lines of constant k_i in the s plane and looking for a cusp formation. At the cusp, both s_i and s_r reach a local extremum ($\partial s_r/\partial k_r=\partial s_i/\partial k_r=0$) as functions of k_r and $ds/dk=0$. This method is shown in Fig 6 for the secondary viscous mode in the thin annular limit. It agrees with the method followed earlier because the frequency at the cusp formation is the same as the frequency at the merging of the secondary and the evanescent modes.

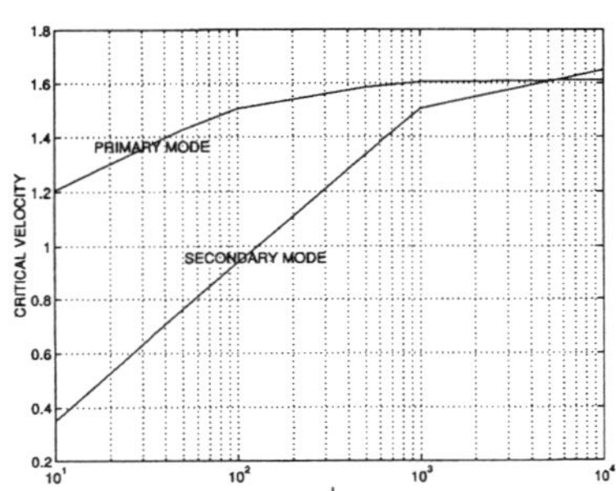

Figure 7. Critical velocities - both absolutely unstable viscous modes as f(J). d=1,γ=2,a=2.

To locate the critical velocity, we solve, ds/dk=g'(k)-iV=0, to obtain the V at which s_r at absolute instability becomes 0. Above this velocity, s_r at the merger of the roots in the k plane is negative and hence the jet is not absolutely unstable. For the inviscid compound jet with β=1, γ=2 and a=2, the critical velocity is V_c=1.77. In the viscous case, the critical velocity depends on the viscosities (on J and m). Fig 7 plots the V_c(J) for β=1, γ=2, a=2 and m=1. Note that even which mode determines the V_c depends on the parameters.

3.4 COMPARISON WITH EXPERIMENT

The theory developed in this paper predicts the breakup of a compound jet resulting in the formation of compound drops and gives the linear theory's estimate of the breakup lengths and of the drop sizes at breakup. Hertz and Hermanrud (1983) did experiments with a compound jet generated using a single nozzle assembly and filmed as the jet moved in the axial direction. Their primary fluid was a water soluble ink and their secondary fluid was a silicone oil. The parameters were: σ_2 =20*10^{-3} N/m, σ_1=50*10^{-3} N/m , ρ_2=ρ_1=1000 kg/m^3, R_1=75x10^{-6}m, R_2=150x10^{-6} m, V=1.98m/s. So, γ=0.4, β=1 and W=2.426.

They observed the compound jet exiting from the orifice and the formation of compound drops. From the pictures one can measure the wavelength(s) of the growing disturbances. The wave number measured from Hertz & Hermanrud's (1983) Fig. 5 changes in the axial direction; it is 0.64 for the first wave and its average is 0.6. In this experiment there is no specific imposed frequency disturbance; so the system selects the dominant wave with the maximum growth rate. Fig. 8 shows the predictions based on the temporal and the spatial analyses of both the viscous and inviscid theories. The spatial analyses predicts that the jet is convectively unstable. The velocity is large but not large enough for the temporal and spatial growth rates to overlap. J is large enough so that the difference between the viscous and inviscid predictions is small. k_m for the inviscid temporal is k=0.66, for the inviscid spatial

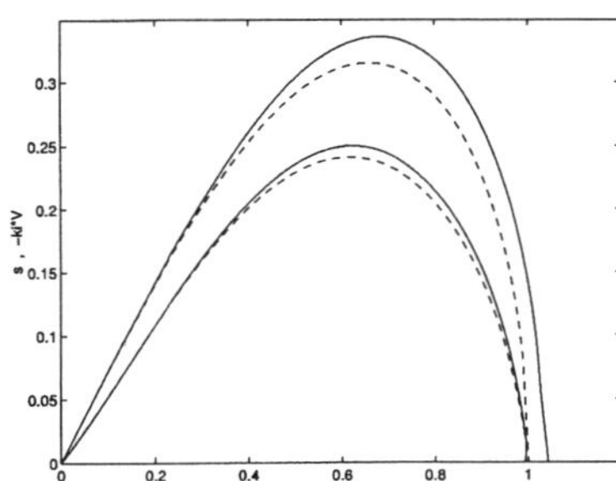

Figure 8. Comparison of the predictions of viscous and inviscid, temporal and spatial theories with experiment performed by Hertz & Hermanrud. d=1,γ=0.4,a=2,V=2.426.

k=0.682, for the viscous temporal k =0.62 and viscous spatial k=0.623, which are all in very good agreement with the experimental findings. Sanz & Meseguer (1985) one dimensional temporal model gave k=0.7.

4. Conclusions

A compound jet, whether viscous or inviscid, is unstable due to capillarity. The difference between the temporal and the spatial instabilities lies in the origin of the disturbance. If the instability is initiated by a disturbance in the initial condition, the situation is one of temporal instability and if the disturbances originate from a localized spatial source, the instability is spatial, with the latter being more appropriate for jet problems. At very high jet velocities, viscous effects are negligible; the disturbances grow spatially, but there is no dispersion and the results of all four analyses agree. At velocities below a critical value which is a function of the jet parameters the jet becomes absolutely unstable and this critical cutoff depends on the jet parameters, including the viscosities.

The spatial analysis predicts that the compound jet is unstable for $z>0$. There are two unstable modes, the primary and the secondary. The growth rate of the primary mode is higher. The primary and the secondary modes are unstable below critical Strouhal numbers that are close to 1 and 1/a, respectively. The exact values depend on the Weber number. The axial growth rate of both the modes increases with a decrease in the jet velocity. The critical velocity below which the jet is absolutely unstable is different for the two modes. At high velocities, the primary and the secondary modes merge into the stretching and the squeezing modes of the temporal analyses, respectively. The breakup size and length can be controlled by manipulating the source frequency or the system parameters. If the source is broad-band, the frequency with the highest axial growth dominates at long lengths if the linear theory still applies. At short lengths, all the frequencies will contribute; the result is an axial variation of the observed wavelength.

Acknowledgements

We thank the Air Force Office for Scientific Research (F49620-94-1-0242 to DP), the National Science Foundation (DMS 9-9-0070 to DP, CTS 8658147 to DR) and the donors of the Petroleum Research Fund (ACS-PRF #27403-AC9 to DR) for supporting this work. DR also thanks the Alexander von Humboldt Foundation for supporting his stay at the Ruhr University of Bochum where a portion of this work was done. Additional support for DTP was provided by the National Aeronautics and Space Administration under Contract No. NAS1-19480 while DTP was at the Institute for Computer Applications in Science and Engineering (ICASE), NASA Langley Research Center, Hampton, VA 23681-0001.

References

Briggs, R.J., *Electron stream interaction with plasmas*, MIT Press, Cambridge, MA., 1964

Chauhan, A., Maldarelli, C., Rumschitzki, D., Papageorgiou, D., Temporal and spatial instability of an inviscid compound jet, *Rheol. Acta*, **35**:567-583, 1996; Temporal instability of a compound thread, *Chem. Eng. Sci.*, in review 1997a; Absolute instability of an inviscid compound jet, *Phys. Fluids*, in review, 1997b; Convective and absolute instabilities of a viscous compound jet, in preparation, 1997c

Felderhof, B.U., Dynamics of fee liquid films., *J. Chem. Phys.*, **49**: 44-51, 1968

Hertz, C.H. & Hermanrud, B., A liquid compound jet, *J. Fluid Mech.*,**131**:271-287, 1983

Keller, J.B., Rubinow, S.L. & Tu, Y.O. (1973), Spatial instability of a jet., *Phys. Fluids*, **16**:2052-2055

Leib, S.J., Goldstein M.E., The generation of capillary instabilities on a liquid jet., *J. Fluid Mech.*, **168**:479-500, 1986a; Convective and absolute instability of a viscous jet, *Phys. Fluids*, **29**(4):952-954, 1986b

Lin, S.P., Lian, Z.W., Absolute instability of a liquid jet in a gas, *Phys. Fluids A*, **1**:490-493, 1989

Radev, S. Shkadov, V., On a stability of two-layer capillary jet, *Theor. Appl. Mech.*, **16**:68-75, 1985

Radev, S., Tchavdarov, B., Linear capillary instability of compound jets, *Int. J. Mulitph. Flow,* **14**:67-69, 1988

Rayleigh, Lord, On the instability of jets., *Proc. Lond. Math. Soc.*, **10**: 4-13, 1879

Sanz, A., Meseguer, J., One-dimensional linear analysis of the compound jet., *J. Fluid Mech.,* **159**:55-68, 1985

Shkadov, V.Y. and Sisoev, G.M., Instability of a two layer capillary jet, *Int. J. Multiph. Flow*, 363-377, 1996

Taylor, G.I., The dynamics of thin sheets of fluid. II. Waves on fluid sheets., *Proc. R. Soc.*, **A253**:296-312, 1959

Tomotika, S., On the instability of a cylindrical thread of a viscous liquid surrounded by another viscous liquid, *Proc. R. Soc.*, **A150**:322-337, 1935

Yarin, A.L., *Free liquid jets and films: Hydrodynamics and rheology*, Longman Sci. and Techn., Essex, 1993

FREE-SURFACE DEFORMATION AND FORMATION OF CUSPS AT LOW REYNOLDS NUMBER FLOW

JAE-TACK JEONG

Department of Mechanical Engineering
Chonnam National University
Kwangju, 500-757, KOREA

Free surface deformation and cusp formation are analyzed by considering a viscous flow due to a vortex dipole below the free surface where the direction and the strength of vortex dipole are given arbitrarily. In the analysis, the Stokes' approximation is used and surface tension effects are included, but gravity is neglected. The solution is obtained analytically by using conformal mapping and complex variable techniques. From the solution, the deformation of the free surface and the formation of a cusp on the free surface are discussed. For some typical directions and strengths of vortex dipole, the deformations of free surface shape are shown. As the capillary number grows, it is shown that the radius of curvature at the converging point on the free surface becomes extremely small, which means that the free surface shape becomes singular and in a real fluid a cusp should form on the free surface above some critical capillary number. In the limit of infinite capillary number, the local structure of the cusp is discussed.

D. Durban and J.R.A. Pearson (eds.),
IUTAM Symposium on Non-Linear Singularities in Deformation and Flow, 283-296.
© 1999 *Kluwer Academic Publishers. Printed in the Netherlands.*

1. Introduction

Some flow visualizations by photographs of free-surface flows of both Newtonian and non-Newtonian fluids at low Reynolds numbers have been carried out (Joseph et al. 1991, Liu et al. 1995). These photographs show compelling evidence for the formation of two-dimensional cusps on the free surface in the regions of convergence of the flow. The dynamical mechanism related to the cusp on the free surface in a Stokes' flow has recently become a subject of theoretical and experimental researches. In a paper published by Jeong & Moffatt (1992), they carried out experiments using a pair of counter-rotating cylinders in a Newtonian fluid with free surface. For very slow rotation rates, there is a stagnation line on the free surface, and in some circumtances a small rounded crest can form in the neighborhood of this stagnation line. When the rotation rate Ω is increased however, the surface dips downwards, and simple visual observation indicates the presence of a very sharp cusp on the free surface. A complete analytical solution of a model problem of this experiments was also obtained, where the full nature of the flow and the detail of the cusp formation procedure were explained. In the model problem, they considered a Stokes' flow induced by a vortex dipole below the otherwise undisturbed free surface, where direction of the vortex dipole was perpendicular to the free surface.

In the present paper, we generalize the above idealized problem allowing the direction of the vortex dipole to be arbitrary. As a low Reynolds number limit, the Stokes' approximation is used and the effect of surface tension is included, but gravity effects are neglected. The solution is obtained analytically by using conformal mapping and complex variable techniques. From the solution, the deformation of the free surface and the formation of a cusp on the free surface are discussed.

2. The method of solution

2.1 MATHEMATICAL FORMULATION

Consider the flow region shown in Fig.1(a), where the undisturbed fluid occupies the half space $y<0$. A vortex dipole of strength B ($B>0$) with direction λ is located at $z(=x+iy)=-id$. Since d is the only length scale in the problem, we may take $d=1$ in what follows, and the vortex dipole singularity is then located at $z=-i$. The free surface Γ is distorted by the resulting flow. Assume that the Reynolds number $Re=B/d\nu$ is small so that stream function Ψ satisfies the biharmonic equation $\nabla^4\Psi=0$.

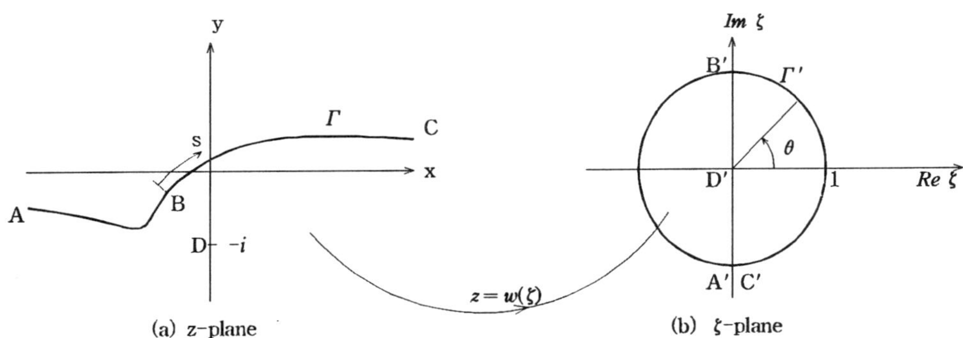

Fig.1 Free surface deformation with a vortex dipole and conformal mapping into the unit circle.

It is well-known that Ψ can then be expressed in the form

$$\Psi = Im[\,f(z) + \overline{z}g(z)\,] \tag{1}$$

where $f(z), g(z)$ are two analytic complex functions at all points z in the fluid domain except at the location of the vortex dipole singularity, $z = -i$. The velocity components (u, v) are then given by

$$u - iv = f'(z) + \bar{z}g'(z) - \overline{g(z)} \, ,\tag{2}$$

and the pressure (p) and vorticity (ω) fields by

$$p - i\mu\omega = 4\mu g'(z)\tag{3}$$

where μ is the viscosity. It is easy to verify that, with these relations, the Stokes equation $\nabla p = \mu \nabla^2 \boldsymbol{u}$ is satisfied in the fluid.

As shown by Richardson (1968), velocity and stress boundary conditions on the free surface Γ take the form

$$f'(z) + \bar{z}g'(z) - \overline{g(z)} = u_0(z) \overline{\left(\frac{dz}{ds}\right)}\tag{4}$$

$$f'(z) + \bar{z}g'(z) + \overline{g(z)} = -i\frac{\gamma}{2\mu} \overline{\left(\frac{dz}{ds}\right)}\tag{5}$$

where s is the arclength on Γ measured from some point (B in Fig.1(a)), γ the surface tension coefficient, and $u_0(z)$ the (real) tangential velocity at an arbitrary point z of the free surface. Manipulation of (4) and (5) yields the equations (for $z \in \Gamma$)

$$f(z) + \bar{z}g(z) = 0,\tag{6}$$

$$Im\left[\overline{\left(\frac{dz}{ds}\right)} g(z)\right] = \frac{\gamma}{4\mu} .\tag{7}$$

Near the vortex dipole singularity at $z=-i$,

$$f(z)\to -\frac{\beta}{z+i} \quad (\beta \equiv Be^{i\lambda}), \qquad g(z)\to finite, \tag{8}$$

where β is a given complex number (B represents strength of vortex dipole and λ direction of the vortex dipole). The conditions $u,v \to 0$ at infinity ($|z|\to\infty$) are satisfied provided

$$f(z)\to kz, \quad g(z)\to \overline{k}, \qquad as \ |z| \to \infty, \tag{9}$$

where k is an arbitrary constant. We can find from (4) and (5) that the choice $k=-i\gamma/4\mu$ is appropriate. Equations (4)–(9) constitute the essential boundary conditions that f(z) and g(z) must satisfy.

2.2 CONFORMAL MAPPING

Now let $z=w(\zeta)$ be the conformal mapping that maps the fluid domain D onto the unit disk $D': |\zeta|<1$, and which places the (imaged) vortex dipole at $\zeta=0$, so that $w(0)=-i$ (Fig. 1(b)). The points A,B,C,D of Fig. 1(a) map to the points A',B',C',D' of Fig. 1(b). Let

$$F(\zeta)=f(w(\zeta))=f(z) \tag{10}$$
$$G(\zeta)=g(w(\zeta))=g(z) \tag{11}$$
$$U(\zeta)=u_0(w(\zeta))=u_0(z) \tag{12}$$

so that

$$f'(z)=\frac{F'(\zeta)}{\omega'(\zeta)}, \quad g'(z)=\frac{G'(\zeta)}{\omega'(\zeta)} \tag{13}$$

By the conformal mapping property, it follows that, for $z\in\Gamma$, i.e. $|\zeta|=1$,

$$\frac{dz}{ds} = -i\zeta \frac{\omega'(\zeta)}{|\omega'(\zeta)|}, \tag{14}$$

$$\overline{\left(\frac{dz}{ds}\right)} = \frac{i}{\zeta}\frac{\omega'(\zeta)}{|\omega'(\zeta)|}. \tag{15}$$

Hence boundary conditions (4) and (5) become, on $|\zeta|=1$,

$$\frac{F'(\zeta)}{\omega'(\zeta)} + \overline{\omega(\zeta)}\frac{G'(\zeta)}{\omega'(\zeta)} - \overline{G(\zeta)} = U(\zeta)\frac{i}{\zeta}\frac{\overline{\omega'(\zeta)}}{|\omega'(\zeta)|}, \tag{16}$$

and

$$\frac{F'(\zeta)}{\omega'(\zeta)} + \overline{\omega(\zeta)}\frac{G'(\zeta)}{\omega'(\zeta)} + \overline{G(\zeta)} = \frac{\gamma}{2\mu\zeta}\frac{\overline{\omega'(\zeta)}}{|\omega'(\zeta)|}. \tag{17}$$

Subtracting (16) from (17) and taking the complex conjugate gives

$$\frac{G(\zeta)}{\zeta\omega'(\zeta)} = \frac{1}{2|\omega'(\zeta)|}\left[\frac{\gamma}{2\mu} + iU(\zeta)\right], \quad \text{on } |\zeta|=1 \tag{18}$$

Since $w(\zeta)$ is a conformal mapping function, $w'(\zeta)$ is analytic and non-zero in $|\zeta|<1$, so that the left-hand side of (18) is analytic in $|\zeta|<1$ except at $\zeta=0$. To remove this singularity at $\zeta=0$, we subtract $G(0)/\zeta w'(0)$ from each side of (18):

$$\frac{G(\zeta)}{\zeta\omega'(\zeta)} - \frac{G(0)}{\zeta\omega'(0)} = \frac{1}{2|\omega'(\zeta)|}\left[\frac{\gamma}{2\mu} + iU(\zeta)\right] - \frac{G(0)}{\zeta\omega'(0)} \quad \text{on } |\zeta|=1. \tag{19}$$

Now, the left-hand side is the boundary value of a function analytic in $|\zeta|<1$. If $w(\zeta)$ and G(0) can be found, then the real part of the right-hand side of (19) will be known. By using Cauchy's integral theorem (Muskhelishvili 1953), the $G(\zeta)$ for $|\zeta|<1$ may then be found from

$$\frac{G(\zeta)}{\zeta\omega'(\zeta)} - \frac{G(0)}{\zeta\omega'(0)} = \frac{1}{2\pi i} \oint_{|\zeta_0|=1} \left[\frac{\gamma}{4\mu|\omega'(\zeta)|} - Re\left\{ \frac{G(0)}{\zeta_0|w'(\zeta_0)|} \right\} \right] \frac{\zeta_0 + \zeta}{\zeta_0(\zeta_0 - \zeta)} d\zeta_0 + ic$$

$$(20)$$

In (20) we have to determine $w(\zeta)$, $G(0)$ and an unknown real constant c. The boundary conditions (6), (8) transform in the ζ-plane to

$$F(\zeta) = -\overline{\omega(\zeta)}G(\zeta), \qquad \text{on } |\zeta| = 1, \tag{21}$$

$$F(\zeta) \rightarrow -\frac{\beta}{\omega'(0)\zeta}, \qquad \text{as } \zeta \rightarrow 0. \tag{22}$$

Now, considering bilinear transformation and observing the boundary conditions (21) and (22) carefully, we are led to an appropriate form of mapping function

$$z = \omega(\zeta) = a(\zeta + i) + (a+1)i\frac{\zeta - i}{\zeta + i} + ib\zeta \tag{23}$$

where a, b are real constants to be determined. Note that (23) satisfies $\omega(0) = -i$ and $Im[\omega(\zeta)] \rightarrow 0$ as $\zeta \rightarrow -i$. For conformality, a, b are constrained to satisfy $\omega'(\zeta) \neq 0$ in $|\zeta| < 1$. By introducing $w(\zeta)$ in the form (23), the boundary condition (22) can be satisfied.

Substituting (23) into (21) and using $\overline{\zeta} = 1/\zeta$ on $|\zeta| = 1$, we get $F(\zeta)$ in $|\zeta| < 1$, by applying Cauchy's integral theorem,

$$F(\zeta) = -\left\{ a(\frac{1}{\zeta} - i) + (a+1)i\frac{\zeta - i}{\zeta + i} - \frac{ib}{\zeta} \right\} G(\zeta). \tag{24}$$

If we compare (24) with (22), we get

$$G(0) = \frac{\beta}{(a-ib)\omega'(0)} = \frac{\beta}{(a-ib)(3a+2+ib)} \qquad (25)$$

Now, with $w(\zeta)$ and $G(0)$ given by (23) and (25), there are only 3 unknown real constants a, b and c. The relations for these 3 unknown constants come from the fact that the velocity should vanish at infinity. From this condition, a, b may be determined from the relation

$$16\pi\, Ca = -(a-ib)(3a+2+ib)^2 \int_0^{2\pi} \frac{\cos\theta_0 - i(1+\sin\theta_0)}{|\omega'(e^{i\theta_0})|\,(1+\sin\theta_0)^2}\, d\theta_0 \,, \qquad (26)$$

where complex capillary number Ca is defined as

$$Ca \equiv \frac{\mu\beta}{\gamma} = \frac{\mu B}{\gamma}\, e^{i\lambda}. \qquad (27)$$

From (26), for given Ca (i.e. for a given vortex dipole), a and b can be determined. The remaining unknown constant c may be easily expressed using (20) in terms of a, b. $G(\zeta)$ can be expressed as

$$G(\zeta) = -\frac{\beta}{32\pi Ca}\, \omega'(\zeta)\frac{(\zeta+i)^2}{\zeta}\, I\!\left(\frac{\zeta-\zeta^{-1}}{2i}; a\right) \qquad (28)$$

where

$$I(\zeta; a) \equiv \int_0^{2\pi} \frac{(\zeta^2+1)(1+\sin\theta_0) - i(\zeta+i)^2\cos\theta_0}{|\omega'(e^{i\theta_0})|(1+\sin\theta_0)^2(\sin\theta_0 - \zeta)}\, d\theta_0$$

Therefore, 2 complex functions $F(\zeta)$ and $G(\zeta)$ in the ζ-plane are obtained as (24) and (28), and hence $f(z)$ and $g(z)$ in z-plane which provide a solution of Stokes' flow. Hence we have obtained the solution of the flow due to the vortex dipole.

3. Results

3.1 FREE SURFACE

From the mapping function (23)

$$z(=x+iy)=\omega(\zeta)=a(\zeta+i)+(a+1)i\frac{\zeta-i}{\zeta+i}+ib\zeta \tag{23a}$$

To make $\omega'(\zeta)\neq0$ in $|\zeta|<1$, a and b should satisfy

$$(3a+1)(a+1)>b^2, \tag{29a}$$

and from the constraint $y\to-\infty$ as $\zeta\to-i$ in $|\zeta|<1$,

$$a>-1. \tag{29b}$$

Hence (a,b) should be in the right side of hyperbola $(3a+1)(a+1)=b^2$ in (a,b) plane as shown in Fig.2. As described in the previous section, a and b can be calculated from the relation (26) for given complex capillarity Ca. Carrying out numerical integration in (26) for various directions of vortex dipole (λ=const.), sets of (a,b) are calculated and shown in Fig.2.

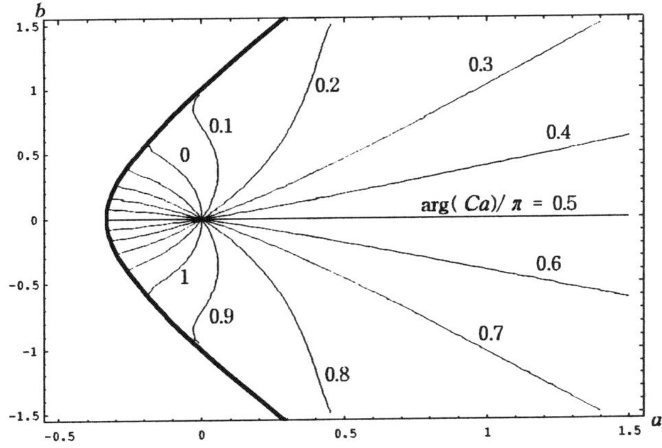

Fig.2 Parameters a and b in the mapping function (23)
related with complex capillary number Ca.

Since $|\zeta|=1$ on free surface, putting $\zeta=e^{i\theta}$, we get a parametric equation of the free surface.

$$\begin{cases} x=-\dfrac{(a+1)\cos\theta}{1+\sin\theta}+a\cos\theta-b\sin\theta \\[2mm] y=a(1+\sin\theta)+b\cos\theta \end{cases} \tag{30}$$

Eliminating θ in (30) we get the equation of free surface,

$$(x+b)^2 y-2b(x+b)(y+a+1)-(2a-y)(y+a+1)^2=0. \tag{31}$$

For constant directions of the vortex dipole ($\beta=Be^{i\lambda}$, $\lambda=const.$) deformations of the free surface are shown in Figs. 3-5 for a few cases. The corerspondences between surface shapes and values of Ca are the obvious ones in all three figures.

For very large surface tension or very weak strength of vortex dipole, deformation of the free surface is very small and free surface shape is horizontal ($y=0$).

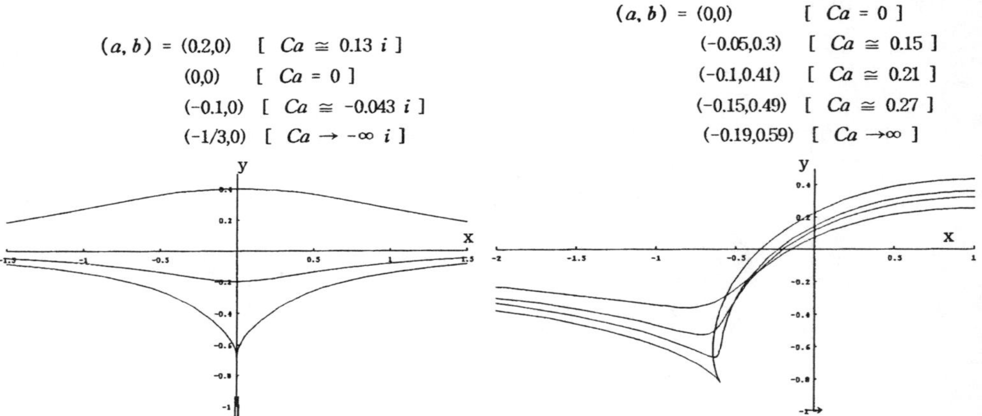

Fig.3 Free surface shapes for vertical direction of vortex dipole ($\lambda=\pm\,\pi/2$).

Fig.4 Free surface shapes for horizontal direction of vortex dipole ($\lambda=0$).

(a,b) = (0.3,0.54)   [$Ca \cong 0.50\nearrow$]
          (0.2,0.24)   [$Ca \cong 0.23\nearrow$]
          (0.1,0.11)   [$Ca \cong 0.086\nearrow$]
          (0,0)   [$Ca = 0$]
          (−0.1,−0.09)   [$Ca \cong 0.058\nearrow$]
          (−0.2,−0.16)   [$Ca \cong 0.095\nearrow$]
          (−0.31,−0.22)   [$Ca \rightarrow \infty\nearrow$]

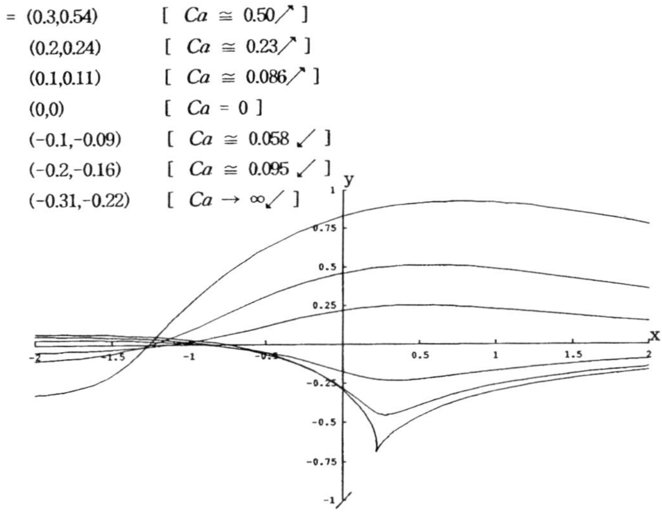

Fig.5   Free surface shapes for vortex dipole with
an inclination angle λ ($\lambda = \pi/4, -3\pi/4$).

3.2 FREE SURFACE CUSP

When $(3a+1)(a+1) = b^2$ or

$$
\begin{cases}
a = -\dfrac{\sin\theta_0}{1+2\sin\theta_0}, \\[2mm]
b = -\dfrac{\cos\theta_0}{1+2\sin\theta_0},
\end{cases}
\qquad \text{for } -\frac{\pi}{6} \le \theta_0 \le \frac{7\pi}{6},
\tag{32}
$$

then $\omega'(e^{i\theta_0}) = 0$ and a cusp occurs on the free surface (at $\zeta = e^{i\theta_0}$). In
this case the equation of the free surface is

$$
\begin{cases}
x = \dfrac{\cos\theta}{1+2\sin\theta_0}\dfrac{1+\sin\theta_0}{1+\sin\theta} + \dfrac{\sin(\theta-\theta_0)}{1+2\sin\theta_0}, \\[3mm]
y = -\dfrac{\sin\theta_0 + \cos(\theta-\theta_0)}{1+2\sin\theta_0},
\end{cases}
\qquad \text{for } -\frac{\pi}{2} < \theta < \frac{3\pi}{2}
\tag{33}
$$

and the cusp point is, at $\theta = \theta_0$,

$$\begin{cases} x_0 = \dfrac{\cos \theta_0}{1 + 2\sin \theta_0} = -b, \\[2mm] y_0 = -\dfrac{1 + \sin \theta_0}{1 + 2\sin \theta_0} = -(a+1). \end{cases} \tag{34}$$

Hence the cusp point (x_0, y_0) is on the curve $x^2 - 3y^2 - 2y = 0$ as shown in Fig.6. Near the cusp point (x_0, y_0) the equation of the free surface is, asymptotically,

$$X^2 = (1 + \sin \theta_0)\sqrt{\frac{1 + \sin \theta_0}{8}}\, Y^3, \tag{35}$$

where (X, Y) is a set of local coordinates shown in Fig.7,

$$X = \frac{(1 + \sin \theta_0)(x - x_0) - \cos \theta_0 (y - y_0)}{\sqrt{2(1 + \sin \theta_0)}} ,$$

$$Y = \frac{\cos \theta_0 (x - x_0) + (1 + \sin \theta_0)(y - y_0)}{\sqrt{2(1 + \sin \theta_0)}} .$$

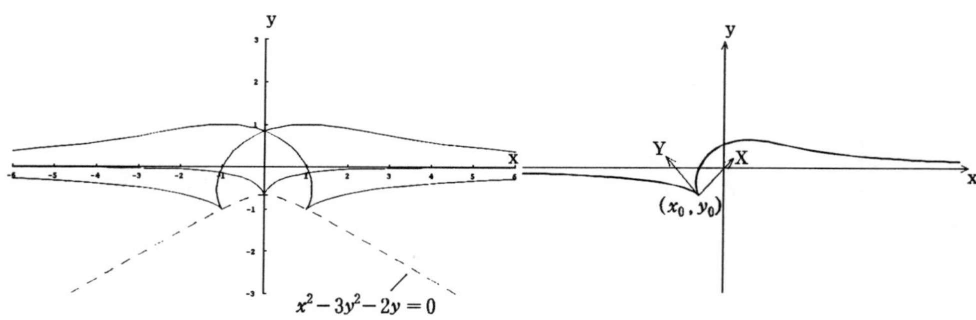

Fig.6 Locations of cusps on the free surface.   Fig.7 Asymptotic equation of free surface near cusp point.

For arbitrary a, b in the region satisfying (29), the radius of curvature of the free surface at the local minimum point (at the point $\theta = \theta_1$ where $a\cos\theta_1 - b\sin\theta_1 = 0$) is

$$R = -\frac{(\dfrac{a+1}{\sin\theta_1+1} + a\sin\theta_1 + b\cos\theta_1)^2}{a\sin\theta_1 + b\cos\theta_1}, \tag{36}$$

As a, b approach nearly to (32), then $\theta_1 \approx \theta_0$ and the radius of curvature R in (36) becomes extremely small. This means that a cusp occurs on the free surface. In the real fluid, when the radius of curvature R at the cusp point on the free surface is so small that the characteristic length scale near the cusp is comparable to the molecular length scale of the fluid, then the continuum hypothesis fails and the effects of molecular interactions become dominant. In this case, the flow near the cusp point may be unstable by molecular interactions of the fluid and air (second fluid outside).

3.3 STREAM FUNCTION

With (25) and (28) $F(\zeta)$, $G(\zeta)$ may be obtained and substituing into (1) we get a stream function

$$\Psi = Im\left[G(\zeta)\left\{ (a - ib)(\bar\zeta - \frac{1}{\zeta}) - 2(a+1)i\frac{\bar\zeta\zeta - 1}{|\zeta + i|^2} \right\} \right] \tag{37}$$

When the complex capillary number $Ca(=\dfrac{\mu\beta}{\gamma})\rightarrow 0,$

$$b + ia \rightarrow 2\,Ca \tag{38}$$

and

$$\omega'(\zeta) \to \frac{-2}{(\zeta+i)^2} \ , \quad G(\zeta) \to \frac{i\gamma}{4\mu} . \tag{39}$$

Hence stream function becomes, from (1),

$$\Psi \to -\frac{\gamma}{\mu} Im \frac{a+bz}{z^2+1} , \tag{40}$$

which is a well known potential flow solution.

References

Jeong, J. & Moffatt, H. K. (1992) Free-surface cusps associated with flow at low Reynolds number *J. Fluid Mech.* **241** 1-22.

Joseph, D. D, Nelson, J., Renardy, M. & Renardy, Y. (1991) Two-dimensional cusped interface, *J. Fluid Mech.* **223**, 383-409.

Liu, Y. J., Liao, T. Y. & Joseph, D. D. (1995) A two dimensional cusp at the trailing edge of an air bubble rising in a viscoelastic liquid *J. Fluid Mech.* **304** 321-342.

Muskhelishvili, N. I. (1953) *Some Basic Problems of the Mathematical Theory of Elasticity*, 3rd Edn. P. Noordhoff.

Richardson, S. (1968) Two dimensional bubbles in slow viscous flows *J. Fluid Mech.* **33** 476-493.

FREE-SURFACE CUSPS AND MOVING CONTACT LINES.
A COMMON APPROACH TO THE PROBLEMS

Y. D. SHIKHMURZAEV
Department of Applied Mathematics,
University of Leeds,
Leeds LS2 9JT, United Kingdom

Abstract. A model is described which allows one to pose and solve mathematical problems associated with viscous flows in domains with piecewise smooth boundaries in a regular way. The singularities inherent in the conventional approach are removed by incorporating the interface formation-disappearance process in the boundary conditions for the Navier-Stokes equations. The model is illustrated by considering the moving contact-line problem and the free-surface cusp problem as particular examples.

1. Introduction

Moving contact lines and steady two-dimensional free-surface cusps are the two examples of singularities which highlight some important general aspects of the mathematical modeling of viscous flows. The term "singularity" is used here, as always in continuum mechanics, to indicate that a conventional way of modelling a certain physical process mathematically leads to consequences which for some reasons cannot be accepted. This definition implies that the assessment is made on the basis of some *physical* criteria associated with what features of the flow should be modelled, i.e. the criteria which lie outside the mathematics of the problem, though in the 'most singular' cases non-existence of a solution makes the formulation of the problem unacceptable also from the mathematical point of view. If a solution exists but some physical arguments indicate a "singularity", one has to find out whether it appeared due to simplifications and/or additional assumptions made in the process of obtaining the solution (see Goldshtik, 1990 for a review of such cases) or is inherent in the very formulation of the problem. In the former case, the singularity can be removed mathematically, without altering the underlying model, while in the latter, one has to look for those physical processes which are not accounted for in the conventional formulation and generalize the model itself.

In the simplest case, the "moving contact-line problem" emerges when one considers displacement of an inviscid gas by a viscous liquid from a smooth homogeneous solid surface. The conventional mathematical formulation of this problem involves the following three basic components:

(i) the Navier-Stokes equations for the flow of the liquid in the bulk,

297

D. Durban and J.R.A. Pearson (eds.),
IUTAM Symposium on Non-Linear Singularities in Deformation and Flow, 297-308.

(ii) a piecewise smooth boundary of the flow domain so that the free surface meets the
solid at a *contact angle,* which is not equal neither to 0 nor to π,

(iii) the classical boundary conditions, which include the no-slip condition on the solid
surface, and balance of normal stresses together with zero tangential stress and normal
relative velocity on the free boundary.

This formulation requires a prescribed value of the contact angle, i.e. the dynamic contact
angle, which, according to experiments, is not equal to the static one, must be prescribed
as a function of the particular conditions of the flow and the properties of contacting
materials, so that from the very beginning the formulation is, in fact, semi-empirical.

An analysis of the conventional model shows that under assumptions (i)–(iii) no solution to the problem exists, and in a simplified formulation, with a prescribed free-surface
shape and the normal-stress boundary condition dropped, the singularity appears in another form as a non-integrable shear stress at the contact line (Dussan V. & Davis, 1974).
The vulnerable point among (i)–(iii) is assumption (iii), i.e. the boundary conditions, simply because assumption (i) *defines* the fluid one is interested in and the equations for which
the boundary conditions must be set up, while assumption (ii) reflects the experimental
fact, which is a part of what should be modelled theoretically. Obviously, non-existence of
a solution to the moving contact-line problem in the general case indicates that the model
should be improved to account for those 'extra' physical processes which come into action
when the contact line starts to move. A recent review of the theories proposed to improve
the conventional formulation of the moving contact-line problem can be found in §9 of
Shikhmurzaev (1997*a*).

The steady two-dimensional free-surface cusp is another example of a singularity in
viscous flow, though the reasons for considering it as such are less straightforward. Free-
surface cusps are produced by convergent viscous flows near free boundaries (Joseph,
Nelson, Renardy & Renardy, 1991), and their appearance corresponds to a change in the
flow kinematics: fluid particles which used to constantly belong to the free surface are now
advected through the cusp into the interior (Jeong & Moffatt, 1992) and become 'ordinary'
bulk particles.

In contrast to the moving contact-line problem, the conventional fluid mechanics formulation allows for solutions to the problems associated with convergent flows near free
boundaries (Richardson, 1968; Jeong & Moffatt, 1992). However, these solutions, being
applied to the real situation, show that the results do not describe correctly the key elements of the flow. For example, according to experiments, cusping occurs at finite capillary
numbers while in this case the conventional model either leads to a genuine cusp but with
an infinite rate of dissipation of energy in the neighbourhood of the cusp line (Richardson,
1968) or predicts that there is no cusp both in the free surface geometry and, what is more
important, in the flow kinematics (Jeong & Moffatt, 1992). It is also worth mentioning
that modelling of the gas-liquid interfaces as geometrical surfaces of zero thickness and a
constant surface tension becomes incorrect as the free-surface curvature goes to infinity,
so that in this case the classical model falls outside its limits of applicability. At the same
time, a genuine cusp implies a capillary force acting on the cusp line, and this force, being
a *concentrated* one, can be balanced by viscous stress, which is a *distributed* force, only if
the latter has a non-integrable density so that the associated rate of dissipation of energy
is infinite (Richardson, 1968). One can find a more detailed ˙discussion of the problem in
Shikhmurzaev (1997*c*).

In the present work we will try to show that both the moving contact-line problem and the cusp problem can be solved on the basis of a model which treats the two flows as particular cases of a more general physical phenomenon.

2. Physical Background

Experiments point out a very essential similarity between the flows with cusps and those with moving contact lines: both the cusp line and the moving contact line are regions where the free surface disappears, and in both cases the fluid particles initially located in the free surface traverse these regions in a finite time and either form another interface (the moving contact-line problem) or become "ordinary" bulk particles (the cusp problem). Hence in both cases the "surface" properties of such particles (e.g. the surface tension) have to relax to new equilibrium values (or disappear). Thus, the two flows are particular cases of the process of interface formation/disappearance, the process which is not accounted for in the classical boundary conditions, where the interfaces are treated as already formed. Therefore it is not surprising that the classical formulation gives rise to physically unacceptable singularities for the flows where this process does take place.

In §§3 and 4 we will describe the simplest mathematical model which incorporates the interface formation-disappearance into the boundary conditions for the Navier-Stokes equations and its application to moving contact lines and steady free-surface cusps. Then, in §5, we will return to the physical background of the model to discuss some general aspects of the problem.

3. Model

The simplest boundary conditions for the Navier-Stokes equations, which generalize the classical ones to account for the process of interface formation-disappearance in a self-consistent way, were formulated in Shikhmurzaev (1993a).

On the liquid-facing side of an interface between a viscous liquid and an inviscid gas the bulk velocity u and stress P satisfy the following conditions:

$$n \cdot P \cdot n + p_g + p^s \kappa = 0, \tag{1}$$

$$(I - nn) \cdot P \cdot n - \nabla p^s = 0, \tag{2}$$

$$p^s = \gamma(\rho^s - \rho_0^s), \tag{3}$$

$$\frac{\partial \rho^s}{\partial t} + \nabla \cdot (\rho^s v^s) = -\frac{\rho^s - \rho_e^s}{\tau}, \tag{4}$$

$$(1 + 4\alpha\beta)\nabla p^s = 4\beta(u - v^s) \cdot (I - nn), \tag{5}$$

$$(u - v^s) \cdot n = 0. \tag{6}$$

Here $n \cdot P \cdot n$ and $n \cdot P \cdot (I - nn)$ are the normal and the tangential projection of stress acting on the interface, respectively (I is a metric tensor; n is a unit vector normal to the interface, which points from the gas to the liquid); p_g is the gas pressure; κ is the free surface curvature. The *surface* pressure p^s, defined as the negative surface tension, is related with the *surface* density by the equation of state (3), which is given in the simplest form applicable for barotropic processes in the surface phase; the surface mass balance equation (4) includes the relaxation term on the right-hand side, where $\rho_e^s = \rho_{1e}^s$ is the equilibrium *surface* density of the liquid-gas interface and τ is the relaxation time.

Equations (5) and (6) relate the velocity of the surface mass transport v^s with the bulk velocity u evaluated on the liquid-facing side of the interface and the surface pressure gradient (α and β are phenomenological constants).

On the liquid-solid interface one has the generalized Navier condition

$$n \cdot P \cdot (I - nn) - \tfrac{1}{2}\nabla p^s = \beta(u - U) \cdot (I - nn) \tag{7}$$

together with equations which describe distributions of the surface parameters along the interface

$$p^s = \gamma(\rho^s - \rho_0^s), \tag{8}$$

$$\frac{\partial \rho^s}{\partial t} + \nabla \cdot (\rho^s v^s) = -\frac{\rho^s - \rho_{2e}^s}{\tau}, \tag{9}$$

$$v^s = \tfrac{1}{2}(u + U) - \alpha \nabla p^s, \tag{10}$$

$$(u - U) \cdot n = 0. \tag{11}$$

Here u and U are the velocities of the liquid and the solid on the opposite sides of the liquid-solid interface; the unit normal vector n points form the solid to the liquid. It is necessary to emphasize that p^s is not the surface tension of the solid; it is associated with the stress in a thin layer of the liquid adjacent to the solid surface, which thickness is neglected in the macroscopic model, and this quantity may be positive or negative.

A generalization of the model for the case of a liquid-liquid interface and a discussion on the physical meanings of parameters can be found elsewhere (Shikhmurzaev, 1993a, b, 1994, 1996[1], 1997a, c).

Boundary conditions (1)–(11) are applicable on smooth interfaces, while at the lines of discontinuity the distributions of the surface parameters must be linked by the surface mass and momentum balance conditions. In hydrodynamic processes, i.e. those where the characteristic time scale is large compared with τ, these conditions give that

- the *surface* mass fluxes into and out of the line of intersection of two interfaces are equal,
- the total force acting on such a line is zero.

These conditions together with some conditions specifying the overall flow provide a closed well-posed mathematical problem for describing the flows associated with intersecting interfaces.

4. Wetting and Cusping

The model described in the previous section can be applied without additional *ad hoc* assumptions in different situations where intersection of interfaces takes place. It can be easily generalized to account for surfactants and thermal effects and give a conceptual frame for modelling the interface formation process in complex fluids. Below, we will consider application of the model to the moving contact-line problem and the free-surface cusp problem.

---

[1]Note a misprint in the sign of ∇p^s in Eq. (6).

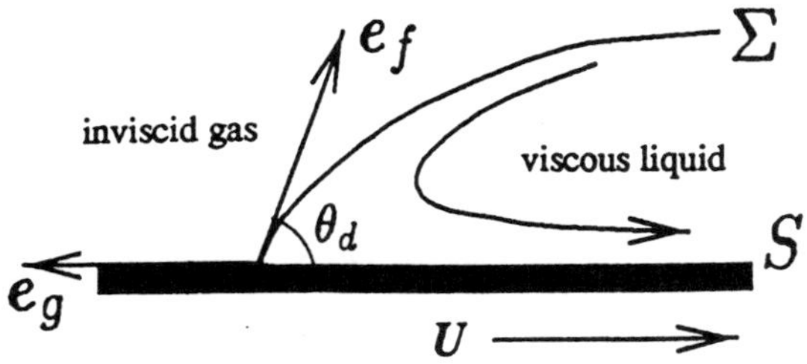

Figure 1.   A definition sketch for the moving contact-line problem

4.1. MOVING CONTACT LINE

The simplest variant of the moving contact-line problem is associated with the intersection of a liquid-gas and a liquid-solid interfaces (Fig. 1). In this case, one may use (1)–(6) and (7)–(11) as boundary conditions for the Navier-Stokes equations on the two interfaces, respectively. The mass and momentum balance conditions at an arbitrary point r_0 of the contact line in a reference frame where the contact-line speed is zero may be written down as follows:

$$(\rho^s v^s)|_{r\to r_0, r\in\Sigma} \cdot e_f = (\rho^s v^s)|_{r\to r_0, r\in S} \cdot e_g, \tag{12}$$

$$p^s_{|r\to r_0, r\in\Sigma} \cos\theta_d = p^s_{SG} - p^s_{|r\to r_0, r\in S}. \tag{13}$$

Here e_f and e_g are unit vectors normal to the contact line and directed along the liquid-gas and gas-solid interface, respectively; the notation $r \to r_0$, $r \in \Sigma$(or S) is used to denote the limit of a function as r tends to r_0 along the liquid-gas interface Σ(or along the liquid-solid interface S). The dynamic contact angle θ_d measured through the liquid is defined by

$$\cos\theta_d = -e_f \cdot e_g.$$

The mass balance equation (12) can be easily generalized to account for a precursor film ahead of the moving contact line (Shikhmurzaev, 1996) or a more complicated structure of the contact line.

As in the Young-Laplace theory, we imply that the projection of the total force acting on the contact line normal to the solid surface is zero, so that the corresponding component of the capillary force from the free surface is balanced by the reaction force from the solid. Equation (13) states that the tangential projection of the total force at $r = r_0$ is zero as well. Besides the surface pressures determined by (3) and (8), equation (13) contains the tangential projection of the force acting from the solid on the contact line p^s_{SG} determined by the nature of the contacting media. As the contact line starts to move, the values of the surface pressures p^s at the contact line begin to deviate from those in equilibrium $p^s(\rho^s_{1e})$ and $p^s(\rho^s_{2e})$, while p^s_{SG} remains a constant not involved in the drama of the surface pressure changes. The result of those changes is that the dynamic contact angle θ_d determined by (13) goes away from the static one θ_s. The latter satisfies the classical Young equation

$$p^s(\rho^s_{1e})\cos\theta_s = p^s_{SG} - p^s(\rho^s_{2e}),$$

which links together the equilibrium values of the surface parameters.

The boundary conditions (1)–(13) allow one to remove the shear-stress singularity and find θ_d as a function of the parameters of the problem. Condition (7) guarantees that the stress is regular in the vicinity of the contact line while (12) preserves the so-called "rolling" motion observed in experiments (Dussan V. & Davis, 1974; Chen, Ramé & Garoff, 1996). The model does not require any auxiliary concepts, such as, for example, the "apparent" contact angle, to interpret experimental data. Since the distributions of the surface parameters are coupled with the bulk flow, one should expect that the dynamic contact angle depend on the flow field in the vicinity of the contact line. Recent experiments (Blake, Clarke & Ruschak, 1994) show that indeed this is the case.

The present model has been investigated in a number of papers (Shikhmurzaev, 1993a, 1994, 1996) and extended to the general case of a fluid/liquid/solid system (Shikhmurzaev, 1993b, 1997a). For small capillary numbers, the formulae can be considerably simplified (Shikhmurzaev, 1993a), and in some cases one can even arrive at an analytical expression for the dynamic contact angle as a function of the flow parameters for both the advancing and the receding contact-line motion (Shikhmurzaev, 1994, 1996). The results are in good agreement with experimental data (Shikhmurzaev, 1993a, 1997a,b).

4.2. TWO-DIMENSIONAL FREE-SURFACE CUSP

To apply the present model to the cusp problem one has to "find" the third interface that meets the two free surfaces at the cusp line. This question requires a few comments.

As the fluid particles which initially form the liquid-gas interface traverse the cusp line and go into the bulk of the liquid, they gradually loose their "surface" properties and become "ordinary" bulk particles thus forming a relaxation "tail", which stretches from the cusp towards the interior. This "tail" may be regarded as a gradually disappearing "internal interface". The length of the relaxation "tail" depends on the flow rate and for sufficiently high flow rates becomes large compared with the interfacial layer thickness so that it may be regarded as macroscopic from the point of view of fluid mechanics. In this case, we may use our model to describe this process. One may say that, in a sense, the flow with a cusp is a specific case of dynamical wetting, where the liquid "wets" a geometrical surface, which passes through the cusp line, with the dynamic contact angle equal to π.

Let us consider a symmetrical cusp formed by an interface between an incompressible Newtonian liquid and an inviscid gas which opens on the negative x-axis of the Cartesian coordinate frame (x, y) with the origin coincident with the cusp (Fig. 2). Due to the symmetry of the problem we may consider the region $y \geq 0$ and in what follows will assume that the flow is steady and the associated Reynolds number is small so that the inertial terms may be neglected. In the bulk in the absence of mass forces one has:

$$\nabla \cdot u = 0, \quad \nabla \cdot P = 0, \tag{14}$$

where u is the fluid velocity and $P = -pI + \mu[\nabla u + (\nabla u)^{\mathrm{T}}]$ is the stress tensor (p and μ are the pressure and viscosity, respectively).

The boundary conditions on the free surface are given by (1)–(6) while the distribution of the surface parameters along the relaxation "tail" ($x \geq 0, y = 0$) are described by (2)–(6). The change in the environment for a fluid particle as it comes from the free surface

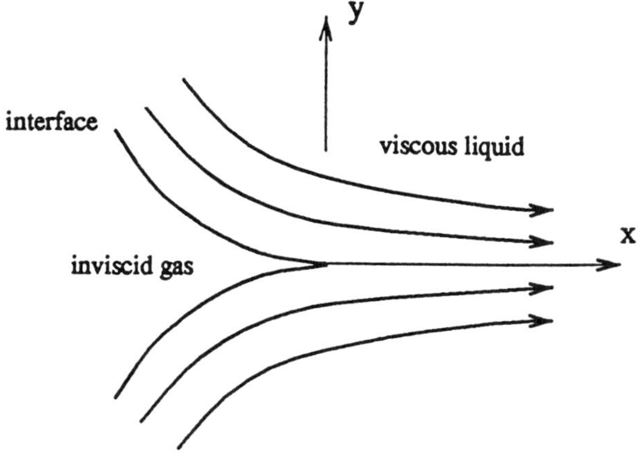

Figure 2.   A definition sketch for the flow near the cusp

to the bulk is given by

$$\rho_e^s = \begin{cases} \rho_{1e}^s, & x < 0 \\ \rho_0^s, & x \geq 0 \end{cases}, \tag{15}$$

so that the equilibrium surface pressure in the bulk is zero.

The mass and momentum balance equations at an arbitrary point r_0 of the cusp line take the form:

$$(\rho^s v^s)|_{r \to r_0, \ r \in \Sigma} = (\rho^s v^s)|_{r \to r_0, \ r \in R}, \qquad p^s|_{r \to r_0, \ r \in \Sigma} = p^s|_{r \to r_0, \ r \in R}, \tag{16}$$

where R denotes the relaxation "tail". With account of (3), conditions (16) require simply the continuity of the surface parameters at the cusp point.

To complete the problem formulation one has to prescribe the values of ρ^s $(= \rho_{1e}^s)$ and v^s $(= u)$ far away from the cusp and specify the boundary conditions for the outer flow which give rise to the cusp formation.

We are going to analyze the asymptotics of the solution near the cusp, where the classical hydrodynamic approach faces the difficulty of principle briefly described in §1. If U is the characteristic value of the flow velocity, then one may use the following scales for lengths, velocities, the pressure difference $p - p_g$ (which will be used instead of p), surface pressures and densities

$$U\tau, \ U, \ \frac{\mu}{\tau}, \ \sigma, \ \rho_0^s$$

to make the equations and boundary conditions non-dimensional $(\sigma = -p^s(\rho_{1e}^s)$ is the equilibrium surface tension). Using the notation $\theta = \pi - g(r)$ for the position of the free surface located above the cusp in the plane polar coordinates (r, θ) and eliminating p^s with the help of (3), one can rewrite (2), (4), (5) and (6) as

$$Ca(\boldsymbol{I} - \boldsymbol{nn}) \cdot \boldsymbol{P} \cdot \boldsymbol{n} = \lambda \nabla \rho^s, \tag{17}$$

$$\nabla \cdot (\rho^s v^s) = -(\rho^s - \rho_e^s(\theta)), \tag{18}$$

$$\nabla \rho^s = 4V^2(\boldsymbol{u} - \boldsymbol{v}^s) \cdot (\boldsymbol{I} - \boldsymbol{nn}), \tag{19}$$

$$\boldsymbol{u} \cdot \boldsymbol{n} = 0, \tag{20}$$

where

$$Ca = \frac{\mu U}{\sigma}, \quad \lambda = \frac{1}{1 - \rho_{1e}^s}, \quad V^2 = \frac{\tau \beta U^2}{(1 + 4\alpha\beta)\sigma\lambda}, \quad \rho_e^s(\theta) = \begin{cases} \rho_{1e}^s, & \theta = \pi - g(r) \\ 1, & \theta = 0 \end{cases}$$

and (1) takes the form

$$\boldsymbol{n} \cdot \boldsymbol{P} \cdot \boldsymbol{n} = \frac{\lambda(\rho^s - 1)}{Ca} \frac{rg'' + 2g' + r^2 g'^3}{(1 + r^2 g'^2)^{3/2}}. \tag{21}$$

We will look for a solution for the flow field near the cusp as a perturbation of uniform flow in the positive x-direction and the free surface as a perturbation of the negative x-semiaxis. If ρ_c^s and v_c^s denote respectively the values of the surface density and the x-component of the surface velocity at the origin, then one can immediately find from (18) and (19) that the leading terms of an asymptotic expansion of ρ^s and v^s, which is a projection of v^s on the tangent to the interface directed along the flow, about the origin are given by

$$\rho^s = \rho_c^s - ar + \dots \quad , \quad v^s = v_c^s + b_1 r + \dots, \quad (\theta = \pi - g(r)), \tag{22}$$

$$\rho^s = \rho_c^s + ar + \dots \quad , \quad v^s = v_c^s + b_2 r + \dots, \quad (\theta = 0), \tag{23}$$

where

$$v_c^s = 1 - \frac{a}{4V^2}, \quad \rho_c^s = \frac{av_c^s - \rho_{1e}^s}{b_1 - 1}, \quad b_2 = \frac{1}{\rho_c^s}(1 - \rho_c^s - av_c^s),$$

and a and b_1 are determined externally. It is noteworthy that the surface density gradient (and hence, due to (3), the surface-pressure gradient) is continuous at the cusp since, as is clear from (19), the second terms on the right-hand sides of (22) and (23) depend on the leading terms of the expansions of $\boldsymbol{u}$ and v^s, which are continuous at the origin.

We will introduce the stream function ψ by

$$u_r = \frac{1}{r}\frac{\partial \psi}{\partial \theta}, \quad u_\theta = -\frac{\partial \psi}{\partial r}$$

and look for the leading terms of an asymptotic series

$$\psi = \psi_1(r, \theta) + \sum_q \psi_q(r, \theta) \equiv r\sin\theta + \sum_q r^q F_q(\theta), \tag{24}$$

where ψ_q are solutions of the biharmonic equation.

Substituting (24) in (20), (17), (21) and using (22) and (23), we get after straightforward calculations that the second term in (24) is associated with the shear flow

$$\psi_2(r, \theta) = \frac{\lambda a}{2Ca} r^2 \sin^2\theta. \tag{25}$$

The third term, which corresponds to

$$q > 2, \tag{26}$$

is the solution of an eigenvalue problem, provided that

$$q < 3, \tag{27}$$

and hence the contribution from the third terms on the right-hand sides of (22) and (23) to the right-hand side of (17) does not appear when the terms of order r^{q-2} are considered.

In other words, conditions (26), (27) correspond to the situation where the flow-induced Marangoni effect caused by the interface disappearance process is taken into account already ($q > 2$) but only in the leading terms ($q < 3$).

The term (25) does not give us the leading term of the free-surface shape expansion about the origin. The cusp "opens" only when we arrive at the third term of (24).

The function F_q satisfies the equation

$$\left(\frac{d^2}{d\theta^2} + q^2\right)\left(\frac{d^2}{d\theta^2} + (q-2)^2\right) F_q = 0, \tag{28}$$

and, using that (20) gives

$$g(r) = -r^{q-1}F_q(\pi)$$

on the free surface, one can rewrite (20), (17) on $\theta = 0$ and (17), (21) on the free surface as

$$F_q(0) \;=\; 0, \tag{29}$$
$$F_q''(0) \;=\; 0, \tag{30}$$
$$F_q''(\pi) - q(q-2)F_q(\pi) \;=\; 0, \tag{31}$$
$$F_q'''(\pi) + (3q^2 - 6q + 4)F_q'(\pi) +$$
$$\frac{\lambda}{Ca}[(1 - \rho_c^s)q + 2a](q - 1)(q - 2)F_q(\pi) \;=\; 0. \tag{32}$$

Equation (28) together with (29)–(32) give the above-mentioned eigenvalue problem.

The solution of (28), which satisfies (29) and (30), has the form

$$\psi_q(r, \theta) = r^q[C_1 \sin(q\theta) + C_2 \sin((q - 2)\theta)], \tag{33}$$

where C_1 and C_2 are nonzero if q satisfies an equation

$$\tan(q\pi) = -\frac{2qCa}{\lambda[(1 - \rho_c^s)q + 2a]}, \tag{34}$$

which can be obtained after substituting (33) in (31), (32) and making the determinant of the corresponding set of linear algebraical equations equal to zero. Conditions (26), (27) provide selection of a unique root of (34). Constants C_1 and C_2 satisfy

$$C_1 q + C_2(q - 2) = 0$$

so that there is only one degree of freedom which has to be determined from external conditions.

Condition (26) implies that the free-surface curvature tends to zero as the cusp is approached, and (24), (25), (33) show that the flow field near the cusp is regular. At the cusp point the angle of the interface has a jump of 2π, but the concentrated capillary force, which appears because of this jump, is balanced by that from the surface-tension relaxation 'tail' stretching from the cusp towards the interior of the liquid and does not require any non-integrable (or even integrable) singularities in the flow field. Equation (22) together with (3) show that, in general, the surface tension begins to disappear *before* the particles belonging to the interface reach the cusp, and (22), (23) give that both the surface tension and its gradient are continuous across the cusp point.

5. Discussion

As was shown in the previous section, the present model does remove the singularities inherent in the conventional approach to the moving contact-line and the cusp problems. It seems interesting to discuss some general aspects of how the interface formation-disappearance process is related with the way of removing the singularities irrespectively of the particular mathematical model, which uses this process as its physical basis. In the rest of this section we will restrict ourselves to the moving contact-line problem simply because, on one hand, it was investigated experimentally more thoroughly than the cusp problem and, on the other hand, the unacceptability of the conventional approach to its modeling is evident and well-known.

For the reason mentioned in §1, in the case of the moving contact-line problem one has to alter the classical boundary conditions in the vicinity of the contact line to make the solution exist. This implies that the length scale on which the boundary conditions deviate from the classical ones has to be macroscopic, i.e. large compared with the thickness of the interfacial layers, which is neglected in macroscopic fluid mechanics. (The opposite statement converted into hydrodynamical terms obviously leads to the classical boundary conditions everywhere and hence to non-existence of the solution.)

On the other hand, the interface formation-disappearance process implies in particular that the surface tensions, being forced to change as a fluid particle belonging to the interface approaches, traverses and goes away from the contact line, differ from their equilibrium values in the vicinity of the moving contact line. The same conclusion follows also from experiments, for example, from the very fact that the dynamic contact angle is not equal to the static one (Zhou & Sheng, 1990) and hence the surface tensions, which act along the contacting interfaces and are related with the contact angle through the Young equation, do deviate from their equilibrium values when the contact line is moving. Obviously, the surface tension, as a macroscopic hydrodynamic quantity, varies only on a macroscopic (hydrodynamic) length scale.

So, we may assert that (a) the length scale on which the classical boundary conditions must be altered is macroscopic and that (b) the interface formation-disappearance process and the surface tension changes do take place, and the associated length scale is also macroscopic. Now the remaining question is whether (a) and (b) are interrelated, i.e. whether the process of interface formation-disappearance is associated with the physical process resulting in alteration of the classical boundary conditions in the neighbourhood of the moving contact line. This question can be reformulated as follows:

- Is there experimental evidence that the surface tension distribution near the moving contact line is influenced by the flow field?
- Is there experimental evidence that the length on which the classical boundary conditions must be abandoned depends on the contact-line speed, i.e. on the rate at which

the interface is forced to pass through the contact line?

The answers to both questions are positive. The first one is illustrated by the so-called "hydrodynamic assist of dynamic wetting" (Blake *et al.*, 1994), i.e. the influence of the flow field in the vicinity of the moving contact line on the dynamic contact angle. In experiments the flow field was varied while the contact-line speed and the contacting media remained the same. This effect has no explanation in the frame of existing theories and is in agreement with the mechanism of interface formation-disappearance. Indeed, the dynamic contact angle is determined by the values of the surface tensions at the contact line (through the Young equation), while the surface tension relaxation process of the fluid particles belonging to the interfaces depends on how they are transported to and from the contact line by the external flow.

The answer to the second question requires a look at the experimental data through the results obtained in the frame of classical fluid mechanics since the boundary conditions are not the object of direct observations. The dependence of the length scale on which the classical boundary conditions must be altered on the contact-line speed was detected experimentally through its influence on the free surface shape (see Chen *et al.*, 1996, p. 123). This length increases with the contact-line speed as should be expected from the mechanism of interface formation-disappearance described in the present paper.

It is also worth mentioning that direct measurements of the dynamic surface tension (Kochurova, Shvechenkov & Rusanov, 1974) give the relaxation times which lead to macroscopic relaxation lengths as well.

Thus we may conclude that the idea of considering dynamical wetting as a particular case of interface formation-disappearance process is in qualitative agreement with experimental observations, and this process may be incorporated in the macroscopic boundary conditions for the Navier-Stokes equations. The model described in §3 is the simplest example of how this can be done. It was derived from first principles in the framework of irreversible thermodynamics (Shikhmurzaev, 1993*a*) with all cross-effects neglected, so that it can be used as a basis for further work in this direction.

6. References

Blake, T. D., Clarke, A. & Ruschak, K. J. (1994) Hydrodynamic assist of dynamic wetting. *AIChE J.* **40**, 229–242.

Chen, Q., Ramé, E. & Garoff, S. (1996) Experimental studies on parametrization of liquid spreading and dynamic contact angles. *Coll. Surf.* **116**, 115–124.

Dussan V., E. B. & Davis, S. H. (1974) On the motion of a fluid-fluid interface along a solid surface. *J. Fluid Mech.* **65**, 71–95.

Goldshtik, M. A. (1990) Viscous-flow paradoxes. *Ann. Rev. Fluid Mech.* **22**, 441–472.

Jeong, J.-T. & Moffatt, H. K. (1992) Free-surface cusps associated with flow at low Reynolds number. *J. Fluid Mech.* **241**, 1–22.

Joseph, D. D., Nelson, J., Renardy, M. & Renardy, Y. (1991) Two-dimensional cusped interfaces. *J. Fluid Mech.* **223**, 383–409.

Kochurova, N. N., Shvechenkov, Yu. A. & Rusanov, A. I. (1974) Determination of the surface tension of water by the oscillation jet method. *Colloid J.* (USSR) **36**, 785–788.

Richardson, S. (1968) Two-dimensional bubbles in slow viscous flows. *J. Fluid Mech.* **33**, 476–493.

Shikhmurzaev, Y. D. (1993*a*) The moving contact line on a smooth solid surface. *Int. J. Multiphase Flow* **19**, 589–610.

Shikhmurzaev, Y. D. (1993*b*) A two-layer model of an interface between immiscible fluids. *Physica* **A192**, 47–62.

Shikhmurzaev, Y. D. (1994) Mathematical modeling of wetting hydrodynamics. *Fluid Dynamics Research* **13**, 45–64.

Shikhmurzaev, Y. D. (1996) Dynamic contact angles in gas/liquid/solid systems and flow in vicinity of moving contact line. *AIChE J.* **42**, 601–612.

Shikhmurzaev, Y. D. (1997*a*) Moving contact lines in liquid/liquid/solid systems. *J. Fluid Mech.* **334**, 211–249.

Shikhmurzaev, Y. D. (1997*b*) Spreading of drops on solid surfaces in a quasi-static regime. *Phys. Fluids* **9**, 266–275.

Shikhmurzaev, Y. D. (1997*c*) On cusped interfaces. *J. Fluid Mech.* (submitted).

Zhou, M. Y. & Sheng, P. (1990) Dynamics of immiscible fluid displacement in a capillary tube. *Phys. Rev. Lett.* **64**, 882–885.

EFFECTS OF TIME-PERIODIC FIELDS ON THE RHEOLOGY OF SUSPENSIONS OF BROWNIAN DIPOLAR SPHERES

I. PUYESKY AND I. FRANKEL
Faculty of Aerospace Engineering
Technion-Israel Institute of Technology
Haifa 32000, Israel

The present contribution studies the rotary motion of a spherical dipolar particle suspended in homogeneous shear in the presence of a time-periodic external field, with the goal of describing the rheology (i.e. the macroscopic stress) of a dilute suspension of such particles in the limit of weak Brownian rotary diffusion. In this singular limit, the macroscopic behaviour of the suspension is largely dependent upon the deterministic rotary motion of the particles. This motion is governed by a nonlinear and non-autonomous system. The analysis reveals two modes of motion: convergence of all particles to a global time-periodic attractor (TPA), and quasi-periodic (QP) motion. The former mode, which is characterized by both frequency and phase locking is shown to result from an appropriate resonance interaction of the respective effects of the fluid shear and external field. The distinction between the two modes of motion is essential in the calculation of the particle contribution to the effective stress. Thus, when TPAs occur, diffusive effects are confined to a narrow domain about the attractor. If, on the other hand, the rotary motion is QP, the (weak) diffusion has a global effect throughout the entire orientation space. A sufficient condition for the occurrence of a global TPA is here established for the particular square-wave oscillation of the external field (and is elsewhere extended to cover more general modes of time variation). Explicit results for the bulk stress are presented for the case of a TPA rotary motion. These results indicate that the particle contribution to the bulk stress may in some cases be negative (i.e. reduce the suspension effective viscosity). These trends are rationalized in terms of the particle deterministic rotary motion.

1. Introduction

When subject to an appropriate external field, dipolar particles experience an orienting torque acting to align their dipole axes with the field direction. This feature significantly affects the macroscopic behaviour (e.g. the rheology and transport phenomena) of suspensions of such particles. These appear in a variety of engineering applications associated with magnetic fluids which are suspensions of ferromagnetic particles, as well as in natural phenomena of bioconvection in sus-

309

D. Durban and J.R.A. Pearson (eds.),
IUTAM Symposium on Non-Linear Singularities in Deformation and Flow, 309-320.

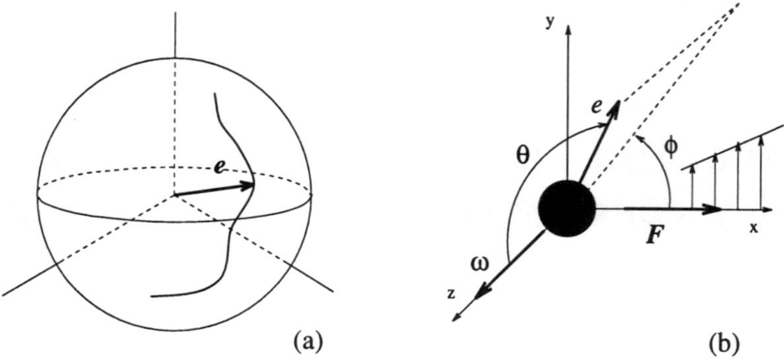

Figure 1. (a) The orientation space. (b) Definition of orientation.

pensions of swimming micro-organisms possessing an asymmetric internal mass distribution.

We study the effects of a zero-mean time-periodic external field $\boldsymbol{F}(t)$ acting parallel to the constant unit vector $\hat{\boldsymbol{F}}$ on the rheology of a dilute suspension of monodisperse Brownian spherical dipolar particles in a steady homogeneous shear flow. Adopting T, the period of external forcing as our basic time scale, we express $\boldsymbol{F}(t) = F f(t)\hat{\boldsymbol{F}}$, wherein $F > 0$ is the amplitude of the external field and $|f(t)| \leq 1$ satisfies

$$f(t) = f(t+1), \quad \text{and} \quad \int_t^{t+1} f(t_1)\, dt_1 = 0.$$

According to the dynamic theory of Batchelor (1970) and Brenner (1970), the difference between $\mathbf{T}$, the suspension bulk stress, and the bulk stress developing in a comparable suspension of *torque-free* particles is

$$\mathbf{T} - 2\mu_0\left(1 + \frac{5}{2}c\right)\mathbf{S} = -\frac{1}{2}n\epsilon \cdot \langle \boldsymbol{L}^e \rangle, \tag{1}$$

in which μ_0 denotes the viscosity of the suspending fluid, c is the volume fraction of particles, $\mathbf{S}$ is the bulk rate of strain, n is the number density of particles, and ϵ is the third-rank alternating pseudo-tensor. We represent the orientation of the particle by the unit vector e attached to its dipole axis (thus the orientation space corresponds to S_2, the surface of the unit sphere, see Fig. 1(a)). The orientation-specific torque on a particle possessing the permanent dipole moment q is

$$\boldsymbol{L}^e(e,t) = q\, e \times \boldsymbol{F}(t). \tag{2}$$

In (1) appears the orientational average

$$\langle \boldsymbol{L}^e \rangle = \int_{S_2} P(e,t)\boldsymbol{L}^e(e,t)\, d^2 e,$$

which introduces as a weighting function $P(e,t)$, the orientational probability density distribution which, in turn, satisfies the dimensionless continuity equation

$$\chi\left[\frac{\partial P}{\partial t} + \nabla_e \cdot (P\dot{e})\right] = \gamma \nabla_e^2 P, \tag{3}$$

which is supplemented by the normalization condition $\int_{S_2} P(e,t)\, \mathrm{d}^2 e = 1$, and
the requirement that $P \geq 0$ be continuous and single-valued throughout S_2. (Subsequent rheological calculations make use of only the long-time limit of $P(e,t)$
which, in turn, is independent of the initial distribution.) In (3) ∇_e denotes
the orientation-space gradient operator and the dimensionless parameter χ is the
Langevin parameter representing the ratio of the orienting effect of the external
torque and the disorienting effect of rotary diffusion

$$\chi = \frac{qF}{kT_0},$$

wherein kT_0 denotes the product of the Boltzmann constant and the absolute
temperature. In the convective term of (3) appears the dimensionless rate of change
of the orientation

$$\dot{e} = \delta\hat{\omega} \times e + \gamma f(t)\,(\mathbf{I} - ee) \cdot \hat{\boldsymbol{F}} \tag{4}$$

consisting of the contributions of fluid shear and external field. In (4) $\hat{\omega}$ is a unit
vector parallel to the fluid vorticity and $\mathbf{I}$ is the unit second-rank isotropic tensor.
The dimensionless parameters are respectively defined as

$$\delta = |\omega|T, \quad \text{and} \quad \gamma = \frac{qFT}{8\pi\mu a^3},$$

in which $|\omega|$ is the angular speed of the undisturbed fluid, and a is the particle
radius. Thus, δ and γ represent the ratios of T and the time-scales respectively
associated with particle rotation by fluid shear and external field. It is interesting
to note that the magnitude of the external-field effect is determined not only by
the field intensity but also by the time span the external field is allowed to act in
a certain direction.

We focus on the limit of weak diffusion. In order to obtain physical insight
into this singular limit, it is essential to study first the deterministic rotary motion
(i.e. in the complete absence of Brownian rotary diffusion). Adopting the standard
parametrization of e in terms of the polar angle, θ, and azimuthal angle, ϕ (Fig.
1(b)), we obtain from (4) the nonlinear and non-autonomous system

$$\dot{\theta} = \gamma f(t) \cos\theta \cos\phi \quad \text{and} \quad \dot{\phi} = \delta - \gamma f(t) \frac{\sin\phi}{\sin\theta}, \tag{5a, b}$$

when the external field acts in the plane of shear, $\hat{\boldsymbol{F}} \perp \hat{\omega}$ (for which case the
interaction of the respective effects of the external field and shear flow is the most
significant).

We start by presenting in §2 some asymptotic results and proceed in §3 to a
general classification of modes of rotary motion appearing in the present problem.
In §4 we obtain the orientational distribution and in §5 present and discuss some
results concerning the bulk stress. Finally, in the Appendix we establish a sufficient
condition for convergence to a global TPA.

2. Deterministic Rotary Motion (Asymptotic Results)

2.1. WEAK FIELD ($\gamma \ll 1$)

In the absence of an external field the particle rotates at a constant rate about
the undisturbed fluid-vorticity vector. In the limit of weak field we thus assume

a regular perturbation expansion and for a time-harmonic external field $f(t) = \sin(2\pi t)$ obtain

$$\phi(t) \cong \phi(0) + t\delta + O(\gamma)$$

$$\theta(t) \cong \theta(0) - \gamma \cos\theta(0)\left\{\frac{\cos((2\pi-\delta)t - \phi(0))}{2(2\pi-\delta)} + \frac{\cos((2\pi+\delta)t + \phi(0))}{2(2\pi+\delta)} - \frac{2\pi\cos\phi(0)}{4\pi^2 - \delta^2}\right\},$$

which indeed shows that the motion consists of a superposition of the previously-mentioned constant-rate rotation together with the small oscillations in the incommensurable frequencies $2\pi \pm \delta$. Thus, the resulting motion is quasi-periodic (QP). When $\delta = 2\pi$, some of the $O(\gamma)$ terms become singular and the expansion is no longer uniform. Thus for $\delta = 2\pi + s\gamma$ ($s = O(1)$), we assume the multiple-scales expansion

$$\theta = \theta_0(t,\tau) + O(\gamma), \quad \text{and} \quad \phi = \phi_0(t,\tau) + O(\gamma), \tag{6a, b}$$

where the slow-time variable is $\tau = 2\gamma t$. In the leading-order we obtain

$$\theta_0 = A(\tau), \quad \text{and} \quad \phi_0 = 2\pi t + B(\tau) - \pi/2.$$

Eliminating the secular terms from the $O(\gamma)$ balance we obtain for $A(\tau)$ and $B(\tau)$ the autonomous system

$$\frac{dA}{d\tau} = \cos A \cos B, \quad \text{and} \quad \frac{dB}{d\tau} = 2s - \frac{\sin B}{\sin A}, \tag{7a, b}$$

which is of exactly the same form as the system governing the rotary motion of a dipolar sphere in the presence of a steady external field. Consequently, from the results of Hall & Busenberg (1969) we conclude that when $|s| > 1/2$, A and B are periodic in τ and thus the rotary motion is QP. However, when $|s| \leq 1/2$ the system (7) possesses a stable node at $A = \pi/2$, $B = \sin^{-1}(2s)$. In terms of the physical variables of the present problem, the rotary motion converges to a time-periodic attractor (TPA) on the unit circle $\theta = \pi/2$ (of S_2). When $\delta = 2\pi$ the external field reverses its sense exactly when the particle has completed one half of a rotation by the fluid shear. Thus, even a weak field may have a significant cumulative effect. This type of resonance interaction lies at the basis of the appearance of TPAs in the present problem. Similar, though possibly weaker, cumulative interactions are anticipated for the above harmonic field whenever the particle completes an odd number of half rotations between successive reversals of the sense of external field.

2.2. WEAK SHEAR ($\delta \ll 1$)

In the absence of shear the particle will oscillate periodically in a plane defined by its initial orientation and the direction of the external field. The introduction of a steady (however weak) shear will have a cumulative effect which will cause the particle to slowly drift out of the previously-mentioned plane. It is convenient in studying the present limit to adopt a somewhat different parametrization of e, where the polar angle is now measured from the direction of the external field. The resulting equations of motion are

$$\dot{\theta} = -\gamma f(t)\sin\theta - \delta\sin\phi, \quad \text{and} \quad \dot{\phi} = -\delta\cos\phi\cot\theta. \tag{8a, b}$$

Assuming a multiple-scales expansion similar to (6) where the slow-time variable is now defined $\tau = t\delta$ and the error terms are presumably $O(\delta)$, we obtain in the leading-order

$$\phi_0 = B(\tau) \quad \text{and} \quad \tan\left(\frac{\theta_0}{2}\right) = \tan\left(\frac{A(\tau)}{2}\right) \exp\left[-\gamma \int_{t_0}^t f(t_1)\,dt_1\right],$$

i.e. the planar periodic (in t) oscillations mentioned before. For $A(\tau)$ and $B(\tau)$ we obtain, via elimination of secular terms in $O(\delta)$, the autonomous system

$$\frac{dA}{d\tau} = -\alpha \sin B, \quad \text{and} \quad \frac{dB}{d\tau} = -\alpha \frac{\cos B}{\tan A}, \qquad (9a, b)$$

in which the value of the constant parameter $\alpha(\geq 1)$ is determined (for a specific functional form of $f(t)$) by γ. The system (9) is in fact equivalent to the original system (8) in the absence of an external field. Thus A and B are periodic in $\alpha\tau$. Returning to the original physical variables we recognize that the resulting motion is QP. It is useful to note that the actual slow-time variable is $\tau^* = \alpha\tau$ rather than τ. For $\gamma \gg 1$ it may be verified that $\alpha \sim \gamma^{-1/2} \exp(\gamma/2\pi)$ and therefore τ^* may no longer qualify as a genuine slow time variable. Indeed, we shall later see that the motion in this limit is not QP but rather converges to a TPA.

3. Classification of Modes of Rotary Motion

The two modes of motion revealed by the asymptotic results of the preceding section are illustrated in Fig. 2 for the case of a harmonic external field and the indicated values of γ and δ. Representative particle trajectories on the surface of the unit sphere (cf. fig 1(a)) are depicted by their respective projections on the plane $\theta = \pi/2$. Also presented (by dots) are the corresponding Poincaré sections obtained via stroboscopic sampling of the motion at $t = n$ ($n = 1, 2, \ldots$). Part (a) of the figure shows the motion of a particle starting at $t = 0$ with the orientation corresponding to point A. Typical of QP motion is the fact that the locus of the Poincaré sections forms in the long time limit a single closed curve. In part (b) of the figure we follow the motion of particles starting at the points B and C. Both of the corresponding Poincaré sections converge in the long time limit to the single periodic point P_S . In this limit all particles, regardless of their respective initial orientations, will exactly follow the periodic motion of P_S along the unit circle.

We now wish to examine how the plane of parameters of the present problem is divided between the two modes of motion. This is motivated by our goal of obtaining the orientational distribution and the bulk stress in the limit of weak diffusion. Thus, if the deterministic motion is characterized by a TPA, diffusive effects are anticipated to be effectively confined to a narrow boundary-layer domain about the time-periodic attractor, whereas, if the motion is quasi-periodic, the weak diffusion will have a global effect throughout orientation space.

In the Appendix we establish a sufficient condition for convergence of the rotary motion to a TPA on $\theta = \pi/2$. We focus on the case of a piecewise-constant external field

$$f(t) = \begin{cases} 1 & n \leq t < n + 1/2 \\ -1 & n + 1/2 \leq t < n + 1 \end{cases} \cdot \quad (n = 0, 1, \ldots) \qquad (10)$$

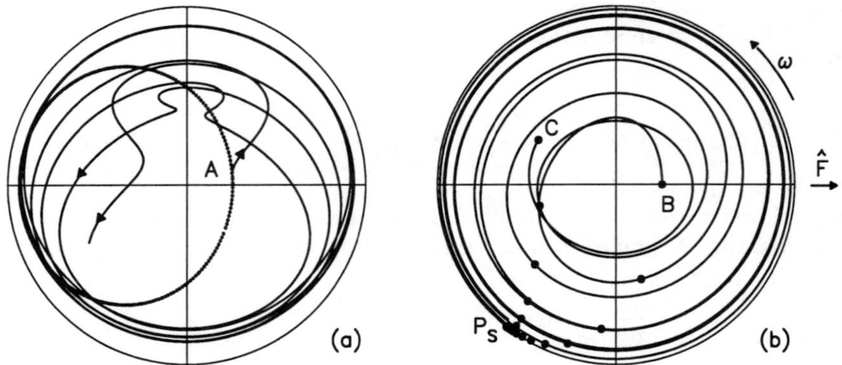

Figure 2. Projections of particle trajectories on the plane of shear for a harmonic field. (a) $\gamma = 2.5, \delta = 4$ quasi-periodic motion (b) $\gamma = 1, \delta = 6$ convergence to time-periodic attractor. • Poincaré sections.

A straightforward calculation shows that for this mode of time variation and for a given $\beta = \delta/\gamma < 1$ (17) is satisfied provided that

$$\gamma > 4 \left(1 - \beta^2\right)^{-1/2} \log \left[\frac{\left(1 - \beta^2\right)^{1/2} + 1 - \beta}{\left(1 - \beta^2\right)^{1/2} - 1 + \beta} \right] . \tag{11}$$

The long-time limit of the motion on the TPA is characterized by the rotation number $\rho = \lim_{t \to \infty} [\phi(t)/2\pi t]$ representing (the long-time average of) the rotation frequency of the dipole about the undisturbed fluid-vorticity vector. It follows from (18) that $\rho = 1$ throughout the domain defined by (11) (frequency-locking).

For $\beta > 1$, it may be established that the same condition (17) insures convergence of the system to a global TPA. The latter condition is satisfied for a given $\beta > 1$ provided that

$$\gamma_L < \gamma < \gamma_U, \tag{12a}$$

where

$$\gamma_U, \gamma_L = 4 \left(\beta^2 - 1\right)^{-1/2} \left[m\pi + 2 \tan^{-1} \left(\frac{\beta \pm 1}{\beta \mp 1} \right)^{1/2} \right], \quad (m = 0, 1, \ldots) \tag{12b}$$

the upper and lower signs respectively corresponding to γ_U and γ_L. Furthermore, throughout each of the domains defined by (12a) $\rho = 2m + 1$. This result supports our earlier intuitive explanation of the appearance of time-periodic attractors in terms of resonance interactions between the effects of fluid shear and external field (cf. the conclusion of §2.1). The division of the plane of parameters of the present problem appears in Fig. 3. The shaded domains correspond to TPA. The values of the rotation number are indicated in the figure.

The result that the same condition (17) applies to both cases $\beta < 1$ and $\beta > 1$ despite the fact that the corresponding rotary motions in the comparable steady problems are so qualitatively different from each other, suggests that this

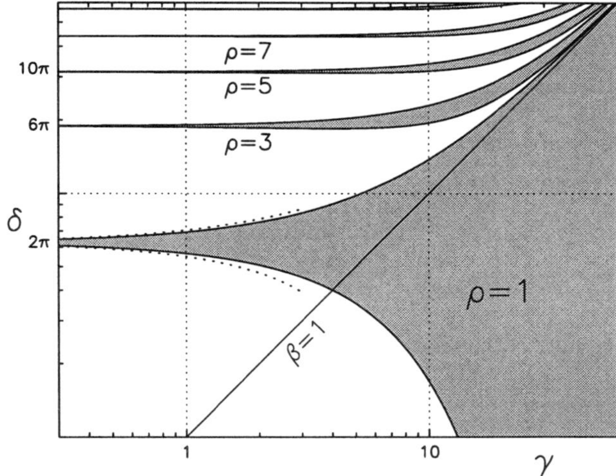

Figure 3. Division of the plane of parameters for the piecewise-constant field (9). Shaded domains correspond to TPA, —— γ_L and γ_U; $\cdots$ weak-field ($\gamma \ll 1$) approximations to γ_L and γ_U for $\rho=1$.

condition may not be a special attribute of the piecewise-constant external field (10). Indeed, it is elsewhere established that (17) readily applies whenever $f(t)$ is antisymmetric with respect to $t = 1/2$, i.e. $f(1 - t) = -f(t)$, and that validity of some of its features extends to a still broader class of time-variations of the external field (cf. Puyesky & Frankel 1998). Finally, (17) has been established as a sufficient condition. However, numerical evidence suggests that it is a necessary condition, as well; no TPAs were found outside the domains marked in Fig. 3.

4. Orientational Distribution

We focus here on the limit of weak diffusion, i.e. $\chi \gg 1$ for fixed values of γ and δ. Assuming that γ, δ correspond to a TPA, we anticipate a boundary-layer distribution about the instantaneous orientation of the TPA, $\theta = \pi/2$ and $\phi = \phi_S(t)$. Accordingly, we define the inner variables

$$\theta = \pi/2 - y\Delta \quad \text{and} \quad \phi = \phi_S(t) + x\Delta.$$

The $O(1/\chi\Delta^2)$ diffusion terms of (3) balance the $O(1)$ convection terms in a domain $\Delta \sim \chi^{-1/2}$. Assuming the perturbation expansion

$$P = P_0 + P_1\Delta + O(\Delta^2),$$

the leading-order term is governed by the equation

$$\gamma^{-1}\frac{\partial P_0}{\partial t} + f(t)\cos\phi_S(t)\left(y\frac{\partial P_0}{\partial y} + x\frac{\partial P_0}{\partial x} + 2P_0\right) = \frac{\partial^2 P_0}{\partial x^2} + \frac{\partial^2 P_0}{\partial y^2} \tag{13}$$

together with the normalization condition $\iint_{R^2} P_0 \, dx dy = \chi$. Equation (13) admits of a solution

$$P_0 = \frac{\chi}{2\pi H(t)} \exp\left(-\frac{x^2 + y^2}{2H(t)}\right) \tag{14}$$

where the function $H(t)$ satisfies the first-order equation

$$\dot{H} + 2\gamma f(t) \cos \phi_S(t) H = 2\gamma.$$

Making use of the deterministic equation of motion (5a), it may be verified that for large t and (γ, δ) corresponding to TPA

$$H \cong 2\gamma \int^t \exp\left(2\gamma \int_t^{t_1} f(t) \cos \phi_S(t) \, dt_2\right) dt_1 \geq 1 \tag{15}$$

is a periodic function of t, and P_0 is thus a time periodic boundary-layer distribution. It is worthwhile to mention that numerical results (Puyesky 1997) indicate that such time-periodicity is a general attribute of the long-time limit of P for arbitrary (γ, δ). This apparent paradox is yet another manifestation of the singularity of the weak-diffusion limit. Thus, while the deterministic motion may not be time periodic, the orientational distribution governed by a linear problem attains a time-periodic limit after diffusive relaxation has been achieved.

5. Bulk Stress

For a macroscopic simple-shear flow, the shear stress $T_{xy} = 2\mu|\omega|$ may be represented in terms of the effective suspension viscosity

$$\mu = \mu_0 \left[1 + c(5/2 + T^a)\right],$$

where the effect of the external field is represented by T^a. Making use of the foregoing asymptotic results for the boundary-layer orientational distribution we have obtained

$$T^a \cong \frac{3}{2}\frac{\gamma}{\delta} f(t) \left[\sin \phi_S(t) + O(\chi^{-1})\right] \tag{16a}$$

up to and including the $O(\chi^{-1})$ term (which has been tabulated by Puyesky 1997). The leading-order term in (16a) is equivalent to the assumption that all the suspended particles exactly follow the deterministic motion of the TPA. Making use of (5b) (for $\theta = \pi/2$), this term may alternatively be written

$$T^a \cong \frac{3}{2}\frac{\delta - \dot{\phi}_S}{\delta} + O(\chi^{-1}). \tag{16b}$$

Figure 4 presents the time variation of T^a during the time interval $n \leq t \leq n + 1/2$ (by symmetry considerations $T^a(t + 1/2) = T^a(t)$) for $\gamma = 20$ and the indicated values of δ and χ. The location of the corresponding points (within the primary TPA domain) in the plane of parameters of the deterministic problem is represented by the inset at the lower right of the figure. Solid thick curves represent exact (numerical) results, solid thin curves correspond to leading-order asymptotic

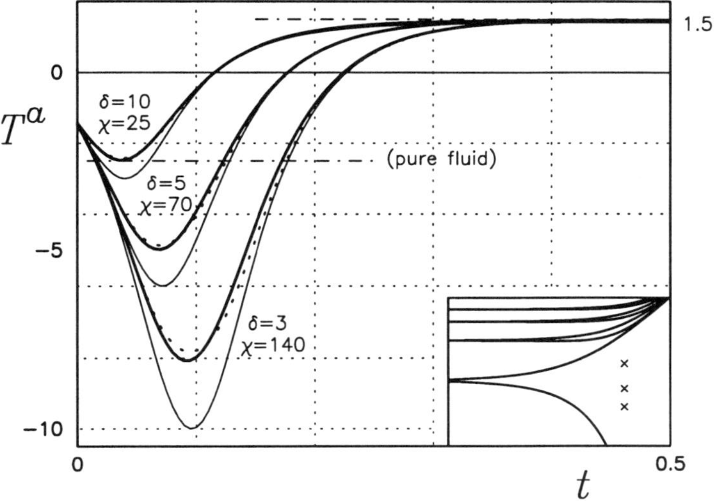

Figure 4. Variation with time of the contribution of the external field to the effective suspension viscosity. Solid curves respectively represent exact numerical (thick) and leading-order asymptotic (thin) results; dotted curves, asymptotic results including $O(\chi^{-1})$ correction.

results and dotted lines (wherever discernible) are based on asymptotic results including first-order correction at the indicated values of the Langevin parameter.

For the specific values selected for the parameters in this illustration, the variation of T^a may be divided into 'transient' and 'steady' parts (roughly two thirds and one third of the entire half-period presented). During the 'transient' interval, T^a assumes negative values, i.e. the external field causes a reduction of the effective suspension viscosity. In fact, since values of $T^a < -5/2$ occur, the effective suspension viscosity may even be smaller than that of the pure suspending fluid. This result may be rationalized by considering the leading-order term on the right-hand side of (16b) which shows that T^a is approximately proportional to the angular speed of the TPA relative to that of the ambient fluid. Initially, and as long as $\sin \phi_S < 0$, the external field makes the particles rotate faster than the ambient fluid. Thus, through viscous stresses, the particles do work on the fluid thereby decreasing the effective viscosity. This effect is maximized when $\phi_S = -\pi/2$ (cf. (16a). At $\phi_S = 0$ the external torque vanishes and so does T^a. Subsequently, when $\sin \phi_S > 0$, the external field inhibits further rotation of the particles, eventually effectively immobilizing them. Thus T^a assumes positive values asymptotically approaching the steady limit (1.5).

The difference between the deterministic, leading-order, approximation and the corresponding exact results (for specific values of χ and δ) varies with the time, the approximation quality being the poorest in the vicinity of the respective minima of the various curves. This observation may qualitatively be explained by recalling that the instantaneous width of the boundary-layer (14) is proportional to $(H(t)/\chi)^{1/2}$ and by relating the time-variation of $H(t)$ to the deterministic rotary motion of particles close to the unit circle $\theta = \pi/2$. Thus, via approximation of the actual $\phi_S(t)$ of the TPA by $\phi(t)$ and use of (5a), we obtain from (15)

$$H \cong \tan^{-2}(\pi/4 + \theta(t)/2) \int^t \tan 2(\pi/4 + \theta(t_1)/2)\, dt_1.$$

The integral factor is a monotonically increasing function of time, hence so is $H(t)$ as long as $\dot\theta < 0$ (i.e. $\phi_S < -\pi/2$, c.f. (5a)). Thus, $H(t)$ attains its maximum at $\phi_S \cong -\pi/2$ which, as mentioned above, corresponds to the respective minima of the T^a curves.

Finally, comparison between the curves corresponding to the different values of δ presented reveals that, with decreasing δ, we need to select larger values of χ in order to achieve quantitative agreement between the asymptotic results including first-order correction and the numerical exact solution. With decreasing δ, we approach the boundary of the TPA domain in the plane of parameters. At the same time, the stable and unstable attractors approach each other until the points P_U and P_S (see Apendix) coalesce just when the boundary of the TPA domain is reached and (17) is no longer satisfied. Now, in order to achieve the requisite quantitative agreement, we need to insure that the distance between the attractors $|\phi_S - \phi_U|$ remains large in comparison with $\chi^{-1/2}$, the width of the prospective boundary-layer. If this condition is not satisfied, Brownian rotations may transfer stray particles back into the U domain of Figs. 5 and 6 thus derailing the (essentially) deterministic process of convergence to the TPA.

This research was supported by the Fund for the Promotion of Research at the Technion.

Appendix: Convergence to TPA

During each half period of the square-wave oscillation (10) the situation is similar to the comparable steady problem which, in turn, is characterized by the single parameter $\beta = \delta/\gamma$. For $\beta < 1$, the entire trajectories emerge from the unstable node N_U^+ and approach the stable node N_S^+ (see Fig. 5(a)). If the sense of external field is reversed the corresponding trajectories are obtainable by symmetry. When $\beta > 1$ the motion is periodic about a centre (cf. Hinch & Leal 1972). We need to study the long-time rotary motion for an arbitrary initial orientation, which is a rather intractable task even for the present relatively simple example. We thus consider instead the Poincaré sections of the motion at constant half-period time intervals.

Figure 5(a) illustrates the motion during a typical first-half period $t \in [n, n + 1/2)$ when $\beta < 1$. Particles approach during this half period the stable node N_S^+, but the interesting question is whether or not they approach the unit circle $\theta = \pi/2$. The answer to this question is somewhat complicated by the fact that (by (5a) and (10)) θ varies non-monotonically, decreasing throughout the motion in the left half of the unit circle while monotonically increasing during the motion in the right half. By symmetry we may find a point K which, during the half period, moves to $K_{1/2}$, its symmetric image relative to the vertical diameter. (Thus for a particle starting at K, $\theta(n + 1/2) = \theta(n)$.) The locus of these points defines the line A^+B^+ which, in turn, divides the unit circle into a pair of simply-connected domains U^+ and $\overline{U}^+$ its complimentary including the origin. Similarly, the locus of the points

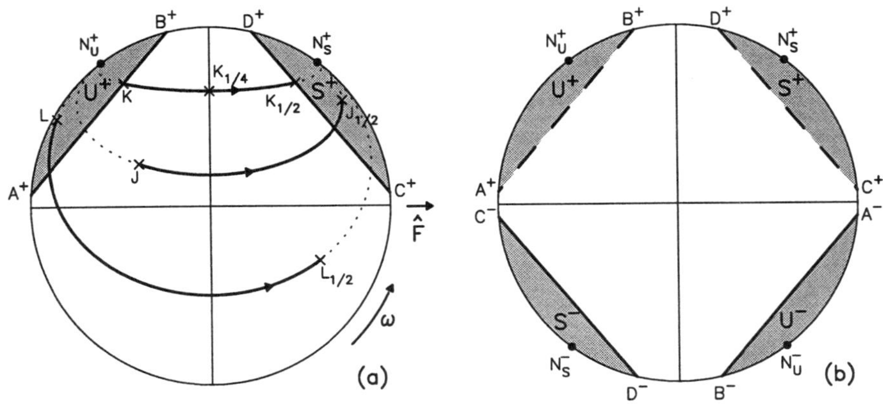

Figure 5. Definition of the domains U, $\overline{U}$, S and $\overline{S}$. $\cdots\cdots$ entire trajectories of corresponding steady problem $(\beta < 1)$.

$K_{1/2}$ defines the line C^+D^+ together with the domains S^+ and $\overline{S}^+$. By these very definitions, for any $J \in \overline{U}^+$, the corresponding $J_{1/2} \in S^+$ with $\theta(n + 1/2) > \theta(n)$ whereas, for any $L \in U^+$, $L_{1/2} \in \overline{S}^+$ with $\theta(n + 1/2) < \theta(n)$. During the following second-half period, $t \in [n + 1/2, n + 1)$, we may similarly define the lines A^-B^- and C^-D^- together with U^-, S^- and their respective complimentary domains $\overline{U}^-$ and $\overline{S}^-$ (see Fig. 5(b)). Again, for any particle starting outside U^- (i.e. within $\overline{U}^-$) $\theta(n+1/2) > \theta(n)$, whereas for particles starting inside U^-, θ decreases during this half period. For future reference we note here that (for $\beta < 1$ fixed) the entire trajectories remain the same while the domains U and S shrink or grow with increasing or decreasing γ, respectively.

We now turn to the question of whether or not $\theta(t)$ converges in the long-time limit. We examine this issue under the assumption that

$$U^- \cap S^+ = \emptyset, \quad \text{or} \quad U^+ \cap S^- = \emptyset . \tag{17a, b}$$

(By symmetry the two assumptions are equivalent.) Figure 6 presents the projections on the plane of shear of a number of particle trajectories. The orbit marked by (1) starts at $t = 0$ within $\overline{U}^-$ and at $t = 1/2$ the particle is located within S^+. From (17a) and the foregoing discussion of Fig. 5 we conclude that $\theta_k = \theta(k/2)$ $(k = 0, 1, 2, \ldots)$ forms in this case a monotonically increasing sequence which is bounded from above by $\theta = \pi/2$ and is thus necessarily convergent. (It is readily verified that the limit is $\theta = \pi/2$.) The pair of other trajectories presented (marked by (2) and (3)) both start within U^+. Accordingly, in both cases θ_k initially decreases until the particles leave the U domain. Subsequently the situation is comparable to that described above for orbit (1). In general, by the very construction of the domains U^+, U^- there exist some $\theta_{min} > 0$ such that for all $e \in U$, $\theta > \theta_{min}$. Thus, whenever we start within U, θ_k initially decreases until $\theta_k < \theta_{min}$ at which instant the particle orientation no longer belongs to either of the U domains. Subsequently, by the above line of reasoning θ_k turns into a monotonically increasing sequence eventually converging to $\theta = \pi/2$.

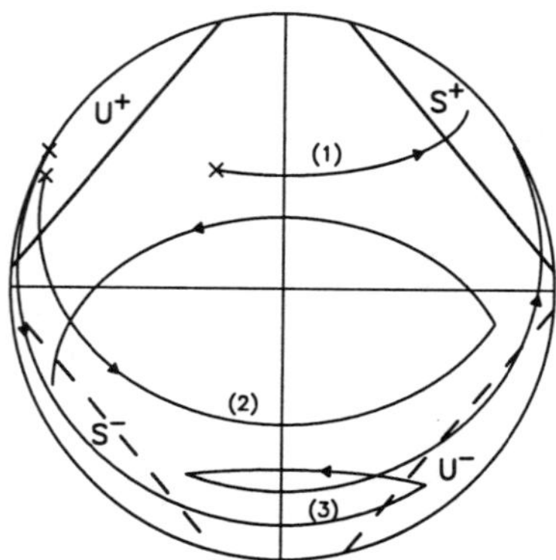

Figure 6.   Projections of typical trajectories starting within $\overline{U}^+$ or U^+.

As regards the motion on the unit circle, by (17), during a first-half period $t \in [n, n+1/2)$, the point A^+ moves to C^+ (cf. Fig. 5(a)) through $\Delta\phi > \pi$ and the point B^+ moves to D^+ through $\Delta\phi < \pi$. Thus, by continuity of the transformation during this half period, there necessarily exists on the arcA^+B^+ a point P_U^- which moves to P_U^+ through exactly $\Delta\phi = \pi$. Similarly, on the complementary arc$\overline{A^+B^+}$ there also exists a point P_S^- which moves to P_S^+ through $\Delta\phi = \pi$. By symmetry, during the following half period, $t \in [n + 1/2, n + 1)$, these two points, P_U^+ and P_S^+, will again move through $\Delta\phi = \pi$, thus returning to their initial respective locations after having completed exactly one full rotation. Thus, P_U and P_S are both periodic points for which

$$\phi_P(n + 1) - \phi_P(n) = 2\pi. \tag{18}$$

By considering the relative motion of pairs of points on the unit circle, it may be established that P_U and P_S are the only periodic points and that, furthermore, P_U is unstable and P_S is stable in the sense that the respective Poincaré sections corresponding to the motion of all points on the unit circle approach P_S. Thus, we have established convergence of the system to a single global TPA.

References

BATCHELOR, G.K. 1970 The stress system in a suspension of force-free particles. *J. Fluid Mech.* **41**, 545-570.

BRENNER, H. 1970 Rheology of two-phase systems. *Ann. Rev. Fluid Mech.* **2**, 137-176.

HALL, W.F. & BUSENBERG, S.N. 1969 Viscosity of magnetic suspensions. *J. Chem. Phys.* **51**, 137-144.

HINCH, E.J. & LEAL, L.G. 1972 Note on the rheology of a dilute suspension of dipolar spheres with weak Brownian couples. *J. Fluid Mech.* **56**, 803-813.

PUYESKY, I. 1997 *Unsteady Shear Flows of Suspensions* PhD Thesis, Technion-I.I.T., Israel.

PUYESKY, I. & FRANKEL, I. 1998 The motion of a dipolar sphere in homogeneous shear and time-periodic fields. *J. Fluid Mech.* (accepted).

A Molecular Theory for Dynamic Contact Angles

Alexandra Indeikina and Hsueh-Chia Chang
Department of Chemical Engineering
University of Notre Dame
Notre Dame, IN 46556

Dynamic contact angle correlations for partially-wetting fluids are derived for $\theta_e < \theta < \pi/2$ by relieving the contact-line stress singularity with attractive van der Waals intermolecular forces. A macroscopic disjoining pressure for an arbitrary interface is derived from the latter forces by a renormalized coarse-graining scheme with different but consistent distinguished limits near and away from the contact point. The resulting dynamical angle condition for glycerine on Plexiglass is reasonable with a critical capillary number of $Ca_c \sim 10^{-2}$ for the onset of rolling motion $(\theta > \pi/2)$ and satisfactory agreement with prior predictions for small $(\theta - \theta_e)$.

1 Introduction

Dynamic-angle conditions for a moving gas-liquid interface on a solid plane, which correlate the apparent contact angle θ near the moving contact line (measured from the liquid side but defined in the literature in a position-dependent manner that seems arbitrary) to the local contact-line speed U (usually expressed in dimensionless form by the capillary number $Ca = \mu U/\sigma$) can only be derived if the apparent stress singularity at the contact line is relieved. Such conditions can be transformed to the classical Navier slip conditions which, in turn, can be used to solve the singularity-free macroscopic flow (Hocking, 1994; Kalliadasis and Chang, 1996). The

D. Durban and J.R.A. Pearson (eds.),
IUTAM Symposium on Non-Linear Singularities in Deformation and Flow, 321-337.

physical mechanism that relieves this singularity must be significant only at microscopic scales since the Navier-Stokes equation, without the additional force provided by the microscopic mechanism, is known to be accurate in the bulk. Several physical mechanisms have been proposed for this purpose and, with appropriate tuning of the model parameters, they can all fit experimental data.

The insensitivity to the candidate mechanism can be partially attributed to two physical facts. At small θ ($\tan\theta \ll 1$), the meniscus becomes a thin wedge and the usual long-wave lubrication arguments, that balance capillary and viscous forces, yield the generic interfacial equation $h^3 h_{xxx} = h$ which has an $h \sim x \ln x$ behavior away from the contact line (Bender & Orszag, 1978). Also in this thin-film limit, the lubrication scaling dictates $h \sim LCa^{2/3}$ and $x \sim LCa^{1/3}$ where L is a macroscopic length scale much larger than the molecular scales (Bretherton, 1961). Except for microgravity conditions or liquid-liquid interfaces, we can take L to be the capillary length scale $H = (\sigma/\rho g)^{1/2}$. Hence, the generic behavior becomes

$$\left(\frac{h}{HCa^{2/3}}\right) \sim \left(\frac{x}{HCa^{1/3}}\right) \ln \left(\frac{x}{HCa^{1/3}}\right) \tag{1}$$

such that

$$\theta \sim h_x \sim Ca^{1/3} \ln \left(\frac{x}{HCa^{1/3}}\right) \tag{2}$$

This Bretherton solution for the meniscus must be matched to the solution closer to the contact line where the candidate mechanism is significant. This is tantamount to replacing x by a molecular length scale and, as such, the dynamic contact angle condition has a generic $\theta \sim Ca^{1/3}$ scaling. The choice of outer length scale L or H, the specific physical mechanism and the model parameters only contribute to the coefficient of the scaling and only through a weak logarithmic dependence. This insensitivity to the outer length scale suggests that the contact angle does not depend on where it is measured and it is a requirement for a well-defined dynamic contact angle. The weak dependence on the system model and parameters, on the other hand, implies that the problem is quite forgiving to poor models at low Ca and θ, especially if there is a surfeit of fitting parameters.

The complete matched asymptotics that support the above scaling arguments have been carried out by Hocking (1994) and Kalliadasis and Chang (1994, 1996) for perfectly wetting fluids with a zero static angle ($\theta_e = 0$) and partially wetting fluids ($\theta_e \neq 0$), using attractive van der Waals intermolecular forces as the singularity-removal physical mechanism. With θ replaced by $\theta - \theta_e$, all expressions are in the form of (2) from simple scaling arguments. While proponents of other mechanisms have yet to carry out the matched asymptotics, it is clear from (2) that the small-angle conditions are insensitive to the proposed physical mechanism or the model parameters.

At larger $Ca \sim 1$, when θ is close or equal to π, the bulk flow and the contact-line speed are insensitive to the contact-line resistance. For all practical purposes, θ can be set to π without any correlation to the speed U. There is hence no need for a stress-relieving mechanism. In a recent report, we showed experimental evidence that when θ exceeds $\pi/2$ at a critical Ca_c at around 10^{-2} for a gravity-driven current on an inclined plane, the interface immediately assumes a multi-valued nose shape which hangs over the contact line and whose lower interface is tangent to the solid. The liquid within this nose rolls around the nose tip (Veretennikov et al., 1997). It is hence quite plausible that the microscopic contact angle jumps from $\pi/2$ to π at Ca_c or at least approaches π very rapidly slightly beyond Ca_c. Stress-relieving mechanisms are essentially unnecessary beyond Ca_c. Equivalently, any reasonable physical mechanism will predict θ close to π beyond Ca_c.

It then seems that physical modeling of the dynamic contact angle condition for a moving triple line is only needed for θ between unit order ($\pi/4$, say) and $\pi/2$, corresponding to a narrow Ca range of 10^{-3} to 10^{-1}. Nevertheless, this represents an important interval since our recent experiments indicate that contact-line instability is often triggered by the jump from a rolling nose to a wedge with $\theta < \pi/2$ (Veretennikov et al., 1997). If there is any hope of discriminating any of the proposed physical mechanisms, it would be in this critical region where θ ranges from 45° to 90°.

In this report, we derive the dynamic contact angle condition for glycerine on Plexiglass over the entire range of $\theta_e < \theta < \pi/2$ by assuming a van der Waals interaction mechanism. Insofar as the pertinent interaction parameters of this model can now be satisfactorily measured, there is little freedom in adjusting system parameter to fit the data. While this may not be the correct mechanism under all conditions, it remains the most aesthetically clean model and certainly one that is most amenable to empirical verification. After it was first proposed by de Gennes (1985) and Teletzke et al. (1988), Joanny (1986), Hocking (1994) and Kalliadasis and Chang (1994, 1996) have applied it to derive small-angle correlations. The extension to the more important large-angle range here is complicated by the need to derive the macroscopic disjoining pressure for an arbitrary interface whose slope is not small and whose curvature is significant. The classical derivation in the long-wave limit (Miller and Ruckenstein, 1974), which assumes a straight wedge, breaks down when curvature is present, as is the case here where the slope varies over a large range over the interface. This weakness is especially acute near the contact point where a high-curvature region appears. Derivation of the general macroscopic disjoining pressure also requires a fundamental examination of the molecular cut-off in the integration of pair-wise interaction terms to yield the net molecular force and of the coarse-graining length scale at various interfacial positions to yield consistent distinguished macroscopic disjoining pressures – one that is independent of the coarse-graining scale. Such considerations are not necessary for the small $(\theta - \theta_e)$ limit when the straight wedge has an angle that is close to θ_e everywhere.

2   Macroscopic Disjoining Pressure

As is consistent with earlier theories, we shall omit retardation forces and consider only a long-range attractive interaction potential between two molecules of the form $W = -A/r^6$. We then assume that the total interaction energy between a single molecule and all the other molecules surrounding it can be obtained by summing all pairwise interactions between the reference molecule and every other molecule. Assuming that all other molecules can be treated as a continuum with a mean density ρ (molecules per unit volume), the summation becomes an integral. For example, the interaction energy of a single liquid molecule with an infinite solid plane across vacuum a distance D away is $W = -(\pi/6)A_{LS}\rho_S/D^3$ where A_{LS} is the liquid-solid interaction constant. When the interaction occurs in a medium, the dependence of interaction energy on D remains the same if only van der Waals forces are important. However, the interaction constant A_{LS} can be different by an order of magnitude and, in some cases, can even change its sign (Israelachvili, 1991). When the interaction is with molecules in contact $(D \to 0)$, the integral becomes singular and a cut-off r_*, representing a distance the order of the mean molecule separation, must be imposed. Selection of this cutoff is of utmost importance at the contact point where the interfacial height also approaches molecular scales. A proper selection of r_* is the key to relieving the stress singularity and determining the dynamic contact angle condition.

Consider a liquid layer which is infinite in the transverse z direction and is bounded by the solid plane at $y = 0$ and by the interface at $y = h(x)$. This interface originates from the contact point at the origin of the $x - y$ plane and extends in the negative x direction. The total interaction potential of the liquid molecule at coordinate (x, y, z) with all other liquid molecules is then, after integrating the van der Waals potential, over the entire volume of liquid up to a sphere of radius r_* around the reference molecule

$$W_{LL} = -\frac{2}{3}\frac{A_{LL}\rho_L}{r_*^3}\psi + \frac{\pi}{8}A_{LL}\rho_L \int \frac{d\psi}{r_\gamma(\psi)^3}$$

The "view angle" ψ defines the sector around the reference molecule which is covered by other liquid molecules – it is equal to 2π in the bulk and $\pi - \kappa r_*$ at the interface with curvature κ. The function $r_\gamma(\psi)$ is the radial coordinate from the reference molecule to the boundary γ, which can be the solid plane at $y = 0$ or the air-liquid interface at $y = h(x)$. The first term originates from the lower limit of integration on the cut-off sphere r_*. The second integral term vanishes as $r_\gamma(\psi) \rightarrow \infty$. As a result, the first term with $\psi = 2\pi$ represents the bulk cohesive energy per unit volume of liquid, $W_{LL}^\infty = -(4\pi/3)A_{LL}\rho_L/r_*^3$. The boundaries give rise to a higher potential energy and reduce the strength of interaction.

An explicit integration of the integral term along the solid boundary yields:

$$W_{LL} = -\frac{2A_{LL}\rho_L}{3r_*^3}\psi + \frac{\pi}{12}A_{LL}\rho_L\frac{1}{y^3}\left[1 - \frac{x}{\sqrt{x^2 + y^2}}\left(1 + \frac{y^2}{2(x^2 + y^2)}\right)\right]$$

$$-\frac{A_{LL}\rho_L\pi}{8}\int_{-\infty}^{0}\frac{h_x(\xi - x) + y - h(\xi)}{[(x - \xi)^2 + (h - y)^2]^{5/2}}d\xi \tag{3}$$

where the condition $x, y \geq r_*$ must hold and the integration in the last term should be carried out by excluding an interval of ξ where $(x - \xi)^2 + (h(\xi) - y)^2 \leq r_*^2$. As the reference molecule approaches the solid plane $(y \rightarrow 0)$ or to the interface $(y \rightarrow h(x))$, the view angle ψ in the first term varies and, since x or y is less than r_*, the second or third terms can also vary. For example, for a reference molecule on a solid plane away from the contact point $(y = 0, x > r_*$ and $h > r_*)$, the view angle is π and the second term vanishes. Hence, the introduction of the cut-off sphere r_* leads to a non-singular interaction potential but W_{LL} can then be discontinuous or have discontinuous derivatives over a distance of r_* near the boundaries.

The interaction potential between this reference liquid molecule with all the solid molecules can be likewise integrated to yield

$$W_{LS} = -\frac{A_{LS}\rho_S\pi}{6(y + D_*)^3} \tag{4}$$

where D_* is the cut-off for the reference liquid molecules adjacent to the solid at $y = 0$.

Away from the solid boundary, D_* can be neglected and both W_{LS} and W_{LL} on the interface scale as h^{-3}. Hence, if one uses these macroscopic versions without the cutoffs, one would be introducing a new singularity at the contact point even if the stress singularity is removed. In the low-θ theory, this new singularity is circumvented by either removing the contact point and replacing it with a precursor film for perfectly wetting fluids (Kalliadasis and Chang, 1992) or, for a partially wetting fluid, removing D_* in (4) and allowing the singularity in (4) to cancel that in (3) by assuming a straight wedge with the static contact angle θ_e at the contact line (Joanny, 1986; Kalliadasis and Chang, 1996). Strictly speaking, a straight wedge is only a static solution and a moving contact line must necessarily assume another shape. However, in the limit of small Ca when $\theta - \theta_e$ is small everywhere, the assumption of a straight wedge anywhere is valid to leading order and one can use it to relieve the singularity. This approximation breaks down in our limit when $\theta - \theta_e$ is not small and another approach must be used. The precursor film relief is also not reasonable for the partially wetting fluids considered here.

Actually, at the microscopic scales where (3) and (4) hold, the limit of $h \to 0$ is an artificial continuum limit. The natural separation of real molecules, of roughly the same order as the cut-off length scale r_* or D_*, can be used to remove the singularity at the contact line. However, W_{LL} and W_{LS} themselves have fluctuations over the same scale and must be smoothed before they can be of use. Such fluctuations occur near the interface and their effect is negligible for thick films and is important only at the tip. In the small θ theory, the tip is assumed to be a precursor film or a static wedge and the fluctuations are removed even at the tip. As a result, smoothing by a coarse graining is unnecessary for small θ_e. For large θ_e, the all-important tip region must be handled differently and smoothing of W is absolutely necessary. It is the breakdown of this coarse-graining scheme at the contact line that requires a different smoothing of the fluctuations near the tip. The contact-line averaging is not a distinguished one and its coarse-grain dimension is actually specified by matching with the coarse-grained field away from the contact line. It is this balance among film dimension, molecular cutoffs and coarse-graining dimension that produces a globally distinguished limit for the macroscopic disjoining pressure for a continuum approach.

While $W_{LL}(x,y) + W_{LS}(x,y)$ represent the total intermolecular potential on a single liquid molecule at (x,y,z), it cannot be used in its raw form. Actually, the reference molecule has a coordinate in three dimensions (x,y,z). However, since we assume invariance in z, the integral over all molecules in three dimension yields an interaction potential that is only a function of x and y. The dependence on z vanishes for all subsequent arguments and we shall, for simplicity, present the arguments only in the $x-y$ plane. To obtain a pressure force, we need to take the gradient of this potential. The jump in ψ from an interfacial molecule to the one immediately below it in (3) indicates that the gradient will be unacceptably large. The same fluctuation occurs for the last integral term in (3) as (x,y) approaches the interface due to the break up of the cut-off sphere. We can, of course, be consistent with the continuum approach and assume the reference molecule position (x,y) is continuous. In this case, the interaction potential is continuous but its gradient is not. Consequently, neither a discrete nor a continuum description can yield a smooth potential. We hence need to smooth out the fluctuations by averaging over a domain that is small compared to the interfacial thickness but large compared to the characteristic lengths of the fluctuations, r_* and D_*. We carry out such coarse graining near a point (x,y) within the liquid surrounded by a circle around (x,y). Near the boundary of the liquid phase, this domain is bracketed by the interface and is not a perfect circle. One can define the coarse grain domain by $x \pm \Delta x_\pm$ and $y \pm \Delta y_\pm$ where Δx_+ may be different form Δx_-. The characteristic domain dimensions are then $\Delta x = \Delta x_+ + \Delta x_-$ and $\Delta y = \Delta y_+ + \Delta y_-$ and their proper scales are

$$r_*, D_* \ll \Delta x, \Delta y \ll x, y \tag{5}$$

We average the potential over such a domain such that the macroscopic force

$$\overline{\underset{\sim}{F}} = \underset{\sim}{\nabla}\overline{W}$$

$$= \underset{\sim}{\nabla} \lim_{\Delta x, \Delta y \to 0} \frac{\rho_L}{\Delta x \Delta y} \int_{x-\Delta x_-}^{x+\Delta x_+} \int_{y-\Delta y_-}^{y+\Delta y_+} (W_{LL}(\xi,\eta) + W_{LS}(\xi,\eta))\, d\xi d\eta \tag{6}$$

has a distinguished limit under the conditions of (5). Whether this is possible often involves a judicious choice of Δx and Δy. These coarse-grain scales have to be smaller than the distance beyond which intermolecular forces are important and larger than the molecular scales. A typical interaction energy $\rho_L^2 A_{LL}$ (or $\rho_L \rho_S A_{LS}$) is of the order of 10^{-12}–10^{-13} erg and the macroscopic capillary pressure is of the order of $\sigma/H \sim 100$ erg/cc. Consequently, the former length scale is about 0.1 to 1 micron while molecular lengths are of the order of 0.2 to 0.5 nm. There is hence a sufficiently large length scale separation to allow an intermediate coarse graining, provided we are not close to the contact line where x and y approach r_* and D_*.

We shall only require $\overline{W}$ at the interface for our derivation of the dynamic contact angle condition. Converting to a polar coordinate ($x = -r\cos\beta$ and $y = r\sin\beta$) and representing the interface by $\beta = \alpha(r)$, the double integral in the averaging can be simplified by allowing that $\alpha(r)$ is a weak function of r away from the contact line at the origin. The distinguished limit for the interfacial potential away from the contact line can then be readily derived with considerable algebra,

$$\overline{W}(\alpha = \alpha(r), r) = -\frac{2\pi A_{LL}\rho_L^2}{3r_*^3} - \frac{\pi A_{LS}\rho_L\rho_S}{6r^3 \sin^3 \alpha(r)}$$

$$+ \frac{A_{LL}\rho_L^2 \pi}{12 r^3 \sin^3 \alpha}\left[1 + \cos\alpha(1 + \frac{1}{2}\sin^2\alpha)\right] + A_{LL}\rho_L^2 \frac{\kappa}{r_*^2}\left(\frac{2}{3} + \frac{\pi}{16}\right)$$

$$- A_{LL}\rho_L^2 \frac{\pi}{32}\left[\frac{\tan(\theta_0 - \alpha)}{r^3} + \frac{\kappa_0}{2r^2 \cos^3(\theta_0 - \alpha)}\right] \tag{7}$$

where κ_0 and θ_0 are the curvature and contact angle at the origin. They are, however, nominal values and not the true curvature and angle at molecular scales since the coarse graining at the contact line breaks down and must be done differently. The first constant term in the interfacial potential (7) can be omitted because it is related only with the choice of zero of the interaction energy.

Several convenient limits of (7) are useful. If the solid is not present, there is no liquid-solid interaction and the third and fifth terms in (7) from the finite liquid phase below the reference interfacial molecule also vanish. As a result, one obtains the van der Waals contribution to the interfacial tension σ

$$\overline{W}_{cap} = \frac{A_{LL}\rho_L^2}{r_*^2}\left[\frac{2}{3} + \frac{\pi}{16}\right]\kappa \tag{8}$$

There are, however, short-range contributions to the interfacial tension σ other than $(2/3 + \pi/16)\,A_{LL}\rho_L^2/r_*^2$ from the long-range van der Waals interaction. We hence include this term in a general interfacial potential $\sigma\kappa$ where σ is determined, instead, from empirical values for interfacial tension. The short-range forces, mostly from the polar groups in the liquid, do not affect the long-range interactions in the other terms significantly.

In the limit of an infinite flat film of thickness h on the solid plane ($\kappa = \kappa_0 = 0, \alpha = 0$ and $h = r\sin\alpha = const$), (7) yields the usual expression for the flat-film disjoining pressure,

$$\overline{W}_{disj} = -\frac{\pi}{6}\left(\rho_{LP}\rho_S A_{LS} - A_{LL}\rho_L^2\right)\frac{1}{h^3} \tag{9}$$

Consequently, the flat-film Hamaker constant A, which is now measurable to high accuracy, can be related to $\pi^2\left(\rho_L\rho_S A_{LS} - A_{LL}\rho_L^2\right)$. Returning to (7), we see that there are three terms that arise due to deviation from the flat-film configuration. The fourth and fifth terms arise from local and tip curvatures. However, the second term within the square bracket of the third term is present when the film remains flat but is slanted. It then corresponds to the h_x^4/h^3 disjoining pressure term in the straight-wedge theory (Miller and Ruckenstein, 1974 and Kalliadasis and Chang, 1994). In the limit of a linear wedge considered by Miller & Ruckenstein ($\kappa = \kappa_0 = 0$ and $\theta = \alpha = \theta_0$ =constant), (7) yields only an implicit relationship for the static contact angle

$$2A_{LS}\rho_L\rho_S = \rho_L^2 A_{LL}\left[1 + \cos\theta_e\left(1 + \frac{\sin^2\theta_e}{2}\right)\right] \tag{10}$$

where $\tan\theta_e = \tan\alpha = -h_x$. In the limit of small θ_e, this yields the explicit formula $\theta_e \sim (2A_{LS}\rho_S/A_{LL}\rho_L - 2)^{1/2}$ which is in agreement with the linear wedge result. Moreover, an expansion of the third term, in the limit of α and θ_e approaching zero, yields the h_x^4/h^3 term of the small-θ theory. However, for the curved wedges here, the more general expression of (7) must be used. With the above convenient limits, it is possible to rewrite the interfacial pressure (including capillary and disjoining pressures) in a more convenient form,

$$P(\alpha(r)) = \sigma\kappa$$

$$-\frac{\pi}{6}\frac{A_{LS}\rho_L\rho_S}{(1 + f(\theta_e))}\frac{1}{r^3\sin^3\alpha}\{f(\theta_e) - f(\alpha)$$

$$+\frac{3}{8}\left[\sin^3\alpha\tan(\theta_0 - \alpha) + \frac{\kappa_0 r\sin^3\alpha}{2\cos^3(\alpha - \theta_0)}\right]\} \tag{11}$$

where $f(\alpha) = 1 + \cos\alpha \left(1 + 1/2\sin^2\alpha\right)$ and we have used (10) to eliminate the liquid-liquid interaction coefficient. Hence, given the Hamaker constant A and the static contact angle θ_e, both of which are readily measurable, $A_{LS}\rho_L\rho_S$ can be determined and the interfacial pressure is fully specified provided θ_0 and κ_0 are known.

If θ_0 and κ_0 are known, one can insert (11) into the equations of motion and use the attractive van der Waals forces to relieve the macroscopic stress singularity. Instead of coming towards the solid at an oblique angle, the interface will curve due the intermolecular forces. When it is sufficiently close to the contact line with curvature κ_0 , viscous effect disappears since the logarithmic stress singularity is relived by the r^{-3} molecular forces and only the capillary and disjoining forces from intermolecular interactions are in play. Such forces will also become infinite if the interfacial height approaches zero. However, since we are at molecular scales here, a simple cut-off can relieve this singular molecular force. Consequently, macroscopic stress singularity is relieved by disjoining pressure at 0.1 to 1 micron while singularity of the disjoining pressure in continuum is relieved at 0.2 to 0.5 nm by molecular separation. The only required quantities for this scheme are θ_0 and κ_0 near the contact point.

However, the coarse graining scheme over domains specified by scaling (5) breaks down near the contact point at the origin, as the reference coordinates x and y within the liquid approach the molecular scales r_* and D_*. We hence need to carry out a different coarse graining. Unlike the current coarse graining, the new domain dimension is necessarily the same order as x and y and it hence appears in the averaged interaction potential $\overline{W}_c$. This implies that $\overline{W}_c$ is not a distinguished limit and is, instead, dependent on the coarse grain dimension ρ. However, since $\overline{W}_c$ has to match into $\overline{W}$ away from the contact point, the matching specifies ρ in terms of r_* and D_*. In anticipation for the matching between $\overline{W}$ and $\overline{W}_c$, we take the limit of $r \to 0$, $\kappa \to \kappa_0$ and $\alpha \to \theta_0$ in (11) to yield.

$$P(r \to 0) = \sigma\kappa_0 \left[1 - \frac{\pi}{16} \left(\frac{A_{LS}\rho_L\rho_S}{\sigma}\right) \frac{1}{r^2(1 + f(\theta_e))}\right]$$

$$-\frac{\pi}{6} \frac{A_{LS}\rho_L\rho_S}{(1 + f(\theta_e))} \frac{1}{r^3\sin^3\alpha} [f(\theta_e) - f(\alpha)] \tag{12}$$

The averaging of $W_{LL} + W_{LS}$ for interfacial molecules near the contact point will be carried out within the liquid portion of the circle of radius ρ

around the molecule where

$$\rho \sim x, y \gg r_*, D_* \tag{13}$$

Since the domain dimension is now the same as the liquid film, ρ appears in the averaged potential. For example, the coarse grained potential of liquid molecules interacting with a solid plane a distance D away transforms from D^{-3} to $(D\rho^2)^{-1}$ to leading order when D is much smaller than ρ (Israelachvili, 1991). To simplify matters further, we notice that this contact-line region contains only a few hundred molecules such that one can approximate the interface to have a constant curvature locally, i.e. a circle that is not centered at the contact point. This is analogous to the other constant curvature surface, a straight wedge, assumed in the small $\theta - \theta_e$ limit. However, as mentioned before, the large $\theta - \theta_e$ limit here dictates a different surface at the tip. This *ad hoc* approximation to neglect molecular dynamics at the tip has also been assumed by Joanny (1986) and it can be justified by a molecular dynamic simulation.

Let this circle have a curvature of κ_0 and let it depart from the origin with an angle θ_0, the above averaging yields the interfacial pressure as the contact point

$$P_c \sim \sigma \kappa_0 \left(1 - \frac{r_*^2}{2r^2} \right)$$
$$- \frac{\pi}{6} \left(\frac{A_{LS}\rho_L\rho_S}{1 + f(\theta_e)} \right) \frac{1}{r \sin \alpha} \left[\frac{1 + f(\theta_e)}{\rho D_* \pi} - \frac{1 + f(\alpha)}{4r_*^2} \right] \tag{14}$$

to leading order in r^{-1} and ρ^{-1}. Note that the coarse graining has reduced the order of singularity from h^{-3} in (12) to h^{-1} in (14). However, as mentioned earlier, this singularity can easily be removed by another cutoff at this tip region.

Away from the tip region, the interface is no longer circular due to viscous effects. Nevertheless, however the interface is distorted by the viscous flow effect, the interfacial pressure must match into (14) at the contact-line tip is approached. Hence, matching (12) to (14), we see that the two agree only if

$$r_* = \left[\frac{\pi A_{LS}\rho_L\rho_S}{8\sigma(1 + f(\theta_e))} \right]^{1/2} \tag{15}$$

$$\rho = \frac{4r_*^2}{\pi D_*} \tag{16}$$

and only at the interfacial height

$$h_m = 2r_*$$ (17)

The choice of r_* in (15) is within the range of molecular lengths appropriate for the cut-off of the singular integral in integrating the liquid-liquid molecular interaction in (3). The cut-off D_* for the liquid-solid interacting is not fully specified but is related to the coarse-graining scale ρ through (16). We hence see that the contact line coarse graining is carried out over only a few molecules, which is also the size of this tip region.

The approximation of the tip by the circle $r\kappa_0 = 2\sin(\theta_0 - \alpha)$ also allows us to relate the matching location (r_m, α_m) and an interfacial pressure at matching point $P_m = P(r_m, \theta_m)$ with θ_0 and κ_0:

$$\kappa_0 = \frac{\sin\alpha_m \sin(\theta_0 - \alpha_m)}{r_*}, \qquad r_m = \frac{2r_*}{\sin\alpha_m}$$ (18)

$$P_m = \sigma\kappa_0 \left(1 - \frac{1}{8}\sin^2\alpha_m\right) - \frac{\sigma}{6r_*}[f(\theta_e) - f(\alpha_m)]$$ (19)

This matching has not specified the tip curvature κ_0 and angle θ_0. However, the former can be determined from matching with macroscopic solution away from contact line. For example, for the static case, matching (11) with a linear wedge at $r \to \infty$ gives $P_m = 0$ which specifies κ_0 through (19). The angle θ_0 may be determined by the molecular dynamics near the contact line. By varying θ_0 from θ_e to $\pi/2$ for the static case, we are able to get from the shooting scheme in the next section an apparent static contact angle far from the contact line ranging from θ_e to $73.8°$ for glycerine on Plexiglas. The reported advancing contact angle θ_a for this system is about $70°$ and we hence take $\theta_0 = \pi/2$ as the appropriate value when the molecules are in motion. Other θ_0 values yield static surfaces with apparent contact angles within the hysteresis range between θ_e and θ_a. One can also associate this θ_0 with surface roughness that rotates the circle at the contact line. Macroscopic motion of the tip begins only when θ_0 is $\pi/2$ in all subsequent analysis.

3   Apparent Angle in the Viscous Outer Region

Having deciphered κ_0 and θ_0 in the interfacial pressure in (11) and determined in (17) when (11) becomes valid, we move further away from the contact line and include macroscopic flow and viscous effects in the thicker film. Nevertheless, in the intermediate overlapping region, the disjoining pressure (11) remains valid. We shall decipher how it sucks liquid onto the solid through attractive van der Waals forces and capillary pressure and in the process determine the asymptotic dynamic contact angle far from the contact line where molecular forces become insignificant.

We assume an interface in the shape of a curved wedge away from the circular tip. To leading order in r, the biharmonic equation for viscous flow within the wedge yields the stream function.

$$\psi = \frac{Ur}{\alpha - \sin \alpha \cos \alpha} [\alpha \sin \beta - \beta \sin \alpha \cos(\beta - \alpha)] \tag{20}$$

where $\alpha(r)$ defines the interface in the (r, β) polar coordinate. The tangential stress balance at the wall has also been satisfied in this curved wedge with negligible $(r\alpha_r)^2$ in the limit of $r \to 0$. The only condition to satisfy is the normal stress condition which describes how the interfacial pressure (11) drives the flow. This is provided by the Cauchy-Riemann condition at the interface

$$\frac{1}{\rho} \frac{\partial P}{\partial r} = \frac{2\nu U \sin \alpha}{r^2(\alpha - \sin \alpha \cos \alpha)} \tag{21}$$

Inserting (11) into (21), invoking (15) and (16) and using the capillary length H to scale r, one obtains

$$\frac{1}{r} \left(\frac{r^2 \alpha_r}{(1 + r^2 \alpha_r^2)^{1/2}} \right)_r = \frac{2Ca \sin \alpha}{r(\alpha - \sin \alpha \cos \alpha)}$$

$$- \frac{4\epsilon^2}{3r^3 \sin^3 \alpha} \left\{ f(\theta_e) - f(\alpha) + \frac{3}{8}(\cos \alpha \sin^2 \alpha + \frac{\kappa_0}{2}r) \right\} \tag{22}$$

where $\epsilon = r_*/H$ is the ratio of molecular cut off to the capillary length which is of the order of 10^{-5} to 10^{-7}. One can best integrate (22) away from the contact point by using the arclength-angle formulation with arclength s measured from the matching point in (17) and the angle $\varphi = -\tan^{-1}(h_x)$ that the interface makes with the horizontal,

$$h_s = \sin \varphi \qquad\qquad \alpha_s = \frac{1}{r} \sin(\varphi - \alpha)$$

$$x_s = -\cos \varphi \qquad\qquad r_s = \cos(\varphi - \alpha) \tag{23}$$

$$\varphi_s = \frac{2Ca\sin^2\alpha}{h(\alpha - \sin\alpha\cos\alpha)}$$

$$-\frac{4\epsilon^2}{3h^3}\left\{f(\theta_e) - f(\alpha) + \frac{3}{8}\left[\frac{r\cos\alpha_m\sin\alpha_m}{2\epsilon} + \cos\alpha\sin^2\alpha\right]\right\}$$

and the integration is initiated at the matching point (17) with curvature κ_0 of the circular tip

$$h(s = 0) = 2\epsilon \quad \alpha(s = 0) = \alpha_m$$

$$x(s = 0) = -\frac{2\epsilon}{\tan\alpha_m} \quad r(s = 0) = \frac{2\epsilon}{\sin\alpha_m} \tag{24}$$

$$\varphi(s = 0) = 2\alpha_m - \pi/2$$

$$\kappa_0 = \frac{\cos\alpha_m\sin\alpha_m}{\epsilon} = -\varphi_s(s = 0)$$

We have integrated (23) from (24) with θ_0 fixed at $\pi/2$ and for glycerine on Plexiglass with $\sigma = 63.4$ dyn/cm, $\pi\rho_L\rho_S A_{LS} = 6.4 \times 10^{-13}$ erg and $\theta_e = 60°$ from literature (Veretennikov et. al., 1997), we obtain $\epsilon^2 = 0.5 \times 10^{-14}$. We find little sensitivity on ϵ. In figure 1, the computed h_x is plotted as a function of $(r/2r_*)$ for Ca ranging from the zero static limit to 2×10^2. For small

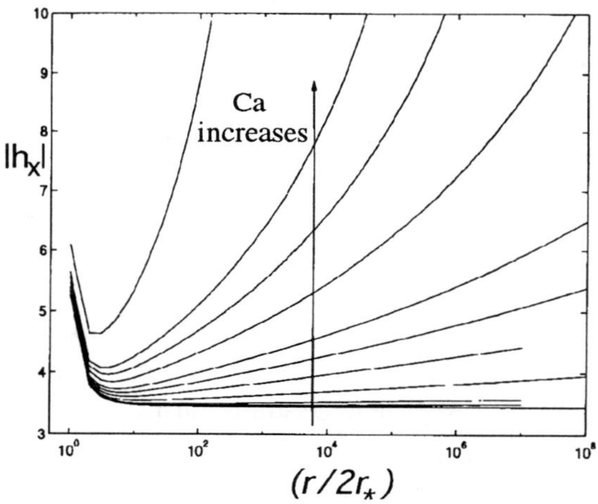

Figure 1: Macroscopic h_x at the interface as a function of r at various Ca values including the static case $Ca = 0$ ($Ca = 0, 10^{-4}, 3 \times 10^{-4}, 10^{-3}, 2 \times 10^{-3}, 3 \times 10^{-3}, 4 \times 10^{-3}, 6 \times 10^{-3}, 8 \times 10^{-3}, 10^{-2}, 2 \times 10^{-2}$)

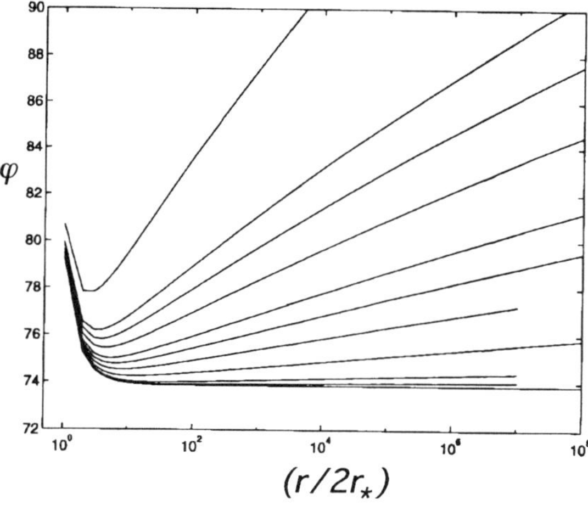

Figure 2: Macroscopic $\varphi = -\tan^{-1}(h_x)$ as a function of r at various Ca. The linear asymptote away from the contact line yields a linear $\ln r$ dependence and insensitivity to outer lengthscale

$Ca < 10^{-3}$, h_x is asymptotically a linear function of $\ln r$ which translates into the $h \sim x \ln x$ behavior of (1) from the small-angle theory. Although we have used a circular tip instead of a straight wedge here, Kalliadasis and Chang (1996) have shown that the result is not sensitive to the tip shape due to the universal scaling (2) and the $\ln$ dependence of the coefficient. The apparent contact angle is then simply a convenient value in the outer limit. Since we have defined the outer length scale as H, this apparent angle should be taken at $r/2r_* = H/2r_* = 1/2\epsilon$. Since h_x has a $\ln r$ dependence, the exact value of ϵ is unimportant - it corresponds to the $\ln \epsilon$ sensitivity of (2). Nevertheless, we are consistent in using the same ϵ we used in the integration. For such small angles, h_x, α and φ are almost identical to each other and one can use any one interchangeably.

For $Ca > 10^{-3}$, h_x is no longer a linear function of $\ln r$ or $\ln x$ in the outer limit. Fortunately, h_x, α and φ are also distinct now and a plot of $\varphi = \tan^{-1}(h_x)$ against $\ln r$ shows linear outer behavior even at large Ca. This implies that , in terms of φ, the insensitive $\ln \epsilon$ dependence is retained and we define the apparent dynamic contact angle θ as φ at $(r/2r_*) = 1/2\epsilon \sim 0.7 \times 10^7$. The estimated contact angle is plotted vs.

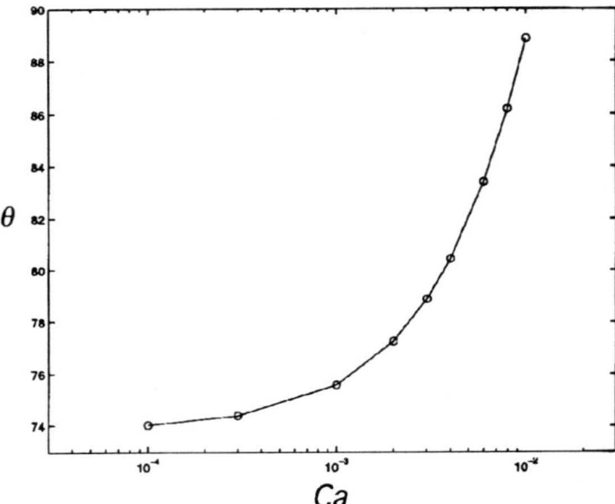

Figure 3: Apparent contact angle evaluated at $r = H$ from Figure 2. There is a little sensitivity to the outer length scale H. The static angle is 73° and the small-angle $Ca^{1/3}$ dependence breaks down beyond $Ca \sim 10^{-3}$ and Ca_c for onset of rolling is at about 2×10^{-2}

Ca in figure 3. It is clear that the $\theta \sim Ca^{1/3}$ universal behavior breaks down beyond $Ca = 10^{-3}$ and the critical Ca_c when rolling begins is about 2.0×10^{-2}. This latter value is consistent with our measured value during our experiment of fingering of glycerine on plexiglass, which we propose is triggered by the transition from rolling to a wedge shape (Veretennikov et. al., 1997). Although we have lost the universal $Ca^{1/3}$ dependence at large Ca, the $\ln \epsilon$ dependence still survives. Such insensitivity to the outer length scale L or H is a prerequisite of a well-defined dynamic contact-angle condition. Here, it also implies that the dynamic contact angle correlations between φ_a and Ca are only parameterized by the static angle θ_e and a proper rescaling with θ_e may collapse all correlations into a master one. We shall examine this postulate with a careful analysis of the solutions to (22) in a separate manuscript.

References

[1] Bender, C.M. and Orszag, S.A. (1978), "Advanced Mathematical Methods for Scientists and Engineers", *McGraw-Hill*, New York.

[2] Bretherton, F.P. (1961),"The Motion of Long Bubbles in Tubes", *J. Fluid Mech.* **10**, 166.

[3] de Gennes, P.G. (1985),"Wetting: Statics and Dynamics", *Rev. of Modem Phys.* **57**, 827.

[4] Hocking, L.M. (1994), "The spreading of Drops with Intermolecular Forces", *Phys. Fluids* **6**, 3224.

[5] Israelachvili, J. N. (1991),"Intermolecular and surface forces.", *Academic Press*, London.

[6] Joanny, J.F. (1986), "Dynamics of Wetting: Interface Profile of a Spreading Liquid", *J. Theor. and App. Mech.* **271**, 249.

[7] Kalliadasis, S. and Chang, H.-C. (1994) "Apparent Dynamic Contact angle of an Advancing Gas-Liquid Meniscus", *Phys. Fluids* **6**, 12.

[8] Kalliadasis, S. and Chang, H.-C. (1996), "Dynamics of Liquid Spreading on Solid Surfaces", *IEC Fund and Res.* **35**, 2860.

[9] Miller, C.A. and Ruckenstein, E. (1974), "The Origin of Flow during Wetting of Solids", *J. Colloid Interface Sci.* **28**, 368.

[10] Teletzke, G.F., Davis, H.T. and Scriven, L.E. (1988), "Wetting Hydrodynamics", *Rev. Phys. App.* **23**, 989.

[11] Veretennikov, I., Indeikina, A. and Chang, H.-C. (1997), "Fingering of a Driven Contact Line on an Inclined Plane *(preprint)*.

REGULARIZATION OF SINGULARITIES IN THE THEORY OF THIN LIQUID FILMS

ALEXANDER ORON
Faculty of Mechanical Engineering,
Technion-Israel Institute of Technology,
Haifa 32000 ISRAEL

AND

S. GEORGE BANKOFF
Department of Chemical Engineering,
Robert R.McCormick School of Engineering and Applied
Science, Northwestern University, Evanston Illinois 60208

1. Introduction

Thin liquid films are often encountered in a variety of applications: as gravity currents under water or as lava flows in geology; as tear films in the eye or as linings of human airways in biology. In engineering thin films are used to limit fluxes and to protect surfaces in heat and mass transfer processes.

One means of studying the dynamics of thin liquid films is to analyze *long-scale* phenomena (Oron *et al.*, 1997) only, in which variations along the film are much more gradual than those normal to it, and in which variations are slow in time. The long-wave theory approach is based on the asymptotic reduction of the governing equations and boundary conditions to a simplified system which consists often, but not always, of a single nonlinear partial differential equation formulated in terms of the local thickness of the film. The rest of the unknowns, i.e. the fluid velocity, pressure, fluid temperature, etc. are then determined via the solution of that differential equation.

The asymptotic reduction of the governing, continuity and Navier-Stokes, equations becomes possible due to the assumption that the aspect ratio of the film is small,

$$\epsilon = h_0/\lambda \ll 1$$

339

D. Durban and J.R.A. Pearson (eds.),
IUTAM Symposium on Non-Linear Singularities in Deformation and Flow, 339-348.
© 1999 *Kluwer Academic Publishers. Printed in the Netherlands.*

where h_0 and λ are the average thickness of the film and the wavelength of a characteristic disturbance of the film interface.

Oron *et al.*(1997) showed using the long-wave approximation that in the two-dimensional case a generic evolution equation describing the spatio-temporal behavior of a non-volatile liquid film placed on a solid planar wall and subjected to the normal interfacial stress Π, shear interfacial stress τ, and body force with potential ϕ is

$$\mu h_t + [\frac{1}{2}h^2(\tau + \sigma_x)]_x - [\frac{1}{3}h^3(\phi|_{z=h} - \sigma h_{xx} - \Pi)_x]_x = 0, \qquad (1)$$

where μ is the liquid viscosity, σ is surface tension, x, t are spatial (along the wall) and temporal variables and $h = h(x,t)$ is the local film thickness. The subscripts denote differentiation with respect to the corresponding variable. If the liquid film is volatile, Eq.(1) contains instead of zero in its right-hand side the term $-j$, where j is the evaporative mass flux.

Equation (1) can be rewritten in non-dimensional form as

$$H_T + [\frac{1}{2}H^2(\tau_0 + \Sigma_\xi)]_\xi - [\frac{1}{3}H^3(\Phi|_{\zeta=H} - SH_{\xi\xi} - \Pi_0)_\xi]_\xi = 0. \qquad (2)$$

written in terms of the dimensionless variables

$$\zeta = z/h_0, \ \xi = \frac{\epsilon x}{h_0}, \ T = \frac{\epsilon U_0 t}{h_0}, \ H = h/h_0,$$

$$(\tau, \Pi, \Phi) = \frac{h_0}{\mu U_0}(\tau_0, \epsilon\Pi_0, \epsilon\phi), \ \Sigma = \frac{\epsilon\sigma}{\mu U_0}, \ S = \frac{\epsilon^3\sigma}{\rho h_0 U_0^2},$$

where U_0 is the characteristic velocity of the problem at hand and S is the non-dimensional Weber number. In the case of a volatile film the right-hand side of Eq.(2) is $-J$, where J is a non-dimensional evaporative mass flux.

At leading order the components of the non-dimensional velocity field U, W are determined by

$$U \equiv \frac{u}{U_0} = (\Phi|_{\zeta=H} - SH_{\xi\xi} - \Pi_0)_\xi(\frac{1}{2}\zeta^2 - H\zeta) + \zeta(\tau_0 + \Sigma_\xi),$$

$$W \equiv \frac{w}{\epsilon U_0} = [(\Phi|_{\zeta=H} - SH_{\xi\xi} - \Pi_0)_\xi(\frac{1}{2}H\zeta^2 - \frac{1}{6}\zeta^3)]_\xi - \frac{\zeta^2}{2}(\tau_0 + \Sigma_\xi)_\xi. \quad (3)$$

Therefore the spatiotemporal evolution of the film interface is found by solving Eq.(2) with appropriate initial and boundary conditions. The latter are most typically periodic. After Eq.(2) is solved the velocity field is found using Eq.(3).

2. Emergence of singularities

In this section several examples of singularities emerging in the theory of thin liquid films are presented. It should be noted that such singularities do not occur in reality and are artifacts of the mathematical model. As it will be shown in the next section, these singularities can be removed by taking into account some of actual physical mechanisms disregarded in previous analyses.

One of the simplest cases considered in the theory of thin liquid films is that of only molecular van der Waals (long-range) forces and constant surface tension present (Williams and Davis, 1982). This case is relevant for films with an average thickness in the range of 100-1000 Angstroms. Thus, if no imposed interfacial stresses are present, $\Sigma_\xi = \Pi_0 = \tau_0 = 0$ and $\Phi = \Phi_r + AH^{-3}$, the evolution equation governing the spatiotemporal behavior of the film interface is obtained from Eq.(2) in the form

$$H_T + A(H^{-1}H_\xi)_\xi + S(H^3 H_{\xi\xi\xi})_\xi = 0, \tag{4}$$

where Φ_r is a reference value for the dimensionless body-force potential, and A is the non-dimensional Hamaker constant

$$A = \frac{\epsilon A'}{6\pi \rho U_0^2 h_0^3},$$

ρ is the fluid density and A' is a dimensional Hamaker constant, $A' \sim \pm 10^{-19} J$. In Eq.(4) the A-term describes the effect of the van der Waals forces, whereas the S-term describes that of the capillary forces. The case of $A > 0$ corresponds to the negative disjoining pressure (attractive van der Waals force), which leads to instabilities of the film interface. On the contrary, the case of $A < 0$ corresponds to the positive disjoining pressure (repulsive van der Waals force) and no interfacial instabilities are present.

The evolution of the film interface as described by Eq.(4) in the domain $0 \leq \xi \leq L$ amended with periodic boundary conditions and the initial condition of the form $H(\xi, T = 0) = const + \delta \cos(2\pi\xi/L)$, $\delta \ll 1$ reveals that the squeeze effect owing to the presence of van der Waals forces leads to rupture of the film in a finite time T_R at some location $\xi = \xi_R$ (Williams and Davis, 1982). This rupture manifests itself by the fact that at a certain time the local thickness of the film becomes zero. Moreover, the rate of film thinning, measured as the rate of decrease of the minimal thickness of the film, increases with time, and becomes much larger than the disturbance growth rate given by the linear theory, see Figure 1. At rupture time $T = T_R$ the rate of the film thinning becomes infinite, thus violating the basic assumptions of the mathematical model, if not more than that. Burelbach *et al.*(1988) and Oron *et al.*(1997) showed that in the neighborhood of the

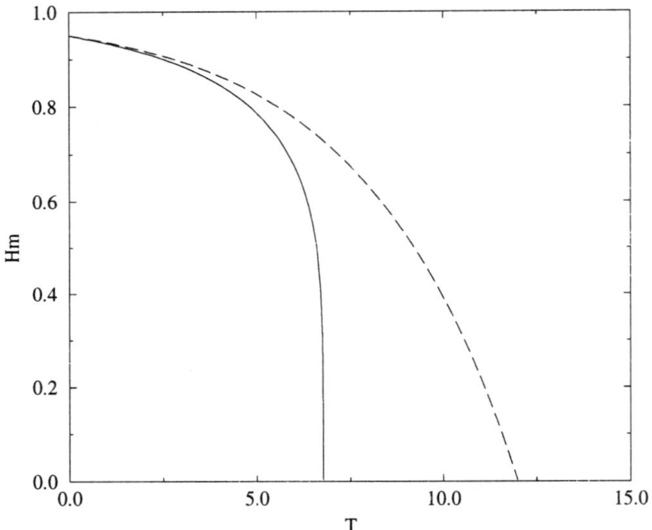

Figure 1. The time evolution of the minimal thickness of the film H_m. Solid line: as given by Eq.(4), for $A = 1, S = 1, L = 2^{3/2}\pi$. The size of the periodic domain here corresponds to the wavelength of the fastest-growing linear mode. Rupture time is $T_R \approx 6.774$. Dashed line: as estimated from the linear stability analysis for the fastest-growing linear mode. Estimated rupture time in the linear case is $T_R \approx 11.983$.

rupture point $\xi = \xi_R$ and close to the moment of rupture $T = T_R$ the solution of Eq.(4) can be approximated by

$$H(\xi, T) = A\frac{\alpha^2}{2}(T_R - T)\sec^2[\frac{\alpha}{2}(\xi - \xi_R)], \qquad (5)$$

where α is a constant that can be found numerically by matching Eq.(5) with the solution away from the rupture point.

The components of the flow field in the case considered are given by

$$U = (AH^{-3} - SH_{\xi\xi})_\xi(\frac{1}{2}\zeta^2 - H\zeta),$$

$$W = [(\frac{1}{2}H\zeta^2 - \frac{1}{6}\zeta^3)(AH^{-3} - SH_{\xi\xi})_\xi]_\xi. \qquad (6)$$

Prior to the moment of rupture, $T \to T_R$ at $\zeta = H$, one obtains from Eqs.(5),(6) that $U \sim (T_R - T)^{-1}$. Thus, by virtue of infinite velocities emerging at rupture the problem becomes singular.

Another example of a singularity occurs, when one considers a heated film of a volatile liquid. Burelbach *et al.*(1988) developed a mathematical model to investigate the evolution of such a film. In the simplest case in which the main effect is evaporation and other mechanisms, such as van der Waals molecular force, the Marangoni effect due to variation of surface tension with temperature, vapor recoil and others vanish, the evolution equation reduces to (Burelbach *et al.*, 1988)

$$H_T + E(H + K)^{-1} = 0, \tag{7}$$

where E is the evaporation number and K is the interfacial resistance to evaporation

$$E = \frac{k_{th}\Delta\vartheta}{\mu\mathcal{L}}, \quad K = \frac{\vartheta_s^{3/2}}{\hat{\alpha}\rho_v\mathcal{L}}\left(\frac{2\pi R_g}{M_w}\right)^{1/2}\frac{k}{h_0\mathcal{L}}.$$

Here k is the thermal conductivity of the liquid, ϑ_s, ϑ_w are the absolute saturation and wall temperatures, respectively, $\vartheta_s \neq \vartheta_w$, $\Delta\vartheta = \vartheta_w - \vartheta_s$ is superheat, $\hat{\alpha}$ is the accommodation coefficient, R_g is the universal gas constant, M_w and ρ_v are the molecular weight and density of the vapor, respectively, and $\mathcal{L}$ is the latent heat of vaporization per unit mass. The E-term in Eq.(7) represents the non-dimensional mass flux due to evaporation

$$J = E(H + K)^{-1}. \tag{8}$$

To the given order of approximation it is found that the dimensionless temperature field is given by

$$\Theta \equiv \frac{\vartheta - \vartheta_s}{\vartheta_w - \vartheta_s} = 1 - \frac{\zeta}{H}, \tag{9}$$

where ϑ denotes the absolute temperature of the film. This means that the heat transfer across the liquid film is approximately conductive.

If the case of quasi-equilibrium evaporation, $K = 0$, is considered (Burelbach *et al.*, 1988) the temperature at the film interface $\zeta = H$ is uniform, $\Theta = 0$, and the film of an initially uniform thickness evolves according to the solution for Eq.(7)

$$H = (1 - 2ET)^{1/2}. \tag{10}$$

Therefore the film disappears at a finite time $T_R = 1/(2E)$ and at the moment of rupture the rate of film thinning

$$\frac{dH}{dT}\Big|_{T=T_R} = -\frac{E}{\sqrt{1 - 2ET_R}}$$

is infinite. At the same time the evaporative mass flux J becomes infinite as well and infinite temperature gradients across the film emerge. These singularities are due to the fact that at rupture two different temperatures

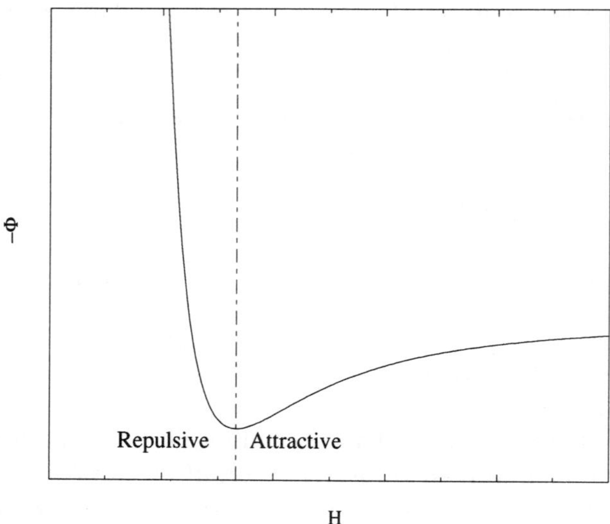

Figure 2.	Sketch of the layered potential $-\Phi$ as a function of H. Domains of repulsive and attractive molecular forces are shown.

ϑ_s and ϑ_w are specified at the same location. To avoid this one must seek for a thermal resistance which separates between those temperatures.

3. Regularization of singularities

In order to regularize the singularities associated with the presence of van der Waals forces a layered potential of the form (Oron *et al.*, 1997; Oron and Bankoff, 1997)

$$\Phi = \Phi_r + AH^{-3} - BH^{-4} \tag{11}$$

is introduced with A, B being dimensionless Hamaker constants. The first two terms represent the potential of attractive van der Waals forces used in Section 2, whereas the last one accounts for repulsive short-range forces based on molecular repelling. The potential (11) is qualitatively similar in its behavior to the Lennard-Jones $6 - 12$ potential (Israelachvili, 1995). However, it is more convenient for use in numerical studies. The potential (11) describes the interaction between the film interface and the wall of the following kind: when the distance between them is sufficiently short they repel each other, while for longer distances they are mutually attracted, see Figure 2.

In the case at hand the evolution equation (2) with $\Sigma_\xi = \tau_0 = \Pi_0 = 0$ reduces to (Oron and Bankoff, 1997)

$$H_T + [(AH^{-1} - BH^{-2})H_\xi]_\xi + S(H^3 H_{\xi\xi\xi})_\xi = 0. \tag{12}$$

A free energy functional

$$F(H) = \int_0^L (\frac{B}{12}H^{-3} - \frac{A}{6}H^{-2} + \frac{S}{2}H_\xi^2)d\xi \equiv \int_0^L (V(H) + \frac{S}{2}H_\xi^2)d\xi \tag{13}$$

is associated with Eq.(12). It has been shown (Oron and Bankoff, 1997) that the functional $F(H)$ is bounded from below and decreases along the trajectories of Eq.(12), i.e. $dF/dT \le 0$. Thus, it follows that $F(H)$ is a Lyapunov functional of the system and the evolution of any initial disturbance of the interface, as governed by Eq.(12), results in a steady state. The minimal thickness of these steady solutions corresponds roughly (see below for a further discussion) to the location of the minimum of the potential part $V(H)$ of the functional F: $H_{min} \approx 3B/4A$. Figure 3 displays a set of steady solutions for Eq.(12) calculated for $A = 1, B = 0.25, S = 1$ in the periodic domain of size $L = 2^{3/2}\pi$ for various values of the mean film thickness H_0. Note that Eq.(12) can always be rescaled to these values for A, B, S by stretching the variables H, T, ξ, while L, H_0 are left as free parameters. Oron and Bankoff(1997) showed also that these steady solutions correspond to static states in which the film interface and the liquid inside the film remain immobile. The entire evolution of the film is free of singularities emerging in the absence of repulsive forces, $B = 0$, as discussed in Section 2.

In order to study steady states of the system the time-independent version of Eq.(12) with $H_T = 0$ is integrated over the periodic domain and next divided by H^3 to obtain

$$(AH^{-4} - BH^{-5})H_\xi + SH_{\xi\xi\xi} = const \cdot H^{-3}. \tag{14}$$

It then follows from the periodicity and positivity of the solution H that $const = 0$. One more integration of Eq.(14) then results in

$$SH_{\xi\xi} + \frac{B}{4}H^{-4} - \frac{A}{3}H^{-3} = \gamma, \tag{15}$$

where γ is a constant of integration that can be found by integrating Eq.(15) over the period as

$$\gamma = \frac{1}{L}\int_0^L (\frac{B}{4}H^{-4} - \frac{A}{3}H^{-3})d\xi. \tag{16}$$

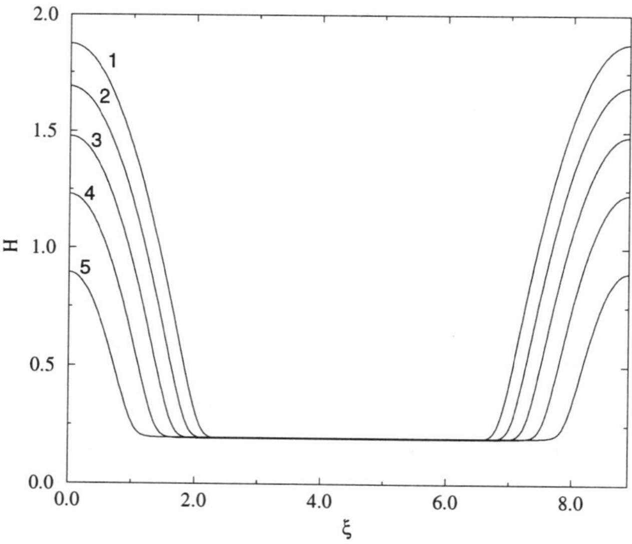

Figure 3. Steady solutions for Eq.(12) for $A = 1, B = 0.25, S = 1, L = 2^{3/2}\pi$ and for various values of the mean film thickness H_0: 1- $H_0 = 0.7$, 2- $H_0 = 0.6$, 3- $H_0 = 0.5$, 4- $H_0 = 0.4$, 5- $H_0 = 0.3$.

Multiplying Eq.(15) by H_ξ and integrating it once again yields

$$\frac{S}{2}H_\xi^2 - V(H) - \gamma H = E \tag{17}$$

with E being a constant of integration.

The minimal thickness H_{min} of the steady solution is found from Eq.(17) by looking for the location of the maximum of the potential $-V(H) - \gamma H$ and solving the equation

$$\frac{B}{4}H_{min}^{-4} - \frac{A}{3}H_{min}^{-3} - \gamma = 0 \tag{18}$$

and thus

$$E = \frac{A}{6}H_{min}^{-2} - \frac{B}{12}H_{min}^{-3} - \gamma H_{min}.$$

It can be shown that $\gamma < 0$, and therefore Eq.(18) has always two real positive and a pair of complex conjugate roots. One of the real roots corresponds to the maximum of $-V(H) - \gamma H$ and the other to its minimum. It follows from the calculations that the term γH is quite small in the vicinity of $H = H_{min}$, and thus causes only a little shift of H_{min} from the minimum of $V(H)$ located at $H = 3B/4A$, which can serve as a good approximation for H_{min}.

The maximal height H_{max} of the steady solution is found by setting $H_\xi = 0$ in Eq.(15) and solving the equation

$$\frac{A}{6} H_{max}^{-2} - \frac{B}{12} H_{max}^{-3} - \gamma H_{max} = E. \qquad (19)$$

It is easy to show that Eq.(19) has always four real roots, one negative which is irrelevant for our problem, and three positive ones. Among the three positive roots, two are double roots corresponding to H_{min}. The fourth root provides us with the value for H_{max}:

$$H_{max} = H_{min} \frac{B - AH_{min} + \sqrt{(AH_{min} + B)^2 - 3B^2}}{4AH_{min} - 3B}, \qquad (20)$$

The values of H_{min} and H_{max} as found from Eqs.(18),(20) are in perfect agreement with the results of numerical computations shown in Figure 3.

The second type of singularity considered in Section 2 and associated with the process of evaporation can be regularized by using the procedure proposed by Oron $et\ al.$(1996). A simple particular case is given here next.

A heated volatile liquid film is again considered when all thermal effects except for evaporation are neglected. However, the film is now assumed to be placed on a "thick" solid planar surface of thickness d such that

$$\epsilon = \max(\frac{h_0}{\lambda}, \frac{d}{\lambda}) \ll 1.$$

It is also assumed that a uniform temperature $\vartheta = \vartheta_b$ is applied to the bottom of the solid surface exposed to the ambient gas phase. In this case one needs to solve a coupled problem of heat transfer in the solid and hydrodynamics with heat transfer in the liquid. These are coupled via the boundary conditions of continuity of both temperature and heat flux at the liquid-solid boundary, cf. Oron $et\ al.$(1996).

The evolution equation describing the spatiotemporal behavior of the film interface in the case considered is (Oron $et\ al.$, 1996)

$$H_T + E(H + K + R)^{-1} = 0, \qquad (21)$$

if van der Waals and capillary forces are neglected, $R = (d/k_w)(h_0/k)^{-1}$ and k_w is the thermal conductivity of the solid. The parameter R represents the ratio between the thermal resistances of the solid and the liquid.

The dimensionless temperatures in the solid, Θ_w, and in the liquid, Θ, are given by

$$\Theta_w = 1 - \kappa(\zeta + \frac{d}{h_0})(H + K + R)^{-1}, \quad \Theta = (H - \zeta)(H + K + R)^{-1}. \qquad (22)$$

where $\kappa = k/k_w$. The dimensionless heat flux in the liquid film and evaporative mass flux are calculated by

$$q = \kappa(H + K + R)^{-1}, \quad J = E(H + K + R)^{-1}. \tag{23}$$

The denominator of the evaporative E-term in Eq.(21) represents a total non-dimensional thermal resistance of the system that includes the relative resistances to conduction in the liquid, H, evaporation, K, and conduction in the solid, R, all in units of h_0/k.

The case of $R = 0$ corresponds to either infinite thermal conductivity or vanishing thickness of the solid wall. Then, if both R and K are zero, one goes back to the singular case discussed earlier. However, if $K = 0$, but $R \neq 0$, one obtains at the gas-liquid (GL) interface $\Theta_{GL} \equiv \Theta(H) = 0$, and at the solid-liquid (SL) interface $\Theta_{SL} \equiv \Theta_w(0) = H(H + R)^{-1}$. Thus, at rupture, $H = 0$, $\Theta_{GL} = \Theta_{SL}$ if $R \neq 0$, and the temperature singularity is removed. At the same time, since the non-dimensional heat and mass fluxes are both proportional to $(H + R)^{-1}$, they are bounded everywhere. including the point of rupture, $q < \kappa R^{-1}$, $j < E R^{-1}$.

If molecular van der Waals and capillary forces were to be included in the analysis of the film dynamics, introduction of short-range repulsive molecular forces would remove the singularities associated with them, as discussed earlier.

Acknowledgments

We thank Prof. S.H. Davis for his very helpful comments. This research was partially supported by the Binational US-Israel Science Foundation through Grant #96-00395 and by S. Faust Research Fund (to A.O.).

References

Burelbach, J. P., Bankoff, S. G. and Davis, S. H. (1988) Nonlinear stability of evaporating/condensing liquid films, *J. of Fluid Mechanics* **195**, 463-488.

Israelachvili, J. N. (1995) *Intermolecular and Surface Forces*, Academic Press, London.

Oron, A., Bankoff, S. G. and Davis, S. H. (1996) Thermal singularities in film rupture, *Physics of Fluids A* **8**, 3433-3435.

Oron, A., and Bankoff, S. G. (1997) Nonlinear dynamics of ultra-thin liquid films, submitted.

Oron, A., Davis, S. H. and Bankoff, S. G. (1997) Long-scale evolution of thin liquid films, *Reviews of Modern Physics* **69**, 931-980.

Williams, M. B., and Davis, S. H. (1982) Nonlinear theory of film rupture, *J. of Colloid Interface Science* **90**, 220-228.

BOUNDS ON THE ENDURANCE LIMIT IN FATIGUE OF DILUTE FIBROUS COMPOSITES BY THE SHAKEDOWN THEOREMS

J. TIROSH

Faculty of Mechanical Engineering

Technion, Haifa, Israel.

Abstract

The intention of this study is to find the 'safe' long term behavior of elasto-plastic materials, reinforced (dilutely) with unidirectional stiff fibers. The composite is subjected to fluctuating load which acts (primarily) transverse to the fibers. The aim is to predict what would be the highest allowable stress amplitude that the composite can endure ('the endurance limit') when undergoing 'infinite' number of cyclic loading.

To reach this goal we employ (a) Melan's static shakedown theorem for formulating the lower bound to the endurance limit, and (b) Koiter's kinematic shakedown theorem for formulating its upper bound. Both theorems are adjusted to capture an isotropic elasto-plastic metal matrix with an embedded non-interactive fibers (i.e. a dilute composite). The fibers/matrix bonding quality is included in a parametric form ranging from no slip condition (m=1) to free-to-rotate condition (m=0).

The solution for the bounds becomes rigorous as the volume fraction of well bonded fibers approaches zero (representing a single fiber in an infinite matrix).

1. Introduction

By-and-large, the phenomenon of fatigue failure in fibrous composites is extremely complex and has been studied hitherto mainly via gathering experimental data for each individual pair of constituents. It has been observed, for example, that in ductile metal-matrices reinforced by fibres plastic strain accumulation is one of the major sources of failure (Dvorak and Johnson, 1980, Jansson and Leckie [1992, 1994]). Under such circumstances, it seems feasible to formulate the bounds on the allowable stress amplitude at which no failure occurs in the long run.

The advantage of using bounds to predict the safe/unsafe stress amplitude is that **no prior information on the actual complex failure mechanisms is required.** The sole hypothesis is that an accumulation of plastic strain during the

349

D. Durban and J.R.A. Pearson (eds.),
IUTAM Symposium on Non-Linear Singularities in Deformation and Flow, 349-360.
© 1999 *Kluwer Academic Publishers. Printed in the Netherlands.*

cyclic loading ('rachetting') in the long run is an unsafe situation . On the other hand, the material is in a safe situation if **eventually** it will respond elastically to the cyclic loading. The border between the safe and unsafe state of the materials is assessed by the shakedown theorems. The deviation between the two bounds represents the 'domain of uncertainty' in the safe/unsafe state of the composites undergoing fluctuating loading. It is anticipated that **all endurance limits of dilute fibrous composites will reside within this domain**. This was evidenced by (some) experiments with metal matrix reinforced composites.

2. The Static Shakedown

Briefly, Melan's theorem (1936) amounts to finding a statically admissible residual state of stress $\sigma_{ij}^{(res)}$ which, when superimposed on the fluctuating 'elastic' stress $\sigma_{ij}^{(e)}(t)$ (not on the actual elasto-plastic stress), does not exceed yielding during further loading. When attaining this condition the material is said to be **safe** for long term behavior.

The outcome of Melan's proof is that the composite structure subjected to cyclic loading will shake down if the sum of the two elastic stress distributions, $\sigma_{ij}^{(res)} + \sigma_{ij}^{(e)}(t)$, nowhere (and at no time) exceeds plastic yielding. Which is

$$f(\sigma_{ij}^{(res)} + \sigma_{ij}^{(e)}(t)) \leq \sigma_0 \qquad \text{in volume V, for all time t } (0 \leq t \leq T). \tag{1}$$

where f(...) is any applicable yield function and σ_0 is the unidirectional yield stress of the matrix. If we define, for convenience, the sum of the elastic stress and the residual stress in the matrix as:

$$\sigma_{ij}^{(sum)}(t) \equiv \sigma_{ij}^{(e)}(t) + \sigma_{ij}^{(res)} \qquad 0 \leq t \leq T \tag{2}$$

then the general expression of (1) is now specified for plane strain situation with Von-Mises yield criterion as:

$$(\sigma_{rr}^{(sum)} - \sigma_{\theta\theta}^{(sum)})^2 + 4\tau_{r\theta}^{(sum)2} = \tilde{\sigma}_0^2$$
$$\text{where} \quad \tilde{\sigma}_0 \equiv \frac{2}{\sqrt{3}}\sigma_0 \quad (\cong 1.15\sigma_0) \qquad\qquad 0 \leq t \leq T \tag{3}$$

3. The Elastic Solution

3.1 THE STRESS FIELD AROUND (AND INSIDE) A DEFORMABLE FIBROUS REINFORCEMENT.

Consider a cylindrical elastic fiber, with Poisson ratio v_0 and shear modulus μ_0. The fiber is embedded in an infinite elastic matrix with properties of v and μ respectively. A remote tensile alternating stress with an amplitude $\sigma_\infty(t)$ is applied transversely to the fibers, as shown in Figure 1.

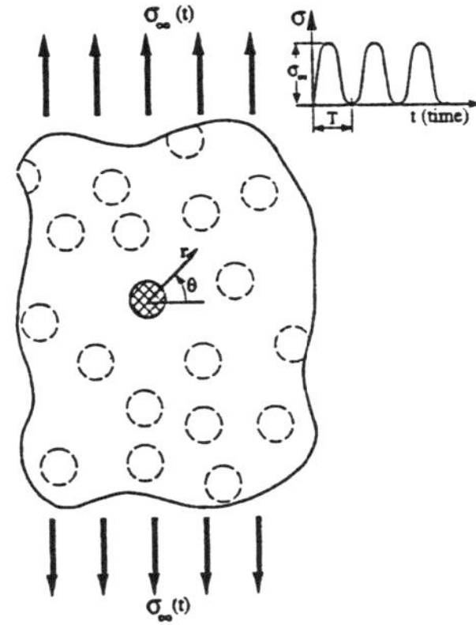

Figure 1. The elasto-plastic metal matrix with dilute stiff fibrous reinforcement . The cyclic zero-tension load, with amplitude $\sigma_\infty(t)$, is prescribed as shown.

The stress distribution in the matrix in terms of the above elastic properties is given (Muskhelishvili, 1963) as:

$$\sigma_{rr}^{(e)} = \frac{\sigma_\infty}{2}\{1 - \gamma\rho^{-2} - [1 - 2\beta\rho^{-2} - 3\delta\rho^{-4}]\cos(2\theta)\}$$

$$\sigma_{\theta\theta}^{(e)} = \frac{\sigma_\infty}{2}\{1 + \gamma\rho^{-2} + [1 - 3\delta\rho^{-4}]\cos(2\theta)\} \qquad \rho \equiv (r/R) \geq 1 \qquad (4)$$

$$\tau_{r\theta}^{(e)} = \frac{\sigma_\infty}{2}[1 + \beta\rho^{-2} + 3\delta\rho^{-4}]\sin(2\theta)$$

The parameters of the matrix β, γ, δ in (4) are

$$\gamma = \frac{\mu(\chi_0 - 1) - \mu_0(\chi - 1)}{2\mu_0 + \mu(\chi_0 - 1)}$$

$$(5)$$

$$\beta = -\frac{2(\mu_0 - \mu)}{\mu + \mu_0\chi} \qquad \delta = \frac{\mu_0 - \mu}{\mu + \mu_0\chi}$$

where

$$\chi = 3 - 4v, \quad \chi_0 = 3 - 4v_0 \quad \text{for plane strain}$$

(v and v_0 are the Poisson ratio of the matrix and the fiber respectively)

3.2 RESIDUAL STRESS DISTRIBUTION (DUE TO THERMAL MISMATCH)

As the composite material is cooled down from its blending temperature, T_c, to the ambient temperature, T_a, residual stresses are produced by the geometrical mismatch along the fiber/matrix interfaces. The mismatch is commonly expressed in terms of the normal strain ε (acting across the interfaces) required to restore the geometrical continuity between the phases. It is given as

$$\varepsilon = (\alpha_m - \alpha_i)(T_a - T_c). \tag{6}$$

where α_m is the thermal expansion coefficient of the matrix and α_0 is that of the fiber. As a result , a residual stress is generated around each fiber with peak magnitude of 'p' along the matrix/fiber interface. This magnitude was found by Muskhelishvili (1963) to yield:

$$p = \frac{4\varepsilon\mu\mu_0}{2\mu_0 + \mu(\chi_0 - 1)} \qquad \text{(at r =R)} \tag{7}$$

When using the data from Table 1 along with (6) and (7), one can see that the magnitude of the residual stress, p, in common pairs of commercial fibrous-matrix composites (B-AL, Be-AL, B-Mg, Be-Ti) are of the order of the matrix yield stress. As an example, the calculated values for the above pairs are: p=2.9 ksi, 2.0 ksi, 2.3 ksi, 0.2 ksi, respectively, provided that the difference between the heat treatment temperature and the ambient temperature is 700 degrees Fahrenheit.

TABLE 1: Mechanical properties of metal constituents (matrix and fibers) used   in common fibrous reinforced composites, (from Tarn et al, 1975).

Material	$\mu \, [or \, \mu_0] \, x10^6 \, psi$	v	$\alpha(10^{-6}/{}^0F)$
Al	3.95	0.329	13.0
B	23.97	0.210	4.5
Be	19.60	0.020	6.5
Mg	2.44	0.332	15.0
Ti	5.59	0.342	5.0

The residual stress distribution is localized around each fiber as follows

$$\sigma_{rr}^{(res)}(r) = \frac{p}{\rho^2} ;$$

$$\sigma_{\theta\theta}^{(res)}(r) = -\frac{p}{\rho^2} \qquad \tau_{r\theta}^{(res)}(r) = 0 \qquad \text{for} \quad \rho \geq 1 \tag{8}$$

but distributed uniformly inside the fiber as:

$$\sigma_{rr}^{(o)(res)}(r) = p$$

$$\sigma_{\theta\theta}^{(o)(res)}(r) = p \qquad \tau_{r\theta}^{(o)(res)}(r) = 0 \qquad \text{for} \quad \rho \leq 1 \tag{9}$$

It is noted that the expressions (8) and (9), plotted in Figure 2, contain a discontinuity in the hoop stress with magnitude of $\sigma_{\theta\theta}^{+} - \sigma_{\theta\theta}^{-} = 2p$. This jump in the stress is physically admissible and does not affect the mechanical integrity of the composite.

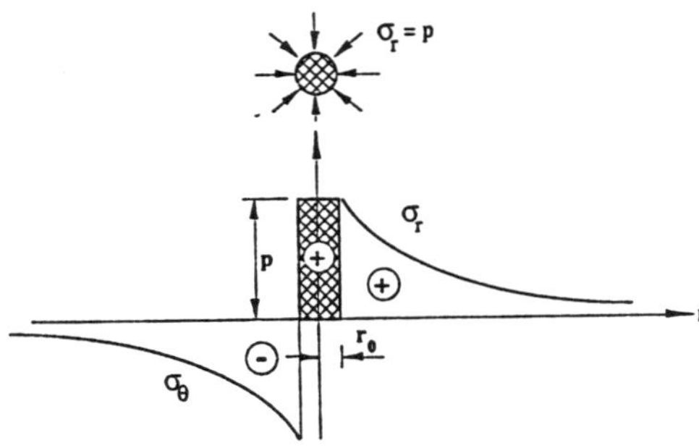

Figure 2. The residual stress distribution near (and inside) a single elastic fiber. The residual stress drawn here is resulted from a thermal fiber/matrix mismatch after cooling-down the metal matrix composite from a certain heat treatment temperature.

4. The Shakedown Condition

By inserting (4) and (8) into (3) one gets the yield function for fibrous composite with residual stress as

$$\frac{\sigma_\infty^2}{\bar{\sigma}_0^2} \{ [(\gamma - \frac{2p}{\sigma_\infty})\rho^{-2} + (1-D)\cos(2\theta)]^2 + [(1+D)\sin(2\theta)]^2 \} = 1, \tag{10}$$

$$\text{for} \quad \rho \geq 1 \text{ and } 0 \leq \theta \leq 2\pi \quad \text{and} \quad D \equiv \beta\rho^{-2} + 3\delta\rho^{-4}$$

From all points in the composite which may yield during the cyclic loading one should look for the point which yields first. This is done by maximizing the expression in the bracket {...} of (10) with respect to the coordinates (ρ, θ). That is

$$\min \frac{\sigma_\infty}{\bar{\sigma}_0} = \frac{1}{\max\{...\}^{1/2}} , \qquad 0 \leq t \leq T(t), \quad 1 \leq \rho, \quad 0 \leq \theta \leq 2\pi \qquad (11)$$

The locus generated from the stress amlitudes of these 'first yielding' points is called the **static shakedown condition.** Examples are plotted in Figure 3.

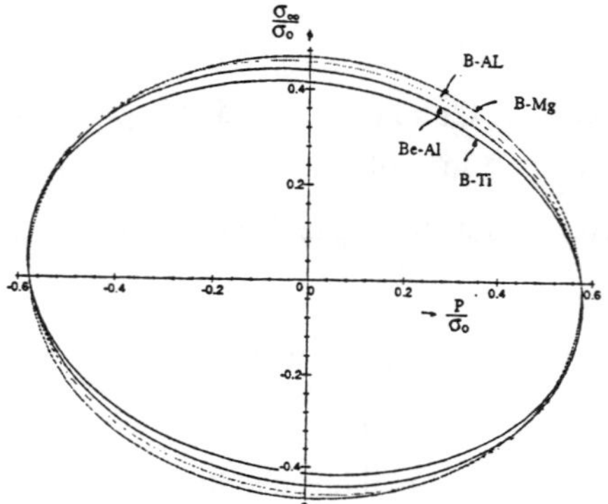

Figure 3. The Melan's shakedown condition (the lower bound solution) for several composite pairs with properties given in Table 1. The composites are subjected to alternating load with amplitude $\sigma_\infty(t)$ and time-independent residual stress p. The area inside each curve is the 'safe working zone' for a long term cyclic loading.

5. Kinematic Shakedown

Koiter (1960) proved that a structure will shakedown under cyclic loading if the rate of plastic dissipation (and the associated frictional loss) exceeds the sum of the work rates due to the applied traction and the internal body force X_i. This inequality reads

$$\int_0^T dt\{\int_V \sigma_{ij}\dot{\varepsilon}_{ij}dV + \int_{s_f} \tau_f|\Delta u|ds\} \geq \int_0^T dt\{\int_V X_i u_i dV + \oint_S T_i u_i ds\} \qquad (12)$$

where the summation convention for repeated subscripts has been employed. $T_i(t)$ is the time-dependent traction, prescribed on S_T and fluctuates between limits. The vector u_i is the velocity field of the plastic flow.

If this inequality is not prevailed, accumulation of plastic strain ('rachetting') is inevitable and the structure is considered as unsafe.

The first term on the left hand side represents the plastic dissipation rate in the deformable volume V through a period T.

$$\int_0^T dt \int_V 2k\sqrt{\tfrac{1}{2}\dot{\varepsilon}_{ij}\dot{\varepsilon}_{ij}}\,dV \qquad (\dot{\varepsilon}_{ij} \text{ is the plastic strain rate}) \qquad (13)$$

The second term on the left hand side is the energy loss due to frictional shear, τ_f, acting along the matrix/fiber interfaces with a relative slip of $|\Delta u|$. The magnitude of the shear stress is considered to be a fraction of the yield stress in shear (k), say

$$\tau_f = mk \qquad (0 \le m \le 1) \qquad (14)$$

The parameter 'm' can be used to assess the quality of the bonding along the fiber/matrix interface. When m=0 the flow around the fiber is friction-less and allows fiber/matrix slippage (debonding). When m=1 the bond is equal to the yield stress of the matrix and no slip is allowed.

In a general problem, the surface which surrounds the volume, $\oint dS = S_u + S_T$, is partially prescribed with a velocity (on s_u) and partially prescribed with traction (on s_T) . Therefore, (12) can be rewritten more specifically as

$$\int_0^T dt\{\int_V X_i u_i dV + \int_{S_u} T_i(t) u_i ds + \int_{S_T} T_i(t) u_i ds\} \le \int_0^T dt\{\int_V 2k\sqrt{\tfrac{1}{2}\dot{\varepsilon}_{ij}\dot{\varepsilon}_{ij}}\,dV + m \int_{s_f} k|\Delta u| ds\} \quad (15)$$

The body force X_i in (15) is the non-uniform stress field generated by the **residual stress** of (8). It is defined as

$$X_i = \frac{\sigma_i^{(res)}}{(perimeter\ of\ a\ fiber)} \qquad (16)\text{a}$$

so that

$$X_r = \frac{p}{2\pi r_0}\frac{1}{\rho^2}, \quad X_\theta = -\frac{p}{2\pi r_0}\frac{1}{\rho^2}, \quad \rho \ge 1 \qquad (16)\text{b}$$

On passing, it is worth recalling that the residual stress acting **inside** the fibers $(\rho \le 1)$ was shown in (16) to be **uniform**. This is in agreement with Eshelby's solution (1957) concerning an elliptical inclusion in a uniform far stress field.

6. The Admissible Velocity Field

We now search for a velocity field, $u_i(r,\theta,t)$, which will describe the kinematics of the plastic flow around a cylindrical circular fiber, of radius r_0. A satisfactory admissible velocity field was derived from the following stream function $\psi(r,\theta,t)$ suggested by Iddan (1996) as

$$\psi(r,\theta,t) = C(t)r_0^2(r^2 - \frac{1}{r^2})\sin(2\theta), \quad 0 \le t \le T, \quad r_0 \le r \le \infty, \quad 0 \le \theta \ge 2\pi \qquad (17)$$

The velocity field which resulting from (17) (according to $u_r = -\dfrac{\partial\psi}{r\partial\theta}$, $u_\theta = \dfrac{\partial\psi}{\partial r}$)

is:

$$u_r(r,\theta,t) = -2C(t)r_0(\rho - \frac{1}{\rho^3})\cos(2\theta), \qquad \rho \equiv \frac{r}{r_0} \ge 1$$

$$u_\theta(r,\theta,t) = 2C(t)r_0(\rho + \frac{1}{\rho^3})\sin(2\theta), \qquad \rho \equiv \frac{r}{r_0} \ge 1 \qquad (18)$$

The associated strain rate components, (using $\dot{\varepsilon}_{ij} = \dfrac{1}{2}(u_{i,j} + u_{j,i})$) yields from (18) the following

$$\dot{\varepsilon}_r = \frac{\partial u_r}{\partial r} = -2C(t)(1 + 3/\rho^4)\cos(2\theta)$$

$$\dot{\varepsilon}_\theta = \frac{1}{r}\frac{\partial u_\theta}{\partial \theta} + \frac{u_r}{r} = 2C(t)(1 + 3/\rho^4)\cos(2\theta) \qquad (19)$$

$$\dot{\varepsilon}_{r\theta} = \frac{1}{2}(\frac{\partial u_\theta}{\partial r} - \frac{u_\theta}{r} + \frac{1}{r}\frac{\partial u_r}{\partial \theta}) = 2C(t)(1 - 3/\rho^4)\sin(2\theta)$$

where 'C(t)' of the above fields is related to the remote strain rate $\dot{\varepsilon}_\infty(t)$ by

$$C(t) = \frac{\dot{\varepsilon}_\infty(t)}{2}$$

In order to reach the **minimal upper bound,** one can rewrite (15) as the following quotient requiring minimal nominator and maximal denominator, namely

$$\frac{\min[\int_0^T dt\{\int_V 2k\sqrt{\frac{1}{2}\dot{\varepsilon}_{ij}\dot{\varepsilon}_{ij}}\,dV + m\int_{s_f} k|\Delta u|ds\}]}{\max[\int_0^T dt\{\int_V X_iu_i dV + \int_{S_*} T_iu_i ds + \int_{S_T} T_iu_i ds\}]} \ge 1 \qquad (20)$$

After inserting (14), (16), (18) and (19) into (20) and performing the algebra for the lowest value of the quotient, one gets the expression of the **kinematic shakedown condition** as

$$\frac{1}{4}\frac{\sigma_\infty}{\sigma_0} + \frac{(1-f)}{\pi^2}\frac{p}{\sigma_0} = \frac{1 - f + 4mf}{\pi\sqrt{3}} \qquad \text{for} \quad (0 < f << 1), \ (0 < m \le 1) \qquad (21)$$

Equation (21) defines the upper bound for a safe stress amplitude for various fiber volume 'f' and bonding quality 'm'.

7. Results

A pre-requisite for employing the shakedown theorems is that the mechanical time-independent load (in our case, the residual stress) will not cause premature yielding of the metal matrix. The residual stress at incipient yielding is readily found from (10) to be

$$\left|\sigma_r^{(res)}\right| = \left|\sigma_\theta^{(res)}\right| = k, \quad (k = \frac{\sigma_0}{\sqrt{3}} \ by \ Von \ Mises \ criterion) \qquad (22)$$

The vertical lines of (22) and the inclined lines of (21), along with the curve (14) exhibit the safe/unsafe bounds for several parameters of Boron/Aluminum fibrous composites, shown in figure 4.

The figures emphasize the variation in the bounds of the fatigue-life limits of composites having different bonding quality (different 'm' values) and different fiber volume fractions, (different 'f 'values). It is seen , for example, that when the fiber/matrix bond is weakened during the cyclic loading (say, from m=0.5 to m=0.1) it lowers the capability of the composite to withstand further loading. The outcome supplies theoretical reasoning to the already known experience [i.e. Dvorak and Johnson, 1980, Jansson and Leckie, 1992] that a **composite with good bonding and higher volume of fibers has a better capability to endure higher stress amplitudes than otherwise** .

In view of the relative insensitivity of the shakedown conditions to the elastic properties of the pair constituents, one can show, for convenience, **the collected data of all pairs** (taken from Table II) **in one figure** (Figure 5) . The ranges of the residual stresses were not recorded in the above references, therefore they were approximated by (11) and added as horizontal lines.

TABLE II Experimental data of endurance limit for some common fibrous composite pairs, measured by their ability to endure at least 10^7 cycles without failure.

Composite	f[%]	endurance limit $[\frac{\sigma_\infty}{\sigma_0}]$	source	symbol
Borsic/titanium	38	0.41	Christian (1975)	⊗
Boron/Al 6061	50	0.53	Christian (1975)	●
E-Glass/epoxy	60	0.62	Hashin and Rotem(1973)	V
Graphite/epoxy	60	0.77	Charewicz and Daniel(1986)	∧

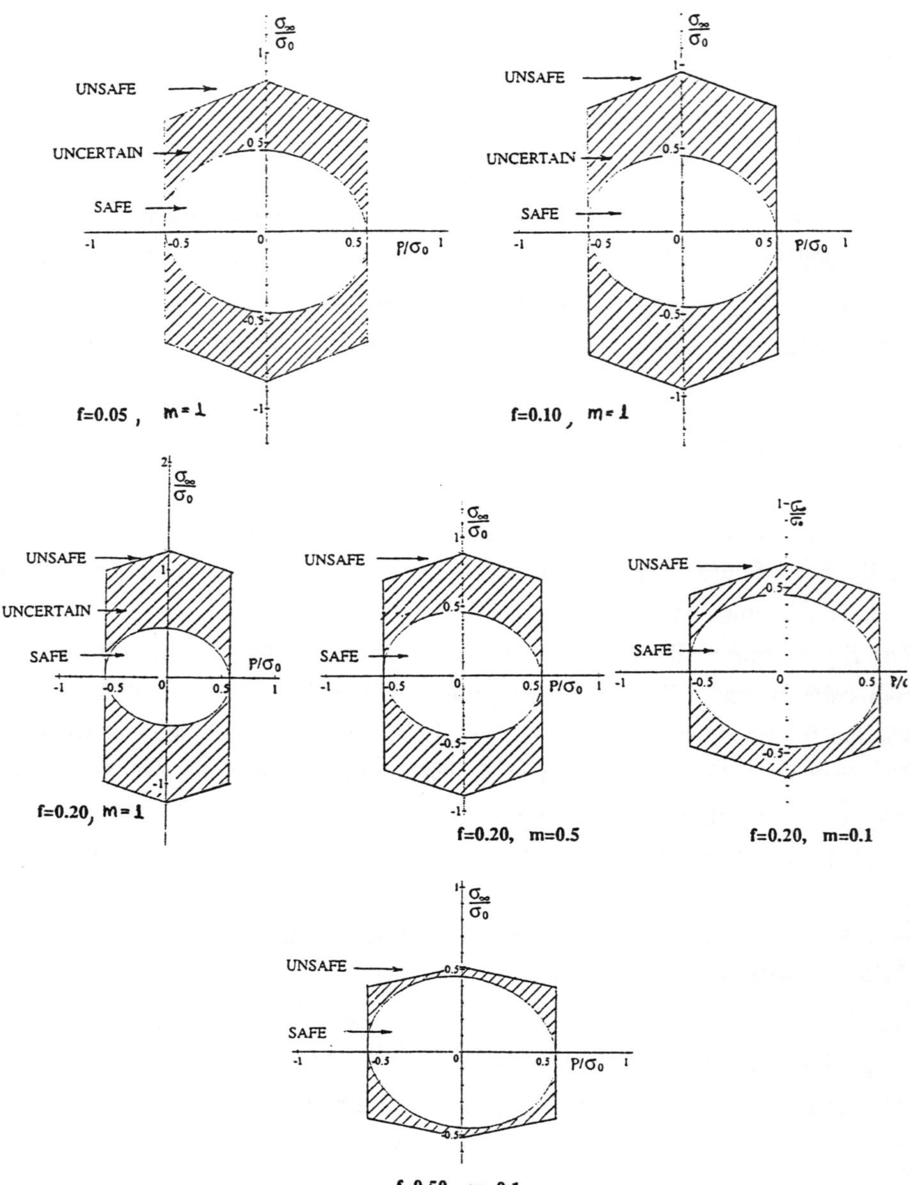

Figure 4: The dual shakedown conditions are plotted from (11) (the lower bound), and from (21), (the upper bound). The zone subtended between the bounds (in dash lines) is interpreted as the zone in which fatigue endurance limit should take place.

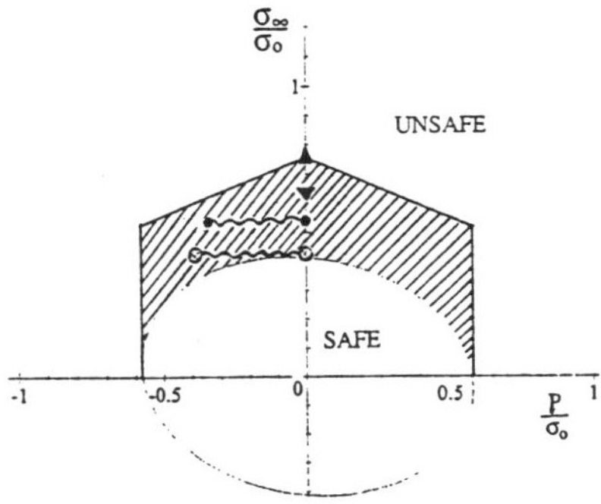

Fig. 5. The plot is a result of the exact shakedown solutions for a single well bonded fiber ($\lim f \to 0$, $m \to 1$). For convenience, several experimental endurance limits (from Table II) are plotted collectively on a single figure. The magnitude of the residual stress in the above composites are estimated [by (10) and (11)] and shown by horizontal lines subtended from each data point.

8. Discussion and Summary

One can assume that the bonding quality 'm' should be related somehow to the sign of the residual stress 'p'. A reasonable relationship between 'm' and 'p' is based on the assumption that the shear resistance to the fiber/matrix slip obeys the Coulomb friction rule, $\tau = \mu p$, where μ is the Coulomb friction coefficient. The shear stress equivalence $\mu p = mk$ leads to the relation

$$m \cong \mu \frac{p}{\sigma_o} \sqrt{3} \qquad \text{(for 'p' in compression)} \qquad (23)$$

From (23) one can conclude that **the higher the residual pressure the higher is the anticipated endurance limit of the composites** (via the increase of 'm'). This, among other issues, wait for a further study.

From (4) and (8) it can be seen that the stress in the matrix, caused by a single fiber, decays away from its center in a rate of the order of $O(\frac{1}{r^2})$. The strain rate, shown in (19) decays by the order of $O(\frac{1}{r^4})$. Accordingly, the energy-rate density (their multiplication) decays steeply by the order of $O(\frac{1}{r^6})$. Therefore, the density of the energy rate is highly localized in the vicinity of each fiber. **This implies (qualitatively) that the interaction between fibers is relatively weak and allows the use of the shakedown bounds to predict endurance limit of composites with substantial fiber volume fraction.**
A quantified assessment of this practical issue is now under consideration.
The outcome of this study is seemingly wide in scope and flexible in use in the sense that:

(i) there is no need to postulate a specific mechanism by which the material fails, beside the common belief that it is associated with accumulation of plastic straining.

(ii) the residual stress in the composite can be taken arbitrarily as an admissible, self-stress   variable.

(ii) the bounds have an analytical form, which makes them accessible to end users.

Acknowledgment

This research was supported by the funds for the promotion of research at the Technion. The assisstance of Dr. Amnon Shirizly from the Technion and the discussions with Prof. Boris Droyanov from the Hebrew University in Jerusalem were very helpful.

References

Charewicz, A. and Daniel I. M, (1986), Damage Mechanisms and Accumulation in Graphite/Epoxy Laminates. Composite Materials- Fatigue and Fracture, ASTM, STP 907 (ed. H. T. Hahn), 274-297.

Christian, J. L., (1975). Axial Fatigue Properties of Metal Matrix composites. Fatigue of Composite Materials, ASTM, STP 569, 280-294.

Dvorak, G. J. and Johnson, W. S., (1980). Fatigue of Metal Matrix Composites. Int. J. Fracture. **16**, 585-607.

Dvorak, G. J. and Tarn, J. Q., (1975). Fatigue and Shakedown in Metal Matrix Composite. Fatigue of Composite Materials, ASTM, STP 569, 145-168.

Eshelby, J. D., (1957).   Proc. Roy. Soc. A 241, 376-386.

Hashin, Z. and A. Rotem , (1973) A fatigue failure criterion for fiber reinforced materials. J. comp. mat., **7**, 448-464.

Iddan , D. (1996), Private communication.

Jansson, S. and Leckie, F. A., (1994). Transverse tensile and inplane shear strength of weakly bonded fiber reinforced MMC's subjected to cyclic thermal loading. Mech. of Materials, **18**, 205-212.

Jansson, S. and Leckie, F. A., (1992). Mechanical behavior of a continuous fiber-reinforced aluminum matrix composite subjected to transverse and thermal loading. J. Mech. Phys. Solids . **40**, No 3. 593-612.

Koiter, W. T., (1960), General theorems for elastic-plastic solids. In: Progress in Solid Mechanics I. § 6, 203-313,Eds.: I. N. Sneddon, R. Hill, North-Holland, Amsterdam.

Melan, E., (1936). Theorie statisch unbestimmter Systeme aus idealplastischem Baustoff. Sitzungsbericht der Akademie der Wissenschaften (Wien) Abt. IIa, **145**, 195-218.

Muskhlishvili, N. I. , (1963). Some Basic Problems of the Mathematical Theory of Elasticity. P. Noordhoff Ltd. Groningen, The Netherlands.

Ponter, A. R. S. and Karadeniz, S., (1985). An Extended Shakedown Theory for Structures that   Suffer Cyclic Thermal Loading, Parts 1, 2. J. Appl. Mech., ASME Trans. **52**, 877-889.

Tarn, J. Q., Dvorak, G. J., and Rao, M. S. M., (1975). Shakedown of Unidirectional Composites. Int. J. of Solids and Structure, **11**, 751-764.